Nitrogen Fixation: Biology and Ecology

Nitrogen Fixation: Biology and Ecology

Editor: Sara Langer

MURPHY & MOORE
www.murphy-moorepublishing.com

Murphy & Moore Publishing,
1 Rockefeller Plaza,
New York City, NY 10020, USA

Visit us on the World Wide Web at:
www.murphy-moorepublishing.com

ISBN: 978-1-63987-397-5 (Hardback)

Cataloging-in-Publication Data

Nitrogen fixation : biology and ecology / edited by Sara Langer.
p. cm.
Includes bibliographical references and index.
ISBN 978-1-63987-397-5
1. Nitrogen--Fixation. 2. Nitrogen-fixing microorganisms. 3. Nitrogen-fixing plants.
4. Biological models. 5. Microbial ecology. I. Langer, Sara.
QR89.7 .N58 2022
589.950 413 3--dc23

Table of Contents

Preface

This book was inspired by the evolution of our times; to answer the curiosity of inquisitive minds. Many developments have occurred across the globe in the recent past which has transformed the progress in the field.

The process of converting atmospheric nitrogen into ammonia and other nitrogenous compounds in soil is termed as nitrogen fixation. Most of the biological nitrogen is fixed by prokaryotic bacteria. A few of these are cyanobacteria, azotobacter, azospirillum, rhizobium and bradyrhizobium. Clover, soybean, peanut, etc., are some of the plants which help in nitrogen fixation process. These processes are part of the nitrogen cycle. They are of utmost significance in agriculture as nitrogen is required for the synthesis of various organic compounds like amino acids, nucleic acids and proteins. Nitrogen is also fixed by natural phenomenon such as lightning. It is also responsible for the formation of nitrogen oxides which may react with water to form nitrous acid or nitric acid. The acids seep into the soil where they are converted into nitrates. This book includes some of the vital pieces of work being conducted across the world, on various topics related to nitrogen fixation. From theories to research to practical applications, case studies related to all contemporary topics of relevance to this field have been included in it. This book is a vital tool for all researching or studying the biological and ecological aspects of nitrogen fixation as it gives incredible insights into emerging trends and concepts.

This book was developed from a mere concept to drafts to chapters and finally compiled together as a complete text to benefit the readers across all nations. To ensure the quality of the content we instilled two significant steps in our procedure. The first was to appoint an editorial team that would verify the data and statistics provided in the book and also select the most appropriate and valuable contributions from the plentiful contributions we received from authors worldwide. The next step was to appoint an expert of the topic as the Editor-in-Chief, who would head the project and finally make the necessary amendments and modifications to make the text reader-friendly. I was then commissioned to examine all the material to present the topics in the most comprehensible and productive format.

I would like to take this opportunity to thank all the contributing authors who were supportive enough to contribute their time and knowledge to this project. I also wish to convey my regards to my family who have been extremely supportive during the entire project.

Editor

1

Systemic Regulation of Root Nodule Formation

Takashi Soyano and Masayoshi Kawaguchi

1. Introduction

Most of legume plants produce *de novo* root lateral organs, root nodules, to accommodate their symbiotic bacteria, collectively called as rhizobia. Production of the *de novo* root organ is initiated by infection of rhizobia on surfaces of plant root tissues [1]. Rhizobia enter host tissues by two major strategies. One is the crack entry from the root epidermis where lateral roots emerge, and the other is the root hair-mediated invasion. Model legume plants, *Lotus japonicus* and *Medicago truncatula,* have adopted the latter mode of the infection pathway. In either case, infected rhizobia invade into host tissues through infection threads, which are tubular paths developed by invagination of the host plasma membrane. Concomitantly with the progression of infection processes from the epidermis into the cortex, a fraction of cortical cells beneath the site of the infection begin to divide and form a root nodule primordium. Invaded bacteria are released into cortical cells in the nodule primordium by an endocytic-like process, and differentiate into bacteroides. Whereas plants usually cannot utilize atmospheric nitrogen as a nutrient, bacteroides convert it to ammonium that is a usable nitrogen source for host plants. Host plants, on the other hand, provide the rhizobia with an energy source from photoassimilates, and compensate components that rhizobia lack for nitrogen fixation [2]. This symbiotic relationship enables legumes to grow under conditions where nitrogen sources are limited.

The nodule symbiosis with rhizobia is a unique feature of legumes among land plants. The infection processes are regulated by interaction with rhizobia. Nodulation factors (Nod factors) that are lipochitin oligosaccharides secreted by rhizobia trigger the symbiotic responses [3,4]. LysM containing receptor proteins, NFR1 and NFR5 in *L. japonicus,* and NFP and HCL in *M. truncatula* are involved in perception of Nod factors [5-9]. Direct binding of Nod factors to NFR1 and NFR5 was recently demonstrated [10]. Nod factor perception leads to activation of transcriptional networks involved in establishing the nodule symbiosis. This process depends on common *SYM* factors, which are required for both the nodule symbiosis and mycorrhizal symbiosis [11-14]. While the mycorrhizal symbiosis is seen in the majority of land plant species, and has been thought to be evolved 400 million years ago, molecular clock data suggest that legumes has acquired the nodule symbiosis about 60 million years ago. Common *SYM*s are involved in transmission of nodulation signals into nuclei. Among common *SYM* factors, CCaMK (for calcium- and calmodulin-depednet protein kinase)/DMI3 plays pivotal roles in regulation of nodulation processes [15-17]. CCaMK localizes to nuclei, and interacts with its phosphorylation substrate, CYCLOPS/IPD3 [14,18,19]. gof-CCaMK spontaneously induces root nodules without rhizobial infection [16,17], indicating this protein substitutes for Nod factor signals in the cortical response for the root nodule organogenesis. Activation of CCaMK is sufficient for onset of gene expression required for the root nodule development. Thus, legumes have invented the nodule symbiosis by recruiting factors from the ancestral mycorrhizal symbiosis, and further have developed nodulation specific factors to acquire the nitrogen-fixing activity.

Although the root nodule symbiosis is beneficial to host plants, excessive nodulation interferes with plant growth, probably, due to a high energy cost against the nitrogen fixation. Therefore, legumes have developed negative feedback pathways that optimize total nodule number and mass in a single plant. A major pathway that regulates nodule number is known as autoregulation of nodulation (AON) [20,21], which is activated by nodulation, and systemically prevents subsequent formation of nodules through root-shoot communications. This systemic effect has been demonstrated by split root experiments. Infection with rhizobia to one part of the root results in reduction in the nodule number in the other part of the root that was inoculated 3-4 days after the first inoculation procedure [20,22]. Reciprocal grafting experiments with mutants defective in AON have further shown that the inhibitory effect seen in the root is mediated by the shoot [23-27]. Two types of long-distance signals that are derived from either the root or shoot have been postulated in AON to explain the root-shoot communication; The root-derived signal that was generated by early nodulation signaling is translocated to the shoot, and activates shoot-acting AON factors to produce the shoot-derived inhibitor (SDI), which, in turn, is transported down to the root, and inhibits nodulation through root-acting AON factors.

In the past decade, many mutants exhibiting impaired nodulation have been isolated, and genes responsible for symbiotic phenotypes have been identified. The accumulating lines of evidence led to basic models for the nodulation signaling pathway. Factors involved in AON have been identified from several legume species. Our knowledge of AON has greatly advanced at molecular levels by efforts of many researchers, although many pieces have remained to understand how AON systemically regulates nodulation. In this review, we highlight past work in this area and describe recent results obtained with forward and reverse genetic approaches as well as with biochemical works, along the basic scheme of AON that has been depicted by pioneer works.

2. Shoot factors involved in AON

Genetic screens have identified loci that affect AON in several legume species. These mutants often display excessive nodules within an enhanced nodulation zone in either the presence of or the absence of nitrate, which possesses inhibitory effects on nodulation (see below). This phenotype is termed hypernodulation or supernodulation [18]. Soybean *nts*, *L. japonicus har1*, *M. truncatula sunn*, and pea *sym29* mutants exhibit the typical hypernodulation phenotype [24,27-29]. Grafting experiments have shown that shoots of these mutants confer excessive nodulation on wild type rootstocks, while mutant rootstocks that were grafted to the wild type scion do not exhibit the hypernodulation phenotype [23-27]. These results indicated that the shoot genotype determined the phenotype in the root, and genetically demonstrated that the root-shoot communication is important for the systemic and negative feedback regulation of the nodule formation.

Positional cloning of genes responsible for the hypernodulation phenotype revealed that AON has diverted factors involved in the shoot apical meristem homeostasis as shoot-acting factors. The causative genes of these mutants encode leucine-rich repeat (LRR) containing receptor protein kinases orthologous to each other [25,26,30,31]. It is expected that ligand-receptor interaction in the shoot regulates nodulation in the root. *In vitro* experiments have shown that the kinase domain of GmNARK/NTS possesses a transphosphorylation activity [32]. This kinase activity is abolished by amino acid substitutions at residues corresponding to sites of missense mutations that were found in loss-of-function *nts* alleles. The kinase activity is required for the GmNARK function, and phosphorylation of its substrates probably leads to the production of the SDI in the shoot. The GmNARK kinase domain phosphorylates two KAPPs (for kinase-associated protein phosphatases) [32]. The phosphorylated KAPPs de-phosphorylated the autophosphorylated GmNARK kinase domain. The LRR receptor protein kinases are closest to Arabidopsis *CLV1* and rice *FON1*, which play a central role in the shoot apical meristem homeostasis, and restrict meristem sizes [33-35]. Multiple receptor proteins are involved in maintenance of meristem sizes in Arabidopsis [36]. Pea and *L. japonicus* genes orthologous to Arabidopsis *CLV2*, which encodes a LRR-containing transmembrane protein that interacts with and stabilizes CLV1 [37], are also involved in AON [38]. Furthermore, *L. japonicus* KLV that acts together with HAR1 is an LRR-receptor protein kinase with a high similarity to Arabidopsis RPK2/TOAD2 [39,40], which regulates the shoot meristem homeo-stasis in parallel with CLV1 [41]. These leguminous receptor proteins probably function as multiple complexes as it is for Arabidopsis counterparts [36,41]. HAR1, indeed, interacts with KLV in a transient expression system using *Nicotiana benthamiana* leaves [40]. Although the function of these leguminous genes in the shoot apical meristem has remained to be analyzed in detail, fasciated stems have been observed in Arabidopsis *klv* mutants and loss-of-function plants of *L. japonicus* and pea *CLV2* genes [38-40]. This phenotypic trait is often observed in mutants, such as *clv* mutants, where the negative feedback system to maintain meristem sizes is abortive and the shoot apical meristem is enlarged [33]. Besides stem fasciation, *klv* mutants further show disconnected vasculatures, indicating defect in cell differentiation or production of stem cells [39]. These observations imply that AON and the shoot meristem-associated pathway may utilize common or similar molecular strategies to maintain either the meristem homeostasis or the nodule number. Alternatively, legumes may have developed pathways that specifically regulate nodule number in the root downstream of the shoot-acting receptor protein kinases.

3. Root-derived CLE peptides

In the shoot meristem regulation, CLV1 restricts meristem sizes through interaction with CLV3 that is a member of CLE (CLV3/ESR-related) small secreted peptide family [42-44]. CLV3 peptide is produced from its primary translational product by proteolytic processing [45]. By analogy to the CLV1-CLV3 pathway, the shoot-acting AON receptor protein kinases were expected to recognize CLE peptides that might act as the root-derived signal. Okamoto et al. have comprehensively analyzed expression of genes that encode CLE-peptide precursors in *L. japonicus*, and found that two *CLE* genes, referred to as *CLE-RS1* and *CLE-RS2*, specifically expressed in roots in response to rhizobial infection [46]. Similar approaches have also been performed in other leguminous species to identify *CLE* genes associated with the nodule development [47-50]. *MtCLE12* and *MtCLE13* from *M. truncatula* and *GmRIC1*, *GmRIC2*, and *GmNIC1* from soybean as well as *L. japonicus CLE-RS1* and *CLE-RS2* have activities to suppress the nodule formation [46,47,49,50]. Ectopic expression of these genes, except *GmNIC1*, in a part of the root system results in systemic inhibition of nodulation in the roots that have not ectopically expressed the *CLE* gene. This effect on nodulation depends on *HAR1* and *KLV* in *L. japonicus* and the corresponding shoot-acting AON genes in *M. truncatula* and soybean [40, 46,48,49,51]. These results are compatible with the idea that these *CLE* gene products are the root-derived signals. However, synthetic CLE peptides corresponding to the putative CLE-RS1, CLE-RS2, MtCLE12, and MtCLE13 failed to suppress nodulation [46,47].

Okamoto et al. have determined the structure of the mature CLE-RS2 [52]. This peptide is composed of 13 amino-acid residues corresponding to the conserved C-terminal CLE domain of its precursor, and the seventh proline residue is hydroxylated, and further posttranslationally modified with three residues of arabinose. This proline residue is conserved in all putative CLE peptides that are encoded by genes whose ectopic expression suppresses nodulation. The same arabinosylation has been also found at the seventh hydroxyproline residue of the CLV3 peptide. This modified CLV3 peptide interacts more strongly with the ectodomain of CLV1 [53]. The synthetic arabinosylated CLE-RS1 and CLE-RS2 bound to HAR1 protein, but not to a loss-of-function HAR1 derivative with an amino acid substitution at a residue critical for ligand binding of the CLV1 family receptor kinases in Arabidopsis [52,54]. The activity of the CLE-RS1 and CLE-RS2 peptides to suppress nodulation has been demonstrated by feeding experiments with the synthetic arabinosylated peptides [52]. The synthetic CLE-RS2 and CLE-RS1 peptides that were fed from cotyledon surfaces suppressed nodulation depending on *HAR1*, but those without arabinosylation failed. The arabinosylation is essential for binding to HAR1 and the activity to inhibit nodulation. These analyses have shown that the arabinosylated CLE-RS2 and CLE-RS1 are ligands for HAR1.

Unlike the short-distance communication between CLV1 and CLV3 (these genes express in restricted region of the shoot apical meristem), long-ranged transport is necessary for the root-derived CLE peptides to interact with HAR1. The xylem often mediates the transport of molecules from the root to the shoot [55]. The arabinosylated CLE-RS2 was detected in xylem sap collected from soybean shoots whose roots have been transformed to express *CLE-RS2* [52]. The secreted mature peptide is thought to be transported to the shoot in the xylem. This is consistent with spatial expression patterns of *HAR1, KLV,* and *NTS/GmNARK*; They express in vascular tissues [40,56]. The arabinosylated CLE-RS2 and CLE-RS1 peptides that were transported in the xylem are thought to bind with HAR1, and exert the inhibitory effect on nodulation.

4. Expression of *CLE* genes that encode the root-derived signal

Transcriptional regulation of the root-derived signal is an important step to regulate AON. The expression in response to rhizobial infection is mediated by the early signaling pathway required for nodulation processes [47]. Transcription factors that are involved in the early nodulation signaling, such as NSP1, NSP2, and NIN, are required for expression of *MtCLE12* and *MtCLE13* in response to rhizobial infection. These transcription factors are essential for formation of the nodule primordium, which is initiated by division of cortical cells beneath epidermal cells infected by rhizobia [57-60]. It has been shown that ectopic expression of *NIN* induces cortical cell division without rhizobial infection [61]. Mortier et al. have shown detailed spatial expression patterns of *MtCLE12* and *MtCLE13* during the course of the nodule meristem formation [47]. These genes begin to express in cortical cells beneath infection sites in the epidermis [47]. The expression is sustained in dividing cortical cells in the incipient nodule primordium, and is restricted in the apical zone of elongated indeterminate type nodules. The apical region corresponds to the meristematic and early infection zones. Expression patterns similar to those of *MtCLE* genes have also shown in the case of *GmRIC2* [49]. In the current model of the nodule meristem formation, perception of Nod factors that are secreted by rhizobia triggers the nodule meristem formation through activation of cytokinin signaling [62-65]. Cytokinin induces expression of *NSP1, NSP2,* and *NIN* [62-64,66], although *NSP2* expression is transient and repressed within 3 hours after the treatment [67]. *MtCLE13* expression in the root treated with exogenous cytokinin is detected in inner cortical cells [68], which is consistent with the site where where *MtCLE13* begins to express when roots were inoculated with rhizobia. These observations indicate that *CLE* gene expression is associated with the nodule primordium or meristem formation.

In addition to rhizobial infection, expression of *CLE-RS2* and *GmNIC1* is induced by nitrate, which also has an inhibitory effect on nodulation [46,50]. *nts* mutants were originally identified as nitrate-tolerant symbiotic mutants [28]. *har1* and *klv* mutants also exhibit the nitrate-tolerance [39]. It is likely that the different inputs, rhizobial infection and nitrate, activate the same *CLE* gene in *L. japonicus,* and interfere with nodulation through AON. It has been shown that NIN-like proteins (NLPs) targets nitrate-responsive elements (NREs) that were found in promoters of nitrate-inducible genes, such as *NIR1,* and activate the gene expression in Arabidopsis [69-71]. All 9 Arabidopsis NLPs are able to bind to NREs and possess potentials to activate gene expression in transient expression assay and *in vitro* experiments. NIN has been thought to be derived from a member of NLPs [72,73]. All of domains that are conserved in NLPs are present in the NIN protein except the GAF-related domain in the N-terminal region, which is necessary for NLPs to sense a nitrate-signal that post-translationally activates the NLP transcriptional activity [70]. NIN-binding nucleotide sequences that were found in promoters of NIN-target genes are similar to those of NREs [61]. NIN and NLPs may target *CLE* genes whose mature translational products act as the root-derived AON signals, and activate the gene expression in response to different stimuli, rhizobial infection and the nitrate-supply, respectively. There may be an evolutionary link between pathways for nitrate-responses and AON. Nitrate also systemically regulates root architecture in Arabidopsis [74,75].

5. Shoot-derived inhibitors

The SDI is thought to be produced depending on activation of the shoot-acting receptor protein kinases. Expression of *GmNARK* in the phloem has suggested that the SDI is produced in the vascular tissue and transported in the phloem. Lin et al. have shown the presence of the SDI activity in leaf extracts from soybean [76]. They developed a feeding system from petioles, and found the activity that suppresses nodulation in aqueous extracts prepared from leaves of inoculated wild type soybean. This activity is generated depending on Nod factors and *GmNARK*, and effective in both wild type plants and *nts* mutants. The generality of the activity was demonstrated with leaf extracts from inoculated *M. truncatula*. These properties of the activity are consistent with that of the presumptive SDI. The factor with the activity is a heat-stable small molecule of < 1 kDa and resistant against either RNase or Proteinase K. It is likely that SDI is neither a protein nor an RNA.

Although the shoot-derived molecule that acts as the SDI has not yet been identified, several phytohormones have shown inhibitory effects on nodulation [77] and associated with AON. Auxin has been postulated to be involved in nodulation [78], and is transported from the shoot to the root by cell-to-cell transport mediated by auxin carriers, and also thought to travel in the phloem. *M. truncatula SUNN* is involved in regulation of this long-distance auxin transport. van Noorden et al. have shown that auxin transport from the shoot to the root is downregulated by inoculation with rhizobia in wild type plants, whereas the rate of the transport is not changed in the *sunn* mutant, and auxin is more abundant in the mutant root compared to the wild type [79]. They have proposed a model of auxin action in AON based on their observations; Downregulation of the long-distance transport of auxin by AON leads to reduction in the auxin level in the roots, resulting in decreased efficiencies of the nodule initiation. Indeed, local application of an auxin transport inhibitor, N-(1-naphthyl)phthalamic acid, to a hypocotyl results in reduction of nodule number in the *sunn* mutant, but not in the wild type plant [79]. Unlike cytokinin, involvement of auxin in the nodule primordium formation has remained obscure. It would be important to elucidate roles of auxin in the root nodule development for understanding relation with AON.

In soybean, on the other hand, it has been reported that metabolic pathways of jasmonic acid in the shoot are altered by *nts* mutations. Kinkema et al. have found that transcripts of genes encoding key enzymes controlling jasmonic acid biosynthesis and of jasmonic acid-responsive genes are accumulated in the shoot of *nts* mutants compared to wild type plants [80]. They have also shown that foliar application of a jasmonic acid biosynthesis inhibitor reduces nodule number. These results suggest that jasmonic acid is a negative regulator in AON. However, other two groups have shown that foliar applied methyl jasmonic acid downregulates nodulation in soybean and *L. japonicus* [81,82]. Involvement of jasmonic acid or methyl jasmonic acid in AON remains controversial. It has been also reported that foliar application of brassinosteroides to loss-of-function *GmNARK* plants suppresses nodule formation [83]. Grafting experiments have shown that shoots of a pea mutant of which causative gene encodes a brassinosteroides biosynthesis enzyme causes decrease in nodule number in wild type rootstacks [84]. It would be required for further analysis on action of brassinosteroides in systemic inhibitory effects.

6. Root-acting factors involved in inhibition of nodulation

According to the scenario described above, the mature CLE peptides whose genes express in response to rhizobial infection act as the root derived signals, and activate shoot-acting receptor protein kinases including HAR1 and KLV to generate the SDI, which in turn inhibits nodulation. Root-acting AON factors involved in production of the root-derived signal or perception of the SDI are required for AON to exert the inhibitory effect. Root-specific hypernodulation mutants, pea *nod3*, *L. japonicus plenty,* and *tml/rdh1*, and *M. truncatula rdn1* and *sickle* have been isolated [85-90]. Reciprocal grafting experiments have shown that root genotypes of these genetic loci influence nodulation.

PLENTY seems to work in a pathway different form theHAR1-mediated AON, since grafting of a *har1* scion resulted in an additive phenotype with respect to number of the root nodules [88]. It has been implied that nodule number is regulated by multiple pathways. Ethylene-insensitive *sickle* mutants have shown that ethylene signaling influence nodule number [91]. The causative gene of this mutant encodes an ortholog of Arabidopsis EIN2 that is involved in ethylene signaling [90]. Similar to *sunn* mutants, downregulation of the long-distance auxin transport in response to rhizobial infection dose not occur in the *sickle* mutant [92]. However, *SICKLE* seems to regulate the nodule number independently of the *SUNN*-mediated pathway, because *sickle sunn* double mutants exhibited a novel hypernodulation phenotype [27]. Nodules densely cover the whole root length in *sunn* mutants, while nodules are formed at the limited region corresponding to the first inoculation zone in *sickle* mutants [91]. Auxin transport inhibition may not be unique to AON.

Pea *NOD3* and *MtRDN1* are orthologous genes, and encode a protein of the endosomal system with unknown function [89]. Similar to *sunn* mutants, *rdn1* mutants show the hypernodulation phenotype that was moderately suppressed by nitrate, and the shorter root phenotype. *RisfixC* locus allelic to *nod3* exhibits an interesting phenotype. A wild type scion that was grafted to a *RisfixC* rootstock produced adventitious roots with the hypernodulation phenotype [93]. Approach-grafting experiments (two plants with intact roots are grafted at stems) have shown that the first inoculation of the *nod3* mutant root fails to suppress nodulation in the wild type root that were inoculated 7 days after the first inoculation, resulting in increase in nodule number compared to the wild type control root [94]. These results suggested that *nod3* mutations systemically affected nodulation. Schnabel et al. have shown that a *GUS* reporter construct for the *MtRND1* promoter intensely expresses in the vascular tissues, in particular, the area of the xylem, throughout the root [89]. NOD3/MtRND1 may be involved in early events of the AON pathway associated with production or transportation of the root-derived signal.

Contrary to NOD3/MtRND1, *L. japonicus* TML is a root-acting AON factor that functions after production of the SDI. *L. japonicus tml* and *rdh1* are allelic mutants [95]. TML has been shown to function downstream of the root-derived CLE peptides [95]. The systemic inhibition of the nodulation by either *CLE-RS1* or *CLE-RS2* is suppressed by *tml* mutations, similar to *har1* and *klv* mutations. Unlike the *plenty* mutant, shoot scions of *har1* and *klv* mutants do not enhance the hypernodulation phenotype of *tml* rootstocks [87,95]. TML is thought to work in the same genetic pathway with HAR1 and KLV. Furthermore, the inverted-Y grafting has shown that a *tml* root that was inserted into a hypocotyl of a wild type seedling with the intact root does not affect efficiencies of nodulation in the wild type root [87]. This result suggested that the *tml* mutation does not systemically affect nodulation, and the *tml* root is insensitive to the SDI.

TML encodes a kelch repeat-containing F-box protein [95]. F-box proteins are subunits of SCF E3 ubiquitin ligase complexes that lead to inactivation of substrate proteins through ubiquitination and subsequent degradation by 26S proteasome. F-box proteins have been thought to confer substrate specificities on SCF E3 ubiquitin ligase complexes. TML protein that was fused to GFP localizes into nuclei. This subcellular localization implies that TML targets transcription factors, similarly to Arabidopsis kelch-repeat F-box protein, FKF1 [96]. Temporal and spatial expression patterns of a GUS reporter construct for the *TML* promoter has suggested that TML suppresses development of the root nodule meristem after initiation of cortical cell division [95]. In addition to the nodule primordium, *TML* constitutively expresses in the root apex transition zone, which is at developmental stages earlier than the susceptible region for rhizobial infection. *tml* mutants produce excess infection threads in the epidermis. One hypothesis is that TML indirectly inhibits formation of infection threads prior to rhizobial infection at the root transition zone. TML may exert inhibition of nodulation with two modes of actions at the site of rhizobial infection and of the nodule meristem formation.

7. Which steps of nodulation are influenced by AON?

Although molecular mechanisms, by which nodule formation is suppressed in the root, have been largely unknown, several results may be able to speculate how AON influences nodulation. Rhizobia usually infect at the root tip region where elongation of root hairs occurs, and nodule formation is initiated at the infection site. Then, nodule formation at the root region that is developmentally younger than the first inoculated region is suppressed. Thereby, nodule formation is limited at the first inoculation zone of the root. The mutants exhibiting impaired AON show increases in number of both infection threads and nodules. Nodule formation in these mutants is deregulated with respect of the nodule density and the region that form nodules. It is suggested that AON influences on early stages of rhizobial infection and local regulation of nodule number. Identification of loss-of-function mutants of a *L. japonicus* cytokinin receptor gene (*LHK1*) as a suppressor of the *har1* hypernodulation phenotype has suggested that AON may act downstream of LHK1 in the nodule initiation [62]. Spatial expression patterns of *TML* suggest that AON targets early nodulation signaling pathway required for rhizobial infection, and inhibits cell division in the cortex [87,95].

Ectopic *MtCLE12* and *MtCLE13* abolished expression of *MtENOD11* expression in the root epidermal cells in response to rhizobial infection [47]. Although *MtENOD11* is unknown function, NSP1-NSP2 transcription factor complex and ERF/AP2 family transcription factors, ERNs, target the *MtENOD11* promoter and activate the gene expression [97-99]. The suppression of *MtENOD11* expression suggests that one of sites AON targets is the early nodulation signaling pathway, including NSP1 and NSP2. It is compatible with expression of *TML* in the root apex transition zone. Murakami et al. have shown that *NSP2* expression is remarkably downregulated at the root tip region within 1 day after rhizobial infection [100]. *MtENOD11* is activated by Nod factor treatment. The expression in the root tip is repressed within 1 day after the treatment, which is consistent with downregulation of *NSP2* expression. Interestingly, this *MtENOD11* repression spatially and temporally regulated is suppressed in *nin* mutants [101]. NIN is required for expression of *MtCLE12* and *MtCLE13*, and NSP2 is required for *NIN* expression in response to rhizobial infection. There may be the negative feedback regulation mediated by *NSP2* and *NIN*. However, it is unlikely that the repression of *NSP2* is mediated by the root-shoot communication of AON, because split-root experiments have shown that it takes 3 days to systemically suppress nodulation in *L. japonicus* [102]. In soybean, on the other hand, *GmNIC1* overexpression locally, but not systemically, suppresses nodulation [50]. This effect of the overexpression depends on GmNARK. Some aspects of local inhibitory effects in soybean may be mediated by LRR-receptor protein kinases.

Saur et al. have shown a strong reduction of *NIN* expression in roots overexpressing *MtCLE12* when compared with controls at 21 days after inoculation [103]. They have proposed a negative feedback regulation that implicates cytokinin signaling and NIN in AON. Type-A response regulators highly express in roots where *MtCLE12* overexpresses at the same time point, although there are no significant differences at 8 days after inoculation. Type-A response regulators are responsible for the negative-feedback regulation of cytokinin signaling [104], and may repress cytokinin-dependent *NIN* expression. Arrested primordia that were composed of cells that originate from the cortex, endodermis, and pericycle were observed in roots overexpressing MtRR9 type-A response regulator [105].

In the shoot apical meristem homeostasis, the CLV1-CLV3 pathway represses a gene encoding a homeobox transcription factor, WUSCHEL, which directly activates *CLV3* gene and represses expression of type-A response regulator genes [106,107]. *WUSCHEL*, itself, is activated by cytokinin. This transcription factor plays a critical role in the feedback loop to maintain the homeostasis of the shoot apical meristem. A WUSCHEL-related homeobox transcription factor, WOX5, expresses in the nodule primordium of *M. truncatula*, [108]. Although a role of

MtWOX5 in the nodule development has not yet been clarified, induction of this gene in response to treatment with auxin, but not cytokinin, suggests that auxin is required for expression of genes that may act in regulation of cortical cell division. An auxin-responsive synthetic promoter, *DR5,* is often used for monitoring distribution of auxin [109,110]. Intense expression of GFP protein whose expression is under the control of the *DR5* promoter is detected in cortical cells beneath the infection site in the epidermis, and sustained in developing nodules [111]. The expression is attenuated in *L. japonicus* roots overexpressing *CLE-RS2*. On the other hand, in *har1* background, the *DR5* construct expresses in broad area of the infected root, suggesting that auxin is more abundantly accumulated in the mutant root. It is consistent with the result obtained by direct measurement of auxin levels in *sunn* mutants [79]. Auxin activates cell division of pericycle cells that is origin of lateral root primordia. Excessive cell division in the root pericycle as well as cortical cell division has been observed in the *har1* mutant [29]. This alteration of the *DR5* expression suggests that AON influences auxin distribution at the site of the nodule initiation and development, leading to inhibition of cell division. In this context, auxin is a positive regulator of the nodule formation. Auxin transport inhibitors elicit pesudonodule formation in *M. sativa* and *M. truncatula* [112,113], indicating alteration of auxin distribution is an important cue for the nodule formation as well as cytokinin.

8. Conclusion

Forward genetic analyses have greatly contributed to identify novel factors that are involved in AON. The series of investigations revealed that AON has diverted factors required for the homeostasis of the shoot apical meristem as the shoot-acting factors. The ligand-receptor relationship between HAR1 and the root-derived CLE peptides has been clearly demonstrated by the reverse genetic approaches and biochemical assays. The similarity of AON to the CLV1-CLV3 pathway suggests that AON may have recruited functions from pathways that generally work in non-leguminous species. How legumes have been molecularly specialized from non-leguminous species is an important question to understand molecular basis on the root nodule symbiosis. Studies on AON would shed light on this fundamental question besides elucidation of the molecular nature of the systemic inhibitory pathway. There remains three major questions to to elucidate AON at molecular levels: (1) How expression of the root-derived signals is regulated, (2) What is a bona-field SDI, (3) Molecular mechanisms, by which nodulation is suppressed in the root. Lin et al. have shown the SDI activity in aqueous extracts prepared from leaves of inoculated wild type soybean. Forward genetics has identified TML as the root-acting AON factor involved in suppression of nodulation downstream of the SDI. These findings as well as data obtained by efforts of many researchers would be clues to resolve these questions.

Author details

Takashi Soyano and Masayoshi Kawaguchi

Division of Symbiotic Systems, National Institute for Basic Biology, Okazaki, Aichi, Japan

References

[1] Oldroyd GE, Downie JA. Coordinating nodule morphogenesis with rhizobial infection in legumes. Annual Review of Plant Biology 2008;59 519-546.

[2] Hakoyama T, Niimi K, Watanabe H, Tabata R, Matsubara J, Sato S, Nakamura Y, Tabata S, Jichun L, Matsumoto T, Tatsumi K, Nomura M, Tajima S, Ishizaka M, Yano K, Imaizumi-Anraku H, Kawaguchi M, Kouchi H, Suganuma N. Host plant genome overcomes the lack of a bacterial gene for symbiotic nitrogen fixation. Nature 2009;462(7272) 514-517.

[3] Ehrhardt DW, Wais R, Long SR. Calcium spiking in plant root hairs responding to Rhizobium nodulation signals. Cell 1996;85(5) 673-681.

[4] Esseling JJ, Lhuissier FGP, Emons AMC. Nod factor-induced root hair curling: continuous polar growth towards the point of Nod factor application. Plant Physiolgy 2003;132(4) 1982-1988.

[5] Madsen EB, Madsen LH, Radutoiu S, Olbryt M, Rakwalska M, Szczyglowski K, Sato S, Kaneko T, Tabata S, Sandal N, Stougaard J. A receptor kinase gene of the LysM type is involved in legume perception of rhizobial signals. Nature 2003;425(6958) 637-640.

[6] Radutoiu S, Madsen LH, Madsen EB, Felle HH, Umehara Y, Grønlund M, Sato S, Nakamura Y, Tabata S, Sandal N, Stougaard J. Plant recognition of symbiotic bacteria requires two LysM receptor-like kinases. Nature 2003;425(6958):585-592.

[7] Limpens E, Franken C, Smit P, Willemse J, Bisseling T, Geurts R. LysM domain receptor kinases regulating rhizobial Nod factor-induced infection. Science 2003;302(5645):630-633.

[8] Arrighi JF, Barre A, Ben Amor B, Bersoult A, Soriano LC, Mirabella R, de Carvalho-Niebel F, Journet EP, Ghérardi M, Huguet T, Geurts R, Dénarié J, Rougé P, Gough C. The *Medicago truncatula* lysin motif-receptor-like kinase gene family includes *NFP* and new nodule-expressed genes. Plant Physiolgy 2006;142(1) 265–279.

[9] Smit P, Limpens E, Geurts R, Fedorova E, Dolgikh E, Gough C, Bisseling T. Medicago LYK3, an entry receptor in rhizobial nodulation factor signaling. Plant Physiolsy 2007;145(1):183-191.

[10] Broghammer A, Krusell L, Blaise M, Sauer J, Sullivan JT, Maolanon N, Vinther M, Lorentzen A, Madsen EB, Jensen KJ, Roepstorff P, Thirup S, Ronson CW, Thygesen MB, Stougaard J. Legume receptors perceive the rhizobial lipochitin oligosaccharide signal molecules by direct binding. Proceedings of the National Academy of Sciences of the United States of America 2012;109(34):13859-13864.

[11] Kistner C, Parniske M. Evolution of signal transduction in intracellular symbiosis. Trends in Plant Science 2002;7(11) 511-518.

[12] Kaló P, Gleason C, Edwards A, Marsh J, Mitra RM, Hirsch S, Jakab J, Sims S, Long SR, Rogers J, Kiss GB, Downie JA, Oldroyd GE. Nodulation signaling in legumes requires NSP2, a member of the GRAS family of transcriptional regulators. Science 2005;308(5729)1786-1789.

[13] Heckmann AB, Lombardo F, Miwa H, Perry JA, Bunnewell S, Parniske M, Wang TL, Downie JA. *Lotus japonicus* nodulation requires two GRAS domain regulators, one of which is functionally conserved in a non-legume. Plant Physiolgy 2006;142(4) 1739-1750.

[14] Messinese E, Mun JH, Yeun LH, Jayaraman D, Rougé P, Barre A, Lougnon G, Schornack S, Bono JJ, Cook DR, Ané JM. A novel nuclear protein interacts with the symbiotic DMI3 calcium- and calmodulin-dependent protein kinase of *Medicago truncatula*. Molecular Plant-Microbe Interactions 2007;20(8) 912-921.

[15] Catoira R, Galera C, de Billy F, Penmetsa RV, Journet EP, Maillet F, Rosenberg C, Cook D, Gough C, Dénarié J. Four genes of *Medicago truncatula* controlling components of a nod factor transduction pathway. Plant Cell 2000;12(9) 1647-1666.

[16] Gleason C, Chaudhuri C, Yang T, Muñoz A, Poovaiah BW, Oldroyd GED. Nodulation independent of rhizobia induced by a calcium-activated kinase lacking autoinhibition. Nature 2006;441(7097) 1149-1152.

[17] Tirichine L, Imaizumi-Anraku H, Yoshida S, Murakami Y, Madsen LH, Miwa H, Nakagawa T, Sandal N, Albrektsen AS, Kawaguchi M, Downie A, Sato S, Tabata S, Kouchi H, Parniske M, Kawasaki S, Stougaard J. Deregulation of a Ca^{2+} calmodulin-dependent kinase leads to spontaneous nodule development. Nature 2006;441(7097) 1153-1156.

[18] Yano K, Yoshida S, Müller J, Singh S, Banba M, Vickers K, Markmann K, White C, Schuller B, Sato S, Asamizu E, Tabata S, Murooka Y, Perry J, Wang TL, Kawaguchi M, Imaizumi-Anraku H, Hayashi M, Parniske M. CYCLOPS, a mediator of symbiotic intracellular accommodation. Proceedings of the National Academy of Sciences of the United States of America 2008;105(51) 20540-20545.

[19] Horváth B, Yeun LH, Domonkos A, Halász G, Gobbato E, Ayaydin F, Miró K, Hirsch S, Sun J, Tadege M, Ratet P, Mysore KS, Ané JM, Oldroyd GED, Kaló P. *Medicago truncatula IPD3* is a member of the common symbiotic signaling pathway required

for rhizobial and mycorrhizal symbioses. Molecular Plant-Microbe Interactions 201;24(11) 1345-1358.

[20] Kosslak, RM, Bohlool BB. Suppression of Nodule Development of One Side of a Split-Root System of Soybeans Caused by Prior Inoculation of the Other Side. Plant Physiology 1984;74(1) 125-130.

[21] Caetanoanolles G, Gresshoff P.M. Early Induction of Feedback Regulatory Responses Governing Nodulation in Soybeen. Plant Science 1990;71(1) 69-81.

[22] Olsson JE, Nakao P, Benbohlool B, Gresshoff PM. Lack of Systemic Suppression of Nodulation in Split Root Systems of Supernodulating Soybean (*Glycine max* [L.] Merr.) Mutants. Plant Physiology 1989;90(4) 1347-1352.

[23] Delves AC, Mathews A, Day DA, Carter AS, Carroll BJ, Gresshoff PM. Regulation of the Soybean-Rhizobium Nodule Symbiosis by Shoot and Root Factors. Plant Physiology 1986;82(2) 588-590.

[24] Sagan M, Duc G. *Sym28* and *Sym29*, two new genes involved in regulation of nodulation in pea (*Pisum sativum L*). Symbiosis 1996;20(3) 229-245.

[25] Krusell L, Madsen LH, Sato S, Aubert G, Genua A, Szczyglowski K, Duc G, Kaneko T, Tabata S, de Bruijn F, Pajuelo E, Sandal N, Stougaard J. Shoot control of root development and nodulation is mediated by a receptor-like kinase. Nature 2002;420(6914) 422-426.

[26] Nishimura R, Hayashi M, Wu GJ, Kouchi H, Imaizumi-Anraku H, Murakami Y, Kawasaki S, Akao S, Ohmori M, Nagasawa M, Harada, K, Kawaguchi M. HAR1 mediates systemic regulation of symbiotic organ development. Nature 2002;420(6914) 426-429.

[27] Penmetsa RV, Frugoli JA, Smith LS, Long SR, Cook DR. Dual genetic pathways controlling nodule number in *Medicago truncatula*. Plant Physiology 2003;131(3) 998-1008.

[28] Carroll BJ, McNeil DL, Gresshoff PM. Isolation and properties of soybean [*Glycine max* (L.) Merr.] mutants that nodulate in the presence of high nitrate concentrations. Proceedings of the National Academy of Sciences of the United States of America 1985;82(12) 4162-4166.

[29] Wopereis J, Pajuelo E, Dazzo FB, Jiang QY, Gresshoff PM, de Bruijn FJ, Stougaard J, Szczyglowski K. Short root mutant of *Lotus japonicus* with a dramatically altered symbiotic phenotype. Plant Journal 2000;23(1) 97-114.

[30] Arai M, Hayashi M, Takahashi M, Shimada S, Harada K. Expression and sequence analysis of systemic regulation gene for symbiosis, *NTS1/GmNARK* in supernodulating soybean cultivar, Sakukei 4. Breeding Science 2005;55(2) 147-152.

[31] Schnabel E, Journet EP, de Carvalho-Niebel F, Duc G, Frugoli J. The *Medicago truncatula SUNN* gene encodes a CLV1-like leucine-rich repeat receptor kinase that regulates nodule number and root length. Plant Molecular Biology 2005;58(6) 809–822.

[32] Miyahara A, Hirani TA, Oakes M, Kereszt A, Kobe B, Djordjevic MA, Gresshoff PM. Soybean Nodule Autoregulation Receptor Kinase Phosphorylates Two Kinase-associated Protein Phosphatases *in Vitro*. Journal of Biological Chemistry 2008;283(37) 25381–25391.

[33] Clark SE, Jacobsen SE, Levin JZ, Meyerowitz EM. The *CLAVATA* and *SHOOT MERISTEMLESS* loci competitively regulate meristem activity in *Arabidopsis*. Development 1996;122(5) 1567-1575.

[34] Clark SE, Williams RW, Meyerowitz EM. The *CLAVATA1* gene encodes a putative receptor kinase that controls shoot and floral meristem size in *Arabidopsis*. Cell 1997;89(4) 575-585.

[35] Suzaki T, Sato M, Ashikari M, Miyoshi M, Nagato Y, Hirano HY. The gene *FLORAL ORGAN NUMBER1* regulates floral meristem size in rice and encodes a leucine-rich repeat receptor kinase orthologous to *Arabidopsis* CLAVATA1. Development 2004;131(22) 5649-5657.

[36] Perales M, Reddy GV. Stem cell maintenance in shoot apical meristems. Current Opinion in Plant Biology 2012;15(1) 10-16.

[37] Jeong S, Trotochaud AE, Clark SE. The *Arabidopsis CLAVATA2* gene encodes a receptor-like protein required for the stability of the CLAVATA1 receptor-like kinase. Plant Cell 1999;11(10) 1925-1933.

[38] Krusell L, Sato N, Fukuhara I, Koch BEV, Grossmann C, Okamoto S, Oka-Kira E, Otsubo Y, Aubert G,Nakagawa T, Sato S, Tabata S, Duc G, Parniske M, Wang TL, Kawaguchi M, Stougaard J. The *Clavata2* genes of pea and *Lotus japonicus* affect autoregulation of nodulation. Plant Journal 2011;65(6) 861-871.

[39] Oka-Kira E, Tateno K, Miura K, Haga T, Hayashi M, Harada K, Sato S, Tabata S, Shikazono N, Tanaka A, Watanabe Y, Fukuhara I, Nagata T, Kawaguchi M. *klavier* (*klv*), a novel hypernodulation mutant of *Lotus japonicus* affected in vascular tissue organization and floral induction. Plant Journal 2005;44(3) 505-515.

[40] Miyazawa H, Oka-Kira E, Sato N, Takahashi H, Wu GJ, Sato S, Hayashi M, Betsuyaku S, Nakazono M, Tabata S, Harada K, Sawa S, Fukuda H, Kawaguchi M. The receptor-like kinase KLAVIER mediates systemic regulation of nodulation and non-symbiotic shoot development in *Lotus japonicus*. Development 2010;137(24) 4317-4325.

[41] Kinoshita A, Betsuyaku S, Osakabe Y, Mizuno S, Nagawa S, Stahl Y, Simon R, Yamaguchi-Shinozaki K, Fukuda H, Sawa S. RPK2 is an essential receptor-like kinase that transmits the CLV3 signal in *Arabidopsis*. Development 2010; 137(22):3911-3920.

[42] Fletcher LC, Brand U, Running MP, Simon R, Meyerowitz EM. Signaling of cell fate decisions by *CLAVATA3* in *Arabidopsis* shoot meristems. Science 1999;283(5409) 1911-1914.

[43] Brand U, Fletcher JC, Hobe M, Meyerowitz EM, Simon R. Dependence of stem cell fate in *Arabidopsis* an a feedback loop regulated by CLV3 activity. Science 2000;289(5479) 617-619.

[44] Cock JM, McCormick S. A large family of genes that share homology with *CLAVATA3*. Plant Physiology 2001;126(3) 939-942.

[45] Matsubayashi Y. Post-Translational Modifications in Secreted Peptide Hormones in Plants. Plant and Cell Physiology 2011; 52(1) 5-13.

[46] Okamoto S, Ohnishi E, Sato S, Takahashi H, Nakazono M, Tabata S, Kawaguchi M. Nod factor/nitrate-induced *CLE* genes that drive HAR1-mediated systemic regulation of nodulation. Plant and Cell Physiology 2009;50(1) 67-77.

[47] Mortier V, Den Herder G, Whitford R, Van de Velde W, Rombauts S, D'Haeseleer K, Holsters M, Goormachtig S. CLE Peptides Control *Medicago truncatula* Nodulation Locally and Systemically. Plant Physiology 2010; 153(1) 222-237.

[48] Mortier V, Fenta BA, Martens C, Rombauts S, Holsters M, Kunert K, Goormachtig S. Search for nodulation-related *CLE* genes in the genome of *Glycine max*. Journal of Experimental Botany 2011;62(8) 2571-2583.

[49] Lim CW, Lee YW, Hwang CH. Soybean nodule-enhanced CLE peptides in roots act as signals in GmNARK-mediated nodulation suppression. Plant and Cell Physiology 2011; 52(9) 1613-1627.

[50] Reid DE, Ferguson BJ, Gresshoff PM. Inoculation- and Nitrate-Induced CLE Peptides of Soybean Control NARK-Dependent Nodule Formation. Molecular Plant-Microbe Interactions 2011;24(5) 606-618.

[51] Saur IM, Oakes M, Djordjevic MA, Imin N. Crosstalk between the nodulation signaling pathway and the autoregulation of nodulation in *Medicago truncatula*. New Phytologist 2011;190(4) 865–874.

[52] Okamoto S, Shinohara H, Mori T, Matsubayashi Y, Kawaguchi M. Root-derived CLE glycopeptides control nodulation by direct binding to HAR1 receptor kinase. Nature communications 2013;4:2191 | DOI: 10.1038/ncomms3191. http://www.nature.com/ncomms/2013/130812/ncomms3191/full/ncomms3191.html (accessed 7 Nov 2013)

[53] Ohyama K, Shinohara H, Ogawa-Ohnishi M, Matsubayashi Y. A glycopeptide regulating stem cell fate in *Arabidopsis* thaliana. Nature Chemical Biology 2009;5(8) 578-580.

[54] Shinohara H, Moriyama Y, Ohyama K, Matsubayashi Y. Biochemical mapping of a ligand-binding domain within *Arabidopsis* BAM1 reveals diversified ligand recognition mechanisms of plant LRR-RKs. Plant Journal 2012;70(5) 845-854.

[55] Schachtman DP, Goodger JQD. Chemical root to shoot signaling under drought. Trends in Plant Science 2008;13(6) 281-287.

[56] Nontachaiyapoom S, Scott PT, Men AE, Kinkema M, Schenk PM, Gresshoff PM. Promoters of orthologous *Glycine max* and *Lotus japonicus* nodulation autoregulation genes interchangeably drive phloem-specific expression in transgenic plants. Molecular Plant-Microbe Interactions 2007;20(7) 769-780.

[57] Schauser L, Roussis A, Stiller J, Stougaard J. A plant regulator controlling development of symbiotic root nodules. Nature 1999; 402(6758) 191-195.

[58] Oldroyd GED, Long SR. Identification and characterization of *nodulation-signaling pathway 2*, a gene of *Medicago truncatula* involved in Nod factor signaling. Plant Physiology 2003;131(3) 1027-1032.

[59] Kaló P, Gleason C, Edwards A, Marsh J, Mitra RM, Hirsch S, Jakab J, Sims S, Long SR, Rogers J, Kiss GB, Downie JA, Oldroyd GED. Nodulation signaling in legumes requires NSP2, a member of the GRAS family of transcriptional regulators. Science 2005;308(5729) 1786-1789.

[60] Smit P, Raedts J, Portyanko V, Debelle F, Gough C, Bisseling T, Geurts R. NSP1 of the GRAS protein family is essential for rhizobial Nod factor-induced transcription. Science 2005;308(5729) 1789-1791.

[61] Soyano T, Kouchi H, Hirota A, Hayashi M. NODULE INCEPTION Directly Targets *NF-Y* Subunit Genes to Regulate Essential Processes of Root Nodule Development in *Lotus japonicus*. Plos Genetics 2013;9(3) e1003352. http://www.plosgenetics.org/article/info%3Adoi%2F10.1371%2Fjournal.pgen.1003352 (accessed 22 July 2013)

[62] Murray JD,Karas BJ, Sato S, Tabata S, Amyot L, Szczyglowski K. A cytokinin perception mutant colonized by Rhizobium in the absence of nodule organogenesis. Science 2007;315(5808) 101-104.

[63] Tirichine L, Sandal N, Madsen LH, Radutoiu S, Albrektsen AS, Sato S, Asamizu E, Tabata S, Stougaard J. A gain-of-function mutation in a cytokinin receptor triggers spontaneous root nodule organogenesis. Science 2007;315(5808) 104-107.

[64] Heckmann AB, Sandal N, Bek AS, Madsen LH, Jurkiewicz A, Nielsen MW, Tirichine L, Stougaard J. Cytokinin induction of root nodule primordia in *Lotus japonicus* is regulated by a mechanism operating in the root cortex. Mol Plant Microbe Interactions 2011;24(11) 1385-1395.

[65] Plet J, Wasson A, Ariel F, Le Signor C, Baker D, Mathesius U, Crespi M, Frugier F. MtCRE1-dependent cytokinin signaling integrates bacterial and plant cues to coordinate symbiotic nodule organogenesis in *Medicago truncatula*. Plant Journal 2011; 65(4) 622-633.

[66] Ariel F, Brault-Hernandez M, Laffont C, Huault E, Brault M, Plet J, Moison M, Blanchet S, Ichanté JL, Chabaud M, Carrere S, Crespi M, Chan RL, Frugier F. Two Direct Targets of Cytokinin Signaling Regulate Symbiotic Nodulation in *Medicago truncatula*. Plant Cell 2012;24(9) 3838-3852.

[67] Maekawa T, Maekawa-Yoshikawa M, Takeda N, Imaizumi-Anraku H, Murooka Y, Hayashi M. Gibberellin controls the nodulation signaling pathway in *Lotus japonicus*. Plant Journal 2009;58(2) 183-194.

[68] Mortier V, De Wever E, Vuylsteke M, Holsters M, Goormachtig S. Nodule numbers are governed by interaction between CLE peptides and cytokinin signaling. Plant Journal 2012;70(3) 367-376.

[69] Konishi M, Yanagisawa S. Identification of a nitrate-responsive cis-element in the *Arabidopsis NIR1* promoter defines the presence of multiple cis-regulatory elements for nitrogen response. Plant Journal 2010;63(2) 269-282.

[70] Konishi M, Yanagisawa S. *Arabidopsis* NIN-like transcription factors have a central role in nitrate signalling. Nuclear retention of the transcription factor NLP7 orchestrates the early response to nitrate in plants. Nature communications 2013;4 1617. http://www.nature.com/ncomms/journal/v4/n3/full/ncomms2621.html (accessed 22 July 2013)

[71] Marchive C, Roudier F, Castaings L, Bréhaut V, Blondet E, Colot V, Meyer C, Krapp A. Nature communications 2013;4 1713. http://www.nature.com/ncomms/journal/v4/n4/full/ncomms2650.html (accessed 22 July 2013)

[72] Schauser L, Wieloch W, Stougaard J. Evolution of NIN-like proteins in *Arabidopsis*, rice, and *Lotus japonicus*. Journal of Molecular Evolution 2005;60(2) 229-237.

[73] Yokota K, Hayashi M. Function and evolution of nodulation genes in legumes. Cellular and Molecular Life Sciences 2011;68(8):1341-1351.

[74] Ruffel S, Krouk G, Ristova D, Shasha D, Birnbaum KD, Coruzzi GM. Nitrogen economics of root foraging: Transitive closure of the nitrate-cytokinin relay and distinct systemic signaling for N supply vs. demand. Proceedings of the National Academy of Sciences of the United States of America 2011;108(45) 18524-18529.

[75] Puig J, Pauluzzi G, Guiderdoni E, Gantet P. Regulation of Shoot and Root Development through Mutual Signaling. Molecular Plant 2012;5(5) 974-983.

[76] Lin YH, Ferguson BJ, Kereszt A, Gresshoff PM. Suppression of hypernodulation in soybean by a leaf-extracted, NARK- and Nod factor-dependent, low molecular mass fraction. New Phytologist 2010;185(4) 1074-1086.

[77] Ferguson BJ, Mathesius U. Signaling interactions during nodule development. Journal of Plant Growth Regulation 2003;22(1) 47-72.

[78] Mathesius U. Auxin: at the root of nodule development? Functional Plant Biology 2008; 35(8) 651-668.

[79] van Noorden GE, Ross JJ, Reid JB, Rolfe BG, Mathesius U. Defective long-distance auxin transport regulation in the *Medicago truncatula* super numeric nodules mutant. Plant Physiology 2006;140(4) 1494-1506.

[80] Kinkema M, Gresshoff PM. Investigation of downstream signals of the soybean autoregulation of nodulation receptor kinase GmNARK. Molecular Plant-Microbe Interactions 2008;21(10) 1337-1348.

[81] Nakagawa T, Kawaguchi M. Shoot-applied MeJA suppresses root nodulation in *Lotus japonicus*. Plant and Cell Physiology 2006;47(1) 176-180.

[82] Seo HS, Li J, Lee SY, Yu JW, Kim KH, Lee SH, Lee IJ, Paek NC. The hypernodulating nts mutation induces jasmonate synthetic pathway in soybean leaves. Molecules and Cells 2007;24(2) 185-193.

[83] Terakado J,Fujihara S, Goto S, Kuratani R, Suzuki Y, Yoshida S, Yoneyama T. Systemic effect of a brassinosteroid on root nodule formation in soybean as revealed by the application of brassinolide and brassinazole. Soil Science and Plant Nutrition 2005;51(3) 389-395.

[84] Ferguson BJ, Ross JJ, Reid JB. Nodulation Phenotypes of Gibberellin and Brassinosteroid Mutants of Pea. Plant Physiology 2005;138(4) 2396-2405.

[85] Postma JG, Jacobsen E, Feenstra WJ. Three pea mutations with an altered nodulation studied by genetic-anlaysis and grafting. Journal of Plant Physiology 1988;132(4) 424-430.

[86] Ishikawa K, Yokota K, Li YY, Wang YX, Liu CT, Suzuki S, Aono T, Oyaizu H. Isolation of a novel root-determined hypernodulation mutant *rdh1* of *Lotus japonicus*. Soil Science and Plant Nutrition 2008;54(2) 259-263.

[87] Magori S, Oka-Kira E, Shibata S, Umehara Y, Kouchi H, Hase Y, Tanaka A, Sato S, Tabata S, Kawaguchi M. *Too much love*, a root regulator associated with the long-distance control of nodulation in *Lotus japonicus*. Molecular Plant-Microbe Interactions 2009;22(3) 259-268.

[88] Yoshida C, Funayama-Noguchi S, Kawaguchi M. *plenty*, a novel hypernodulation mutant in *Lotus japonicus*. Plant and Cell Physiology 2010;51(9) 1425-1435.

[89] Schnabel EL, Kassaw TK, Smith LS, Marsh JF, Oldroyd GE, Long SR, Frugoli JA. The *ROOT DETERMINED NODULATION1* gene regulates nodule number in roots of *Medicago truncatula* and defines a highly conserved, uncharacterized plant gene family. Plant Physiology 2011;157(1) 328-340.

[90] Penmetsa RV, Uribe P, Anderson J, Lichtenzveig J, Gish JC, Nam YW, Engstrom E, Xu K, Sckisel G, Pereira M, Baek JM, Lopez-Meyer M, Long SR, Harrison MJ, Singh KB, Kiss GB, Cook DR. The *Medicago truncatula* ortholog of *Arabidopsis* EIN2, sickle, is a negative regulator of symbiotic and pathogenic microbial associations. Plant Journal 2008;55(4) 580-595.

[91] Penmetsa RV, Cook DR. A Legume Ethylene-Insensitive Mutant Hyperinfected by Its Rhizobial Symbiont. Science 1997;275(5299) 527-530.

[92] Prayitno J, Rolfe BG, Mathesius U. The Ethylene-insensitive sickle mutant of *Medicago truncatula* shows altered auxin transport regulation during nodulation. Plant Physiology 2006;142(1) 168-180.

[93] Novak K. Early action of pea symbiotic gene *NOD3* is confirmed by adventitious root phenotype. Plant Science 2010;179(5) 472-478.

[94] Li DX, Kinkema M, Gresshoff PM. Autoregulation of nodulation (AON) in *Pisum sativum* (pea) involves signalling events associated with both nodule primordia development and nitrogen fixation. Journal of Plant Physiology 2009;166(9) 955-967.

[95] Takahara M, Magori S, Soyano T, Okamoto S, Yoshida C, Yano K, Sato S, Tabata S, Yamaguchi K, Shigenobu S, Takeda N, Suzaki T, Kawaguchi M. Too much love, a novel Kelch repeat-containing F-box protein, functions in the long-distance regulation of the legume-Rhizobium symbiosis. Plant and Cell Physiology 2013;54(4) 433-447.

[96] Imaizumi T, Schultz TF, Harmon FG, Ho LA, Kay SA. FKF1 F-BOX protein mediates cyclic degradation of a repressor of CONSTANS in *Arabidopsis*. Science 2005;309(5732) 293-297.

[97] Andriankaja A, Boisson-Demier A, Frances L, Sauviac L, Jauneau A, Barker DG, de Carvalho-Niebel F. AP2-ERF transcription factors mediate Nod factor-dependent Mt *ENOD11* activation in root hairs via a novel cis-regulatory motif. Plant Cell 2007;19(9) 2866-2885.

[98] Hirsch S, Kim J, Munoz A, Heckmann AB, Downie JA, Oldroyd GED. GRAS Proteins Form a DNA Binding Complex to Induce Gene Expression during Nodulation Signaling in *Medicago truncatula.* Plant Cell 2009;21(2) 545-557.

[99] Cerri MR, Frances L, Laloum T, Auriac MC, Niebel A, Oldroyd GED, Barker DG, Fournier J, de Carvalho-Niebel F. *Medicago truncatula* ERN Transcription Factors: Regulatory Interplay with NSP1/NSP2 GRAS Factors and Expression Dynamics throughout Rhizobial Infection. Plant Physiology 2012;160(4) 2155-2172.

[100] Murakami Y, Yokoyama H, Fukui R, Kawaguchi M. Down-Regulation of NSP2 Expression in Developmentally Young Regions of *Lotus japonicus* Roots in Response to Rhizobial Inoculation. Plant and Cell Physiology 2013;54(4) 518-527.

[101] Marsh JF, Rakocevic A, Mitra RM, Brocard L, Sun J, Eschstruth A, Long SR, Schultze M, Ratet P, Oldroyd GED. *Medicago truncatula NIN* is essential for rhizobial-independent nodule organogenesis induced by autoactive calcium/calmodulin-dependent protein kinase. Plant Physiology 2007;144(1) 324-335.

[102] Suzuki A, Hara H, Kinoue T, Abe M, Uchiumi T, Kucho K, Higashi S, Hirsch AM, Arima S. Split-root study of autoregulation of nodulation in the model legume *Lotus japonicus*. Journal of Plant Research 2008;121(2) 245-249.

[103] Saur IML, Oakes M, Djordjevic MA, Imin, N. Crosstalk between the nodulation signaling pathway and the autoregulation of nodulation in *Medicago truncatula*. New Phytologist 2011;190(4) 865-874.

[104] Hwang I, Sheen J, Muller B. Cytokinin Signaling Networks. Annual Review of Plant Biology 2012;63 353-380.

[105] Op den Camp RH, De Mita S, Lillo A, Cao Q, Limpens E, Bisseling T, Geurts R. A Phylogenetic Strategy Based on a Legume-Specific Whole Genome Duplication Yields Symbiotic Cytokinin Type-A Response Regulators. Plant Physiology 2011;157(4) 2013-2022.

[106] Leibfried A, To JP, Busch W, Stehling S, Kehle A, Demar M, Kieber JJ, Lohmann JU. WUSCHEL controls meristem function by direct regulation of cytokinin-inducible response regulators. Nature 2005;438(7071) 1172-1175.

[107] Yadav RK, Perales M, Gruel J, Girke T, Jönsson H, Reddy GV. WUSCHEL protein movement mediates stem cell homeostasis in the *Arabidopsis* shoot apex. Genes & Development 2011;25(19) 2025-2030.

[108] Osipova MA, Mortier V, Demchenko KN, Tsyganov VE, Tikhonovich IA, Lutova LA, Dolgikh EA, Goormachtig S. *WUSCHEL-RELATED HOMEOBOX5* Gene Expression and Interaction of CLE Peptides with Components of the Systemic Control Add Two Pieces to the Puzzle of Autoregulation of Nodulation. Plant Physiology 2012;158(3) 1329-1341.

[109] Ulmasov T, Murfett J, Hagen G, Guilfoyle TJ. Aux/IAA proteins repress expression of reporter genes containing natural and highly active synthetic auxin response elements. Plant Cell 1997;11(9) 1963-1971.

[110] Sabatini S, Beis D, Wolkenfelt H, Murfett J, Guilfoyle T, Malamy J, Benfey P, Leyser O, Bechtold N, Weisbeek P, Scheres B. An auxin-dependent distal organizer of pattern and polarity in the *Arabidopsis* root. Cell 1999;99(5): 463-472.

[111] Suzaki T, Yano K, Ito M, Umehara Y, Suganuma N, Kawaguchi M. Positive and negative regulation of cortical cell division during root nodule development in *Lotus japonicus* is accompanied by auxin response. Development 2012;139(21) 3997-4006.

[112] Hirsch AM, Bhuvaneswari TV, Torrey JG, Bisseling T. Early Nodulin Genes are Induced in Alfalfa Root Outgrowths Elicited by Auxin Transport Inhibitors. Proceedings of the National Academy of Sciences of the United States of America 1989;86(4) 1244-1248.

[113] Rightmyer AP, Long SR. Pseudonodule Formation by Wild-Type and Symbiotic Mutant *Medicago truncatula* in Response to Auxin Transport Inhibitors. Molecular Plant-Microbe Interactions 2011;24(11) 1372-1384.

2

Effects of Phytohormones on Nodulation and Nitrogen Fixation in Leguminous Plants

Maki Nagata and Akihiro Suzuki

1. Introduction

Leguminous plants form root nodules, in which symbiotic rhizobia fix atmospheric nitrogen. The nodulation process in the legume-rhizobium symbiosis consists of a series of events initiated by an exchange of specific signaling compounds between the two partners. The roots of leguminous plants secrete flavonoids, which trigger the synthesis of lipochitin-oligosaccharide signaling molecules, Nod factors (NFs), by rhizobia. NFs activate nodule organogenesis in the roots by stimulating the division of cortical cells. Many nodule development stages resemble other plant organ development such as cell divisions and differentiations. Because phytohormones are signal molecules involved in most plant physiological activities, they are likely to positively or negatively regulate nodulation and nitrogen fixation in the legume-rhizobium symbiosis.

In this chapter, we review the roles of several key phytohormones, namely auxins, cytokinins, gibberellins, ethylene, brassinosteroids, abscisic acid, salicylic acid, jasmonic acid and strigolactones in nodulation and nitrogen fixation in leguminous plants.

2. Auxins

Auxins were the first class of plant hormones discovered, and play a central role on the regulation of germination, plant growth, flower bud formation and flowering, and other developmental processes. In addition, auxins are involved in responses to environmental stimuli such as temperature, light and gravity. The most important member of the auxin family is indole-3-acetic acid (IAA), a native auxin in plants. The highest auxin levels are found in the cells undergoing cell division, elongation, differentiation and vascular bundle formation.

Therefore, auxin may play a significant role in nodulation. The possible involvement of auxins in nodule formation was first reported by Thimann in 1936. Thimann reported that the nodules of *Pisum sativum* contained auxin and that the auxin content increased during root nodule development [1]. Since then, this indirect evidence for an involvement of auxin in nodulation has been supported by various experiments. Hirsch et al. (1989) showed that auxin transport inhibitors, such as N-(1-naphthyl) phthalamic acid (NPA) and 2,3,5-triiodobenzoic acid (TIBA), caused the formation of nodule-like structures in *Medicago sativa* [2]. In *Trifolium repens*, transgenic plants carrying an auxin-responsive promoter (GH3) fused to the GUS reporter gene have been used to visualize the presence of auxins inside roots during nodule organogenesis [3]. The authors found that rhizobia cause a localized, temporary and early inhibition of auxin transport, which subsequently leads to an accumulation of auxins at the site of nodule initiation. A similar expression pattern was also observed in *Medicago truncatula* by using a DR5::GUS auxin-responsive promoter. In addition, *MtPIN*-silencing plants had significantly fewer nodules than control plants [4]. van Noorden et al. (2006) showed that the supernodulating *sunn* mutant had increased auxin transport and auxin content and that long-distance auxin transport regulation by rhizobia was defective [5]. de Billy et al. (2001) showed that genes in *M. truncatula* related to the auxin import carrier gene *AtAUX1*,which they named *MtLAX*, are predominantly expressed in regions of the root tips and nodule primordia where the vasculature arises (i.e., in the center of the lateral roots and at the peripheral region of the nodules) [6]. These results suggest that auxins are required during the development and differentiation of nodule primordia and of the vasculature within the nodules [6].

These reports also suggest that the auxin transport system is an important control on the number of indeterminate-type nodules. The effects of treatment with auxin transport inhibitors (NPA and TIBA) and with an auxin antagonist, α-(phenylethyl-2-one)-indole-3-acetic acid (PEO-IAA), on determinate-type nodulation were investigated by using *Lotus japonicus*. Both the nodule number and nodule development decreased and the formation of lenticels, which normally develop on the root surface and originate from the root outer cortex, was also inhibited by the treatment [7, 8]. The GH3:GUS transformant of *L. japonicus* showed auxin responses during nodule development. In rhizobia-inoculated roots, *GH3*-driven expression started to increase in the outer cortical cells where cell divisions occurred in the nodule primordia. *GH3*-driven expression was connected to the main root vascular tissues. These results suggest that auxins play an important role in the development of nodule vasculature, regardless of the nodule type [7-9]. In indeterminate-type nodule-forming plants, auxins accumulate at the site of rhizobia inoculation. This is caused by the inhibition of polar auxin transport by an accumulation of flavonoids, which are known as to be auxin transport regulators, around the infection site. In contrast, flavonoid-regulated auxin transport inhibition is not crucial during root nodule formation in *Glycine max*, which produces determinate-type nodules [10]. In the determinate-type nodules of *L. japonicus*, no inhibition of auxin transport was observed [9]. These differences in auxin distribution and transport inhibition between plants with determinate and indeterminate nodules have been attributed to a difference in the developmental pattern of the two nodule types [3, 11].

The role of auxins in nodulation is linked to the development of other root structures, such as lateral roots. Both lateral roots and root nodules development are known to be regulated by the auxin-to-cytokinin ratio. An increase in the auxin concentration stimulated lateral root formation, whereas an increase of in the cytokinin concentration or an inhibition of auxin transport induced the development of pseudo-nodules. Previously, it has been assumed that nodule initiation in plants that form indeterminate-type nodules is stimulated by a low auxin-to-cytokinin ratio [2]. Recently, Suzaki et al. (2012) investigated auxin distribution during root nodule development by using a DR5::GFP transformant of *L. japonicus* [12]. The accumulation of auxin in the dividing cortical cells was positively regulated by the nodule inception (NIN), transcription factor in nodule development and was inhibited by a negative systemic regulatory mechanism called autoregulation of nodulation. Moreover, auxin accumulation was observed in uninoculated roots of the *L. japonicus* mutant *spontaneous nodule formation* 2 (*snf*2), which has a gain-of-function mutation in *LHK1* encoding the putative cytokinin receptor LOTUS HISTIDINE KINASE 1. Therefore, it appears that auxins are involved in the division of cortical cells and acts downstream of cytokinin signaling [12, 13] (Figure 1).

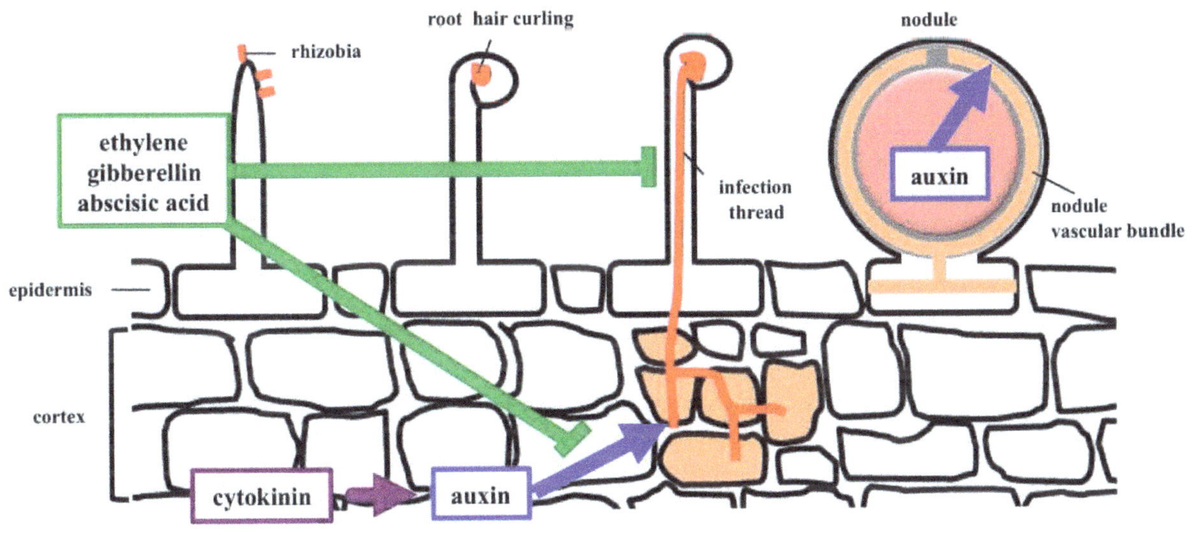

Figure 1. Effects of phytohormones on nodulation process in leguminous plants. Effects of phytohormones on nodulation process in leguminous plants. Auxin and cytokinin are required for cortical cell division and auxin acts downstream of cytokinin signaling. Ethylene, gibberellin, and abscisic acid inhibit the cortical cell divisions and rhizobial infection.

3. Cytokinins

Cytokinins have been found in all living cells of intact higher plants. Cytokinins play roles ranging from minor to major throughout development, from germination to leaf and plant senescence and modulate physiological processes important throughout the life of the plant, including photosynthesis and respiration [14, 15]. Cytokinins are also involved in as the control of root architecture development, including root nodulation.

In *Medicago polymorpha,* the synthetic cytokinin 6-benzyl-amino-purine (BAP) significantly increased the number of nodules [16]. In addition, BAP had a dose-dependent effect on nodulation in *P. sativum*. High levels of BAP results in few, flat, pale nodules and abnormal infection threads in wild-type *P. sativum* [17]. Expression of the alfalfa nodule-sepecific marker gene *MsENOD2* was induced by the exogenous application of a cytokinin in *M. sativa* [18]. The exogenous application of BAP on *M. sativa* induced cortical cell divisions and the expression of the early nodulin *ENOD40* gene [19]. In *G. max,* the cytokinin *trans*-zeatin stimulated nodulation at concentrations lower than 2.5×10^{-8} M, whereas at higher concentrations (5.0×10^{-8} M) it inhibited the formation of nodules [20]. Lohar et al. (2004) developed cytokinin-resistant transgenic *L. japonicus* hairy roots by using cytokinin oxidase (*CKX*) genes from *Arabidopsis thaliana* (*AtCKX3*) and maize (*ZmCKX1*), and showed that these roots had significantly reduced numbers of root nodules compared to control roots [21]. These reports suggest that cytokinins play a significant role in nodule organogenesis.

Recent research has confirmed that cytokinin receptors also play a significant role in nodule organogenesis [22-24]. The gain-of-function *snf2* mutant of the *L. japonicus* cytokinin receptor gene *(LHK1)* can form spontaneous nodules without rhizobial infection. This result indicates that cytokinin signaling is indispensable for cell division and for initiating nodule development [24]. In addition, *L. japonicus* plants homozygous for a mutation in the *hyperinfected 1* (*hit1*) locus exhibit abundant infection thread formation but fail to initiate cortical cell division in response to rhizobial signaling [23]. In addition, RNA interference of the cytokinin receptor homolog *cytokinin response 1* (*MtCRE1*) led to the development of cytokinin-insensitive roots, which showed an increased number of lateral roots and a strong reduction in nodulation [22]. Based on these reports, it has become clear that cytokinins act downstream of early NF signaling to mediate nodule formation. Cytokinins may therefore be the most important differentiation signal for cortical cell division and differentiation and for nodule organogenesis (Figure 1).

Exogenous application of cytokinins enhanced nitrogenase activity in nodules at all stages [25]. Rao et al. (1984) reported that the stimulatory effect of cytokinins increased the efficiency of nitrogen fixation. The application of cytokinins stimulated nitrate-induced nitrate reductase activity in the dark. Stimulation of nitrate reductase by cytokinins was inhibited significantly 6-methylpurine and cycloheximide, suggesting a requirement of RNA and protein synthesis [26].

4. Gibberellins

Gibberellins (GAs) are another important family of growth regulators in higher plants. GAs are also involved in root nodule symbiosis. Mutants deficient in GA biosynthesis or signaling develop dwarf phenotypes [27]. The GA-deficient mutants *of P. sativum* also developed significantly fewer nodules than wild-type plants. The application of an exogenous GA restored the nodule number in these mutants, although the addition of higher concentrations of GAs no longer restored nodule formation in these mutants. These results suggest that reduced levels of root GAs significantly decrease the number of nodules, and that nodule formation is considered to be strictly controlled by the GA concentration [28]. On the other hand, GAs are downstream signals of NFs for root hair curling process and for formation of infection pockets and infection threads at lateral root bases, and are essential for nodule primordium formation and differentiation in *Sesbania rostrata* [29]. The application of GA_3 at concentrations from 10^{-7} to 10^{-4} M results in the formation of nodule-like structures in the roots, and this response is sensitive to nitrogen levels [30]. Maekawa et al. (2009) also reported details of the effects of GAs on root nodulation in *L. japonicus*. Exogenous application of GA_3 (at more than 10^{-8} M) inhibited the number of infection threads and nodules. In contrast, the formation of both infection threads and nodules were stimulated by the application of Uniconazole-P, an inhibitor of GA_3 biosynthesis. Moreover, the degree of NF-induced root hair deformation was attenuated by the application of GA_3. The GA_3-treated *snf1* (a gain-of-function mutation of calcium/calmodulin-dependent kinase, CCaMK) and *snf2* (a gain-of-function mutation of a cytokinin receptor) showed a significant reduction in the number of spontaneous nodules. The cytokinin-dependent induction of *NIN* was suppressed by GA_3 treatment. These results suggest that GAs are involved in the cytokinin signaling pathway for nodulation in *L. japonicus* [31] (Figure 1).

The F-box containing protein, SLEEPY 1 (SLY1), functions as a positive regulator in GA signaling. In the presence of GAs, SLY1 interacts with negative regulators of GA signaling, leading to the degradation of these negative regulators [32, 33]. The over-expression of *L. japonicus* SLY1 carrying a gain-of-function mutation resulted in a reduced number of nodules, though the number of lateral roots and the degree of root growth were not significantly affected [31]. The constitutive GA signaling mutants of *P. sativum*, *la cry-s* mutants, also form significantly fewer nodules than wild-type plants. However, GA deficiency that results from the *na* mutation in *P. sativum* causes a reduction in nodulation. These results suggest that there is an optimal degree of GA signaling required for nodule formation and that the GA signal, and not the concentration of bioactive GA is important for nodulation [34].

The application of a GA decreased nodulation and nitrogen fixation under optimal growth conditions [35]. However, in *G. max* under low soil temperature conditions, nodulation and nitrogen fixation were decreased by GA_3 treatment during early plant development, but were increased during later development. The application of GA_3 increased nodulation and nitrogen accumulation after the early pod filling stage. These results suggest that GAs applied to soybean seeds at the time of planting did not influence final grain and protein yield [36].

5. Ethylene

Ethylene is a gaseous phytohormone involved in fruit ripening, leaf and fruit abscission, germination, seedling morphogenesis, root emergence, root hair elongation, promotion of flowering, senescence and stress response. Several studies have shown that ethylene production can have a negative effect on nodule formation. For example, ethylene production significantly increased in roots infected by rhizobia, and added exogenous ethylene can decrease the number of nodules [37, 38]. In *P. sativum*, exogenous ethylene did not decrease the number of infections per lateral root, but nearly all of the infections were blocked when the infection thread was in the basal epidermal cell or in the outer cortical cells [38]. In addition, ethylene inhibited all of the early plant responses that were tested, including the initiation of calcium spiking in *M. truncatula*. This finding suggests that ethylene acts upstream or at the point of calcium spiking in the NF signal transduction pathway [39]. Nodule formation can be stimulated by treatment of *M. truncatula* roots with aminoethoxyvinyl glycine (AVG) or Ag^+, which are inhibitors of ethylene synthesis and perception, respectively [40, 41]. The hypernodulation phenotype of the *sickle* mutant of *M. truncatula* has been attributed to a mutation causing ethylene insensitivity [42]. This mutant is defective in the ortholog of Arabidopsis *ETHYLENE-INSENSITIVE 2* (*EIN2*), involved in ethylene signaling [43]. In *L. japonicus*, root nodule formation was suppressed by 1-aminocyclopropane-1-carboxylic acid (ACC), a precursor of ethylene, but was enhanced by AVG and silver thiosulfate [44]. These results suggest that an ethylene-mediated signaling pathway is involved in the nodulation process, regardless of the nodule type. Nukui et al. (2004) produced transgenic *L. japonicus* carried the mutated melon ethylene receptor gene *Cm-ERS1/H70A*, which confers ethylene insensitivity [45]. When inoculated with *Mesorhizobium loti*, transgenic plants showed markedly higher numbers of infection threads and nodule primordia in their roots than did control plants. This result is consistent with the result reported for the *sickle* mutant of *M. truncatula* [42]. In addition, transcripts of *NIN* increased in the inoculated transgenic plants as compared with levels in the wild-type plants. In leguminous plants, the early stage of nodule development, including infection thread formation and the emergence of nodule primordia, are likely to be negatively regulated by ethylene signaling (Figure 1). In *snf* mutants, ethylene inhibits spontaneous nodulation. Therefore, ethylene plays a role in nodule formation downstream of the cytokinin signaling pathway [46]. However, ethylene may not play a significant role in nodule formation in all species. In soybean plants that form determinate-type nodules, the application of exogenous ethylene did not inhibit nodulation, and treatment with AVG or Ag^+ did not increase nodule number [38, 47-49]. In addition, ethylene leads to the formation of infection pockets and the initiation of nodule primordia in *S. rostrata* [50].

ACC deaminase catalyzes the degradation of ACC into ammonium and α-ketobutyrate. The ACC deaminase gene (*acdS*) has been found in many rhizosphere bacteria [51, 52]. Through the action of this enzyme, ACC deaminase-containing bacteria can reduce ethylene biosynthesis in plants. In *M. loti*, *acdS* was found in the symbiosis island, and the enhancing effect of this gene on enhancing the nodulation of *L. japonicus* was demonstrated by using an *M. loti acdS* disruption mutant [53]. ACC diaminase in *Rhizobium leguminosarium* bv. *viciae* has been confirmed to enhance nodulation of *P. sativm* [54]. Reports concerning rhizobial strategies to reduce the amount of ethylene synthesized in the host leguminous plant suggest the importance of ethylene-mediated interactions in the establishment of symbiosis between the partners, by decreasing the negative effect of ethylene on nodulation [55].

6. Brassinosteroids

Brassinosteroids (BRs) are a group of plant steroid hormones that regulate a wide range of physiological responses, including cell elongation, photomorphogenesis, xylem differentiation, and seed germination. BRs are present in the plants in extremely low concentrations, but they are highly mobile within the plant. BRs supplied via the root system remarkably promoted the elongation of cotyledon petioles and the hypocotyls of young radish and tomato plants [56]. This study clearly demonstrated the mobility of BRs in the plant system.

In addition, application of BRs affected nodulation and nitrogen fixation in groundnut (*Arachis hypogaea*), pea and soybean [57-59]. In groundnuts, BRs enhanced the growth and yield of the plants, and the growth promotion was associated with enhanced levels of nucleic acids, soluble proteins and carbohydrates [60]. The effect of BRs on nodulation and nitrogen fixation was also investigated. Exogenous application of BR increased in nodulation. Foliar application of BRs also increased the nitrogenase activity [57]. The application of 24-epibrassinosteroid also increased the nodule number, nodule fresh and dry mass, and nitrogenase activity in relation to those of the control in pea [59]. In the hypernodulationg En6500 mutant of the 'Enrei' soybean cultivar, the application of BR to the leaves not only induced stem elongation but also repressed root nodule formation, depending on the dose. However, this effect was not observed in the wild-type. On the other hand, foliar treatment with brassinazole, an inhibitor of BR biosynthesis, increased the nodule number and significantly reduced stem elongation in wild-type 'Enrei'. These results suggest that BRs in the shoots may contribute mainly to the regulation of nodule formation and that brassinazole transferred to the shoot from the culture medium subsequently reduced the level of endogenous BRs in the leaves [58]. From these results, BRs are clearly involved in nodulation and nitrogen fixation. The BR synthesis mutants *lk* and *lkb* exhibit a severe reduction in level of bioactive BRs in the shoot [61, 62]. A reduction in BR levels in the roots has also been confirmed for *lkb* [63]. These mutants *lk* and *lkb* and the BR response mutant *lka* also have fewer nodules than wild-type plants. These mutants also have fewer and shorter lateral roots. However, the average nodule dry weight increased significantly. Thus, although the root system dry weight decreased, the average nodules dry weight increased. This finding illustrated that nodule size is not simply a reflection of root system dry weight. Ferguson et al. (2005) also suggested that BRs affect the nodulation mechanism of the shoot that is involved in regulating the nodule numbers of the root [28]. They found that uniconazole-P, which is a GA_3 biosynthesis inhibitor, partially inhibits BR biosynthesis. In *L. japnicus*, the application of BR combined with Uniconazole-P significantly decreased the number of infection threads, compared with a treatment with uniconazole-P alone, but the decrease was less significant than the decrease in GA levels. However, no significant effects on the nodule number, and on shoot and root lengths, were observed. These results suggest that uniconazole-P inhibits BR biosynthesis in the root hairs, and thus lowers the number of infection threads that develop [31] (Figure 1).

7. Abcisic acid

Abcisic acid (ABA) plays crucial roles in plant growth, development and responses to environment stresses such as cold, drought and high salinity. ABA has been reported to play negative roles at different stages of nodule development. The application of ABA inhibited nodulation in *P. sativum* [64], *G. max* [65, 66], *L. japonicus* [67], *T. repens* [67] and *M. truncatula* [68]. ABA application to wild-type *G. max* and to a hypernodulation mutant, NOD1-3, reduced both nodule numbers and isoflavonoid accumulation. It has been shown that isoflavonoids in *G. max* are responsible for the activation of nodulation, thus ABA could have an indirect role in nodule organogenesis through its effects on isoflavonoid synthesis. The effect of ABA on *T. repens* and *L. japonicus*, which form indeterminate and determinate type nodules respectively, was examined. Both leguminous plants showed a decrease in nodule numbers in response to ABA application. Similarly, after the application of abamine, an ABA biosynthesis inhibitor, nodule number increased in *L. japonicus*. The application of ABA on *T. repens* blocked root hair deformation at the stage between root hair swelling and curling [67]. In addition, ABA treatment inhibited infection thread formation in *L. japonicus* and *M. truncatula* [68, 69]. Moreover, calcium spiking after NF perception was inhibited by ABA treatment in *M. truncatula* [68]. Phillips (1971) found that exogenous ABA inhibited root nodule formation by inhibiting the cytokinin-induced cortical cell divisions required for nodule initiation [64]. These results suggest that ABA controls root nodulation by regulating root hair deformation, infection thread formation, and cytokinin-induced cortical cell division in leguminous plants (Figure 1).

The nitrogenase activity of nodules treated with ABA was lower than in untreated wild-type in *Phaseolus vulgaris* [70] and *P. sativum* [71]. In *P. sativum*, ABA application stimulated an abrupt stress situation of severe drought which led to leghemoglobin reduction. Thus, an effect of ABA on nodule oxygen diffusion might also be involved in the decline of nitrogen fixation. The *enhanced nitrogen fixation1* (*enf1*) mutant of *L. japonicus* was isolated by screening *L. japonicus* seedlings for survival on an agar medium containing 70 μM ABA. The *enf1* mutants showed both increased root nodule numbers and enhanced nitrogen fixation activity. The low ABA sensitivity of the *enf1* mutants was caused by lower endogenous ABA concentration. Moreover, nitrogen fixation activity in the *enf1* mutants increased as a result of decreased nitric oxide production in the nodules [72].

The role of ABA in autoregulation of nodulation was investigated in the *G. max* hypernodulation mutant *nts382*. The basal levels of ABA in the roots of wild-type *G. max* cv. 'Bragg' were higher than those in *nts382*, regardless of *Bradyrhizobium* inoculation. The ABA concentration in the shoot increased at the onset of autoregulation in 'Bragg' but not in *nts382*. The ABA-to-cytokinin ratio in the roots was also consistently higher in 'Bragg' than in *nts382*. This phytohormone ratio had been suggested to be involved in root-to-shoot signaling and photosynthetic gas exchange in *M. sativa* [73]. A model was proposed to explain the possible influence of the ABA-to-cytokinin ratio in autoregulation of nodulation. However, Biswas et al. (2009) proposed that ABA was not directly involved in the systemic autoregulation of nodulation, because an ABA insensitive mutant of *L. japonicus* cv. 'Beyma' did not exhibit altered autoregulation of nodulation, and the application of ABA on one side of the roots inhibited nodulation locally but not systemically in a split-root experiment [74].

8. Salicylic acid

Salicylic acid (SA) is involved in plant responses to pathogen, and its mode of action has been well characterized. SA is an inducer of systemic acquired resistance in the defense responses to pathogen attacks. In terms of the interaction between plants and rhizobia, several reports have shown that SA strongly inhibits nodulation and nodule development, leading to decreased nitrogen fixation activity. When *M. sativa* was inoculated with compatible *Rhizobium meliloti*, SA levels in the roots either decreased or remained close to their basal levels. However, when *M. sativa* was inoculated with incompatible *Rhizobium leguminosarum* or the *nod* mutant of *R. meliloti*, that was defective in NF biosynthesis, SA accumulated in roots. These results suggest the involvement of NFs produced by compatible rhizobia in the inhibition of the SA-mediated defense in leguminous plants [75]. In another study, inoculation of *P. sativum sym30* mutant (nod$^-$) with compatible *R. leguminosarum* increased SA levels in the roots. Similarly, SA accumulation in the roots was found in *P. sativum* inoculated with a *NodC*$^-$ mutant. However, SA levels in roots either remained at the basal level or decreased when *P. sativum* plants were inoculated with compatible *R. leguminosarum*. These results suggest that the *sym30* gene could be involved in a common pathway that leads to the suppression of an SA-dependent defense mechanism in leguminous plants against compatible rhizobia, thus allowing establishment of the symbiosis [76]. van Spronsen et al. (2003) found that SA application completely inhibited the formation of indeterminate-type nodules in *Vicia sativa* subsp. *nigra* and *P. sativum* [77]. However, SA application did not inhibit the formation of determinate-type nodules in *L. japonicus, Glycine soja, G. max* and *P. vulgaris*. On the other hand, SA application at higher concentrations decreased the number of determinate-type nodules and the dry mass of *G. max* seedlings, leading to a low photosynthetic rate and decreased nitrogen fixation [78]. The inhibitory effect of SA on the nodulation of hypernodulating soybean mutants NOD1-3 and NOD2-4 was significantly less pronounced than that in wild-type soybean. These results indicate that SA is directly involved in signal transmission in the autoregulation [79]. In addition, when endogenous SA levels were modulated through the transgenic expression of salicylate hydroxylase (*NahG*) in both *L. japonicus* and *M. truncatula*, a marked reduction in SA levels was correlated with an increase in the number of infections and the number of nodules [80]. These results suggest that endogenous SA levels affect nodulation in both determinate and indeterminate-type nodule-forming species.

9. Jasmonic acid

Jasmonic acid (JA) is also involved in plant defenses against pathogens and in wound responses. JA has been reported to be a negative regulator of nodulation. In *L. japonicus*, shoot-applied methyl jasmonate (MeJA) strongly suppressed nodulation, including infection thread formation and *NIN* gene expression in wild-type plants and even in the *har1* hypernodulation mutant [81]. In *M. truncatula*, nodulation was strongly inhibited in a growth medium containing JA. A high JA concentration significantly decreased the plant's responsiveness to NFs, resulting in a lower number of root hairs that exhibited calcium spiking. In addition, JA inhibited the expression of the early rhizobium-responsive genes, *RIP1* and *ENOD11* [82]. In contrast, the JA concentration in leaves of a *G. max* hypernodulating *nts* mutant was higher than in those of wild-type *G. max* under natural growth conditions. In addition, transcription levels of JA responsive genes increased in the hypernodulating *nts* mutant, which suggests that the *nts* mutation induces changes in certain pathways, including JA synthetic metabolism, resulting in the activation of JA-responsive genes [83]. Kinkema and Gresshoff (2008) also showed that the expression of JA biosynthetic and responsive genes in the leaves of wild-type *G. max* is normally suppressed by inoculation with rhizobia, but not suppression was seen in a hypernodulating *nts* mutant. Furthermore, foliar application of n-propyl gallate, a JA biosynthesis inhibitor, significantly decreased nodulation specifically in the hypernodulating *nts* mutant [84]. Recently, Suzuki et al. (2011) reported that nodulation was enhanced by treatment with a low concentration of JA. Both infection thread formation and nodulation were increased by JA treatment of wild-type *L. japonicus* grown under low red/far-red (R/FR) light condition. These results indicate that nodulation is photomorphogenetically controlled by sensing the R/FR ratio through JA signaling [85]. Therefore, JA functions as a positive regulator of nodulation over a certain range of concentrations in these plant species. Previous studies have shown that JA can act as a signal molecule in the early stages of the development of leguminous plants-rhizobia symbioses; for example, it induces the expression of the *nod* genes in *B. japonicum* [86] and *Rhizobium* [87]. However, it is unclear whether the plant responses, rhizobial responses or combination of both responses are responsible for the positive effects of JA on nodulation. Suzuki et al. (2011) also found that nodulation was suppressed in the roots of a *phytochrome B* (*phyB*) mutant of *L. japonicus* that had not only decreased levels of photo-assimilates but also a reduced concentration of JA-Ile (the active JA derivative). In fact, the number of root nodules in the *phyB* mutant was restored by JA treatment, providing further evidence that JA can act as a positive regulator of nodulation in leguminous plants [85].

10. Strigolactones

Strigolactones (SLs) have been identified as phytohormones that are involved in the regulation of shoot branching in plants and thus have been suggested to be ubiquitous in the plant kingdom [88]. SLs are released by roots into the rhizosphere, and appear to stimulate germination of the seeds of parasitic plants such as *Striga* spp. and *Orobanche* spp. [89]. Recently, they have been shown to play key roles as signaling compounds in the interaction between plants and arbuscular mycorrhizal fungi [90]. Interestingly, SLs also increase nodulation in *M. sativa, P. sativum* and *L. japonicus* [91-93]. In *M. sativa*, treatment with the synthetic SL analogue GR24 clearly increased nodulation. When *M. sativa* plants were treated with GR24, the biosynthesis and the metabolism of the SLs increased, thus, resulting in an enhanced formation of indeterminate-type nodules [91]. The SL deficient *rms1* mutants of *P. sativum* also produce fewer nodules than wild-type plants. Treatment with GR24 elevated nodule number in wild-type *P. sativum* and also elevated nodule number in *rms1* mutant to a level similar to that seen in untreated wild-type plants. These results indicated that endogenous SLs increase nodulation in *P. sativum* [92]. The role of SLs in *L. japonicus* was studied by using transgenic lines in which *CAROTENOID CLEAVAGE DIOXYGENASE 7* (*LjCCD7*), an orthologue of *Arabidopsis More Axillary Growth 3*, was silenced. Silencing of *LjCCD7* is expected to reduce SL levels. The plants with silenced LjCCD7 produced fewer nodules than control plants; this suggests that SLs have a slight positive effect on the formation of determinate nodules [93].

Author details

Maki Nagata[1] and Akihiro Suzuki[1,2*]

*Address all correspondence to: azuki@cc.saga-u.ac.jp

1 Department of Agricultural Sciences, Faculty of Agriculture, Saga University, Honjyo-machi, Saga, Saga, Japan

2 United Graduate School of Agricultural Sciences, Kagoshima University, Korimoto, Kagoshima, Kagoshima, Japan

References

[1] Thimann KV (1936) On the physiology of the formation of nodules on legume roots. Proc Natl Acad Sci USA 22: 511-514

[2] Hirsch AM, Bhuvaneswari TV, Torrey JG and Bisseling T (1989) Early nodulin genes are induced in alfalfa root outgrowths elicited by auxin transport inhibitors. Proc Natl Acad Sci USA 86: 1244-1248

[3] Mathesius U, Schlaman HRM, Spaink HP, Sautter C, Rolfe BG and Djordjevic MA (1998) Auxin transport inhibition precedes root nodule formation in white clover roots and is regulated by flavonoids and derivatives of chitin oligosaccharides. Plant J 14: 23-34

[4] Huo X, Schnabel E, Hughes K and Frugoli J (2006) RNAi phenotypes and the localization of a Protein::GUS Fusion imply a role for Medicago truncatula PIN genes in nodulation. J Plant Growth Regul 25: 156-165

[5] van Noorden GE, Ross JJ, Reid JB, Rolfe BG and Mathesius U (2006) Defective long-distance auxin transport regulation in the Medicago truncatula super numeric nodules mutant. Plant physiol 140: 1494-1506

[6] de Billy F, Grosjean C, May S, Bennett M and Cullimore JV (2001) Expression studies on AUX1-like genes in Medicago truncatula suggest that auxin is required at two steps in early nodule development. Mol Plant Microbe Interact 14:267-277

[7] Takanashi K, Sugiyama A, Yazaki K (2011) Involvement of auxin distribution in root nodule development of Lotus japonicus. Planta 234: 73-81

[8] Takanashi K, Sugiyama A, Yazaki K (2011) Auxin distribution and lenticel formation in determinate nodule of Lotus japonicus. Plant Signal Behav 6: 1405-1407

[9] Pacios-Bras C, Schlaman HRM, Boot K, Admiraal P, Langerak JM, Stougaard J, Spaink HP (2003) Auxin distribution in Lotus japonicus during root nodule development. Plant Mol Biol 52: 1169-1180

[10] Subramanian S, Stacey G and Yu O (2006) Endogenous isoflavones are essential for the establishment of symbiosis between soybean and Bradyrhizobium japonicum. Plant J 48: 261-273

[11] Mathesius U (2008) Auxin: at the root of nodule development? Funct Plant Biol 35: 651-668

[12] Suzaki T, Yano K, Ito M, Umehara Y, Suganuma N and Kawaguchi M (2012) Positive and negative regulation of cortical cell division during root nodule development in Lotus japonicus is accompanied by auxin response. Development 139: 3997-4006

[13] Suzaki T, Ito M, Kawaguchi M (2013) Genetic basis of cytokinin and auxin functions during root nodule development. Front Plant Sci 4: 1-6

[14] Arshad M, and Frankenberger WT (1993) Microbial production of plant growth regulators. In Soil microbial ecology. Applications in agricultural and environmental management. F.B. Metting, Jr. (edit.), Marcel Dekker, Inc., New York; 307-343

[15] Mok MC (1994) Cytokinins and plant development: An overview. In Cytokinins: Chemistry, Activity and Function, D.W.S Mok and M.C. Mok, eds (Boca Raton, FL: CRC Press); 155–156.

[16] Yahalom E, Okon Y, Dovrat A (1990) Possible mode of action of Azospirillum brasilense strain Cd on the root morphology and nodule formation in burr medic (Medicago polymorpha). Can J Microbiol 36: 10-14

[17] Lorteau MA, Ferguson BJ and Guinel FC (2001) Effects of cytokinin on ethylene production and nodulation in pea (Pisum sativum) cv. Sparkle. Physiol Plant 112: 421-428

[18] Cooper JB and Long SR (1994) Morphogenetic rescue of Rhizobium meliloti nodulation mutants by trans-Zeatin secretion. Plant Cell 6: 215-225

[19] Fang Y and Hirsch AM (1998) Studying early nodulin gene ENOD40 expression and induction by nodulation factor and cytokinin in transgenic alfalfa. Plant Physiol 116: 53-68

[20] Taller BJ, Sturtevant DB (1991) Cytokinin production by rhizobia. In Hennecke H, Verma DPS (eds) Advances in Molecular Genetics of Plant-Microbe Interactions 1: 215-221

[21] Lohar DP, Schaff JE, Laskey JG, Kieber JJ, Bilyeu KD and Bird DM (2004) Cytokinins play opposite roles in lateral root formation, and nematode and rhizobial symbioses. Plant J 38: 203-214

[22] Gonzalez-Rizzo S, Crespi M and Frugier F (2006) The Medicago truncatula CRE1 cytokinin receptor regulates lateral root development and early symbiotic interaction with Sinorhizobium meliloti. Plant Cell 18: 2680-2693

[23] Murray JD, Karas BJ, Sato S, Tabata S, Amyot L, Szczyglowski K (2007) A cytokinin perception mutant colonized by Rhizobium in the absence of nodule organogenesis. Science 315: 101-104

[24] Tirichine L, Sandal N, Madsen LH, Radutoiu S, Albrektsen AS, Sato S, Asamizu E, Tabata S, Stougaard J (2007) A gain-of-function mutation in a cytokinin receptor triggers spontaneous root nodule organogenesis. Science 315: 104-107

[25] Fatima Z, Bano A, Sial R and Aslam M (2008) Response of chickpea to plant growth regulators on nitrogen fixation and yield. Pak. J. Bot 40: 2005-2008

[26] Rao LVM, Datta N, Mahadevan M, Guha-Mukherjee S and Sopory SK (1984) Influence of cytokinins and phytochrome on nitrate reductase activity in etiolated leaves of maize. Phytochemistry 23: 1875-1879

[27] Hooley R (1994) Gibberellins: perception, transduction and responses. Plant Mol Biol 26: 1529-1555

[28] Ferguson BJ, Ross JJ and Reid JB (2005) Nodulation phenotypes of giberellin and brassinosteroid mutants of pea. Plant physiol 138: 2396-2405

[29] Lievens S, Goormachtig S, Herder JD, Capoen W, Mathis R, Hedden P, Holsters M (2005) Gibberellins are involved in nodulation of Sesbania rostrata. Plant Physiol 139: 1366-1379

[30] Kawaguchi M, Imaizumi-Anraku H, Fukai S, Syōno K (1996) Unusual branching in the seedling of Lotus japonicus-gibberellins reveal the nitrogen-sensitive cell divisions within the pericycle on roots. Plant Cell Physiol 37: 461-470

[31] Maekawa T, Maekawa-Yoshikawa M, Takeda N, Imaizumi-Anraku H, Murooka Y and Hayashi M (2009) Gibberellin controls the nodulation signaling pathway in Lotus japonicus. Plant J 58: 183-194

[32] McGinnis KM, Thomas SG, Soule JD, Strader LC, Zale JM, Sun T and Steber CM (2003) The Arabidopsis SLEEPY1 gene encodes a putative F-box subunit of an SCF E3 ubiquitin ligase. Plant Cell 15: 1120-1130

[33] Dill A, Thomas SG, Hu J, Steber CM, Sun T (2004) The Arabidopsis F-box protein SLEEPY1 targets gibberellin signaling repressors for giberellin-induced degradation. Plant Cell 16: 1392-1405

[34] Ferguson BJ, Foo E, Ross JJ and Reid JB (2011) Relationship between giberellin, ethylene and nodulation in Pisum sativum. New Phytologist 189: 829-842

[35] Williams PM and Sicardi de Mallorca M (1984) Effect of gibberellins and the growth retardant CCC on the nodulation of soya. Plant Soil 77: 53-60

[36] Zhang F, Pan B and Smith DL (1997) Application of gibberellic acid to the surface of soybean seed (Glycine max (L.) Merr.) and symbiotic nodulation, plant development, final grain and protein yield under short season conditions. Plant Soil 188: 329-335

[37] Ligero F, Lluch C, Olivares J (1987) Evolution of ethylene from roots and nodulation rate of alfalfa (Medicago sativa L.) plants inoculated with Rhizobium meliloti as affected by the presence of nitrate. J Plant Physiol 129: 461–467

[38] Lee KH and LaRue TA (1992) Exogenous ethylene inhibits nodulation of Pisum sativum L. cv Sparkle. Plant Physiol 100: 1759-1763

[39] Oldroyd GED, Engstrom EM and Long SR (2001) Ethylene inhibits the Nod factor signal transduction pathway of Medicago truncatula. Plant Cell 13: 1835-1849

[40] Peters NK and Crist-Estes DK (1989) Nodule formation is stimulated by the ethylene inhibitor aminoethoxyvinylglycine. Plant physiol 91: 690-693

[41] Caba JM, Recalde L and Ligero F (1998) Nitrate-induced ethylene biosynthesis and the control of nodulation in alfalfa. Plant Cell Environ 21: 87-93

[42] Penmetsa RV and Cook DR (1997) A legume ethylene-insensitive mutant hyperinfected by its rhizobial symbiont. Science 275: 527-530

[43] Penmetsa RV, Uribe P, Anderson J, Lichtenzveig J, Gish JC, Nam YW, Engstrom E, Xu K, Sckisel G, Pereira M, Baek JM, Lopez-Meyer M, Long SR, Harrison MJ, Singh KB, Kiss GB, Cook DR (2008) The Medicago truncatula ortholog of Arabidopsis EIN2, sickle, is a negative regulator of symbiotic and pathogenic microbial associations. Plant J 55: 580-595

[44] Nukui N, Ezura H, Yuhashi K, Yasuta T and Minamisawa K (2000) Effects of ethylene precursor and inhibitors for ethylene biosynthesis and perception on nodulation in Lotus japonicus and Macroptilium atropurpureum. Plant Cell Physiol 41: 893-897

[45] Nukui N, Ezura H, Minamisawa K (2004) Transgenic Lotus japonicus with an ethylene receptor gene Cm-ERS1/H70A enhances formation of infection threads and nodule primordia. Plant Cell Physiol 45:427–435

[46] Tirichine L, James EK, Sandal N and Stougaard J (2006) Spontaneous root-nodule formation in the model legume Lotus japonicus: A novel class of mutants nodulates in the absence of rhizobia. Mol Plant Microbe Interact 19: 373-382

[47] Hunter WJ (1993) Ethylene production by root nodules and effect of ethylene on nodulation in Glycine max. Appl Environ Microbiol 59: 1947-1950

[48] Suganuma N, Yamauchi H, Yamamoto K (1995) Enhanced production of ethylene by soybean roots after inoculation with Bradyrhizobium japonicum. Plant Sci 111: 163-168

[49] Schmidt JS, Harper JE, Hoffman TK and Bent AF (1999) Regulation of soybean nodulation independent of ethylene signaling. Plant physiol 119: 951-959

[50] D'Haeze W, Rycke RD, Mathis R, Goormachtig S, Pagnotta S, Verplancke C, Capoen W and Holsters M (2003) Reactive oxygen species and ethylene play a positive role in lateral root base nodulation of a semiaquatic legume. Proc Natl Acad Sci USA 100: 11789-11794

[51] Honma M and Shimomura T (1978) Metabolism of 1-aminocyclopropane-1-carboxylic acid. Agric Biol Chem 42: 1825-1831

[52] Blaha D, Prigent-Combaret C, Mirza MS and Moënne-Loccoz (2006) Phylogeny of the 1-aminocyclopropane-1-carboxylic acid deaminase-encoding gene acdS in phytobeneficial and pathogenic Proteobacteria and relation with strain biogeography. FEMS Microbiol Ecol 56: 455-470

[53] Uchiumi T, Ohwada T, Itakura M, Mitsui H, Nukui N, Dawadi P, Kaneko T, Tabata S, Yokoyama T, Tejima K, Saeki K, Omori H, Hayashi M, Maekawa T, Sriprang R, Murooka Y, Tajima S, Simomura K, Nomura M, Suzuki A, Shimoda Y, Sioya K, Abe M, Minamisawa K (2004) Expression islands clustered on the symbiosis island of the Mesorhizobium loti genome. J Bactriol 186: 2439-2448

[54] Ma W, Guinel FC and Glick BR (2003) Rhizobium leguminosarum biovar viciae 1-aminocyclopropane-1-carboxylate deaminase promotes nodulation of pea plants. Appl Environ Microbiol 69: 4396-4402

[55] Ma W, Penrose DM, Glick BR (2002) Strategies used by rhizobia to lower plant ethylene levels and increase nodulation. Can J Microbiol 48: 947-954

[56] Takatsuto S, Yazawa N, Ikekawa N, Takematsu T, Takeuchu Y and Koguchi M (1983) Structure activity relationship of brassinosteroids. Phytochemistry 22: 2437-2441

[57] Vardhini BV and Rao SSR (1999) Effect of brassinosteroids on nodulation and nitrogenase activity in groundnut (Arachis hypogaea L.). Plant Growth Regul 28: 165-167

[58] Terakado J, Fujihara S, Goto S, Kuratani R, Suzuki Y, Yoshida S and Yoneyama T (2005) Systemic effect of a brassinosteroid on root nodule formation in soybean as revealed by the application of brassinolide and brassinazole. Soil Sci Plant Nutr 51: 389-395

[59] Shahid MA, Pervez MA, Balal RM, Mattson NS, Rashid A, Ahmad R, Ayyub CM and Abbas T (2011) Brassinosteroid (24-epibrassinolide) enhances growth and alleviates

the deleterious effects induced by salt stress in pea (Pisum sativum L.). Aust J Crop Sci 5: 500-510

[60] Vardhini BV and Rao SSR (1998) Effect of brassinosteroids on growth, metabolite content and yield of Arachis hypogaea. Phytochem 48: 927-930

[61] Nomura T, Nakayama M, Reid JB, Takeuchi Y, Yokota T (1997) Blockage of brassinosteroid biosynthesis and sensitivity causes dwarfism in garden pea. Plant Physiol 113: 31–37

[62] Nomura T, Jager CE, Kitasaka Y, Takeuchi K, Fukami M, Yoneyama K, Matsushita Y, Nyunoya H, Takatsuto S, Fujioka S, Smith JJ, Kerckhoffs LHJ, Reid JB, Yokota T (2004) Brassinosteroid deficiency due to truncated steroid 5α-reductase causes dwarfism in the lk mutant of pea. Plant Physiol 135: 2220–2229

[63] Symons GM and Reid JB (2004) Brassinosteroids do not undergo long distance transport in pea. Implications for the regulation of endogenous brassinosteroid levels. Plant Physiol 135: 2196–2206

[64] Phillips, DA (1971) Abscisic acid inhibition of root nodule initiation in Pisum sativum. Planta 100:181-190

[65] Cho M-J, Harper JE (1993) Effect of abscisic acid application on root isoflavonoid concentration and nodulation of wild-type and nodulation- mutant soybean plants. Plant and Soil 152:145-149

[66] Bano A, Harper JE (2002) Plant growth regulators and phloem exudates modulate root nodulation of soybean. Funct Plant Biol 29:1299-1307

[67] Suzuki A, Akune M, Kogiso M, Imagama Y, Osuki K, Uchiumi T, Higashi S, Han SY, Yoshida S, Asami T, Abe M (2004) Control of nodule number by the phytohormone abscisic acid in the roots of two leguminous species. Plant Cell Physiol 45:914-922

[68] Ding Y, Kalo P, Yendrek C, Sun J, Liang Y, Marsh JF, Harris JM, Oldroyd GED(2008) Abscisic acid coordinates Nod factor and cytokinin signaling during the regulation of nodulation in Medicago truncatula. Plant Cell 20:2681-2695

[69] Nakatsukasa-Akune M, Yamashita K, Shimoda Y, Uchiumi T, Abe M, Aoki T, Kamizawa A, Ayabe S, Higashi S, Suzuki A (2005) Suppression of root nodule formation by artificial expression of the TrEnodDR1 (coat protein of white clover cryptic virus 1) gene in Lotus japonicus. Mol Plant Microbe Interact 18:1069-1080

[70] Khadri M, Tejera NA and Lluch C (2006) Alleviation of salt stress in common bean (Phaseolus vulgaris) by exogenous abscisic acid supply. J Plant Growth Regul 25: 110-119

[71] González EM, Galvez L, Arrese-Igor C (2001) Abscisic acid induces a decline in nitrogen fixation that involves leghemoglobin, but is independent of sucrose synthase activity. J Exp Bot 52: 285-293

[72] Tominaga A, Nagata M, Futsuki K, Abe H, Uchiumi T, Abe M, Kucho K, Hashiguchi M, Akashi R, Hirsch AM, Arima S, Suzuki A (2009) Enhanced nodulation and nitrogen fixation in the abscisic acid low-sensitive mutant enhanced nitrogen fixation1 of Lotus japonicus. Plant physiol 151: 1965-1976

[73] Goicoechea N, Antolín MC, Sánchez-Díaz M (1997) Gas exchange is related to hormone balance in mycorrhizal or nitrogen-fixing alfalfa subjected to drought. Physiol Plant 100: 989-997

[74] Biswas B, Chan PK, Gresshoff PM (2009) A novel ABA insensitive mutant of Lotus japonicus with a wilty phenotype displays unaltered nodulation regulation. Mol Plant 2: 487-499

[75] Martínez-Abarca F, Herrera-Cervera JA, Bueno P, Sanjuan J, Bisseling T, Olivares J (1998) Involvement of salicylic acid in the establishment of the Rhizobium meliloti-alfalfa symbiosis. Mol Plant Microbe Interact 11:153-155

[76] Blilou I, Ocampo JA, García-Garrido JM (1999) Resistance of pea roots to endomycorrhizal fungus or Rhizobium correlates with enhanced levels of endogenous salicylic acid. J Exp Bot 50: 1663-1668

[77] van Spronsen PC, Tak T, Rood AM, van Brussel AA, Kijne JW, Boot Kj (2003) Salicylic acid inhibits indeterminate-type nodulation but not determinate-type nodulation. Mol Plant Microbe Interact 16: 83-91

[78] Lian B, Zhou X, Miransari M, Smith DL (2000) Effects of salicylic acid on the development and root nodulation of soybean seedlings. J Agron Crop Sci 185: 187-192

[79] Sato T, Fujikake H, Ohtake N, Sueyoshi K, Takahashi T, Sato A and Ohyama T (2002) Effect of exogenous salicylic acid supply on nodule formation of hypernodulating mutant and wild type of soybean. Soil Sci Plant Nutr 48: 413-420

[80] Stacey G, McAlvin CB, Kim SY, Olivares J, Soto MJ (2006) Effects of endogenous salicylic acid on nodulation in the model legumes Lotus japonicus and Medicago truncatula. Plant physiol 141: 1473-1481

[81] Nakagawa T and Kawaguchi M (2006) Shoot applied MeJA suppresses root nodulation in Lotus japonicus. Plant Cell Physiol 47: 176-180

[82] Sun J, Cardoza V, Mitchell DM, Bright L, Oldroyd G, Harris JM (2006) Crosstalk between jasmonic acid, ethylene and Nod factor signaling allow integration of diverse inputs for regulation of nodulation. Plant J 46: 961-970

[83] Seo HS, Li J, Lee SY, Yu JW, Kim KH, Lee SH, Lee IJ, Paek NC (2006) The hypernodulating nts mutation induces jasmonate synthetic pathway in soybean leaves. Mol Cells 24: 185-193

[84] Kinkema M and Gresshoff PM (2008) Investigation of downstream signals of the soybean autoregulation of nodulation receptor kinase GmNARK. Mol Plant Microbe Interact 21:1337-1348

[85] Suzuki A, Suriyagoda L, Shigeyama T, Tominaga A, Sasaki M, Hiratsuka Y, Yoshinaga A, Arima S, Agarie S, Sakai T, Inada S, Jikumaru Y, kamiya Y, Uchiumi T, Abe M, Hashiguchi M, Akashi R, Sato S, Kaneko T, Tabata S, Hirsch AM (2011) Lotus japonicus nodulation is photomorphogenetically controlled by sensing the red/far red (R/FR) ratio through jasmonic acid (JA) signaling. Proc Natl Acad Sci USA 108: 16837-16842

[86] Mabood F, Smith DL (2005) Pre-inoculation of Bradyrhizobium japonicum with jasmonates accelerates nodulation and nitrogen fixation in soybean (Glycine max) at optimal and suboptimal root zone temperatures. Physiol Plant 125: 311-323

[87] Rosas S, Soria R, Correa N, Abdala G (1998) Jasmonic acid stimulates the expression of nod genes in Rhizobium. Plant Mol Biol 38: 1161-1168

[88] Umehara M, Hanada A, Yoshida S, Akiyama K, Arita T, Takeda-Kamiya N, Magome H, Kamiya Y, Shirasu K, Yoneyama K, Kyozuka J, Yamaguchi S (2008) Inhibition of shoot branching by new terpenoid plant hormones. Nature 455: 195-200

[89] Bouwmeester HJ, Roux C, López-Ráez JA, Bécard G (2007) Rhizosphere communication of plants, parasitic plants and AM fungi. Trends Plant Sci 12: 224-230

[90] Akiyama K, Matsuzaki K, Hayashi H (2005) Plant sesquiterpenes induce hyphal branching in arbuscular mycorrhizal fungi. Nature 435: 824-827

[91] Soto MJ, Fernández-Aparicio M, Castellanos-Morales V, García-Garrido JM, Ocampo JA, Delgado MJ, Vierheilig H (2010) First indications for the involvement of strigolactones on nodule formation in alfalfa (Medicago sativa). Soil Biol Biochem 42: 383-385

[92] Foo E and Davies NW (2011) Strigolactones promote nodulation in pea. Planta 234: 1073-1081

[93] Liu J, Novero M, Charnikhova T, Ferrandino A, Schubert A, Ruyter-Spira C, Bonfante P, Lovisolo C, Bouwmeester H, Cardinale F (2013) CAROTENOID CLEAVAGE DIOXYGENASE 7 modulates plant growth, reproduction, senescence, and determinate nodulation in the model Lotus japonicus. J Exp Bot 64: 1967-1981

3

Role of Boron Nutrient in Nodules Growth and Nitrogen Fixation in Soybean Genotypes Under Water Stress Conditions

Nacer Bellaloui, Alemu Mengistu,
My Abdelmajid Kassem, Craig A. Abel and
L.H.S. Zobiole

1. Introduction

Boron is an essential nutrient for plant growth, development, and seed quality [1-4]. Previous research indicated the involvement of B in cell wall structure [5,6]; cell membrane integrity [2, 7]; sugar metabolism [2], especially sugar alcohols [3,8]; nitrogen assimilation and fixation [9, 10]; nodules [11,12], nodullin protein (ENOD2) and malfunction of oxygen diffusion barrier [13]; phenolic metabolism [2,14,15]; ion uptake [2,16]; and plasma membrane-bount H+ ATPase [7,17,18].

Boron is required for nodules growth and nitrogen fixation [9,11-13], and boron deficiency can occur under certain environmental stress factors even when boron level in soil is adequate [19], leading to yield loss. Among the environmental stress factors that can lead to boron deficiency in plants is drought or water stress. Drought is a major environmental stress factor limiting crop yields worldwide [20], and maintaining boron levels within plants under drought conditions is critical. Boron has low mobility in the phloem [21], although boron mobility in the phloem depends on plant species [3, 22]. Under water stress conditions, plant increases abscisic acid (ABA) production [23], possibly affecting photosynthetic rate in drought-stressed plants, leading to stomata closure and transpiration rate reduction. Under these conditions boron uptake and translocation, and boron movement from leaves to seed is reduced, decreasing seed boron concentration [10]. Although soybean nodule growth and symbiotic N_2 fixation are sensitive to drought [20,24,25], we hypothesized that drought can result in boron deficiency within the plant, impacting nodule growth, N_2 fixation, and CO_2 accumulation. The objective of this research was to investigate the effects of foliar boron on nodules growth and nitrogen fixation rates in several soybean genotypes under water stress. To avoid the confounding effects of multi-environmental factors in the field on boron application effects, the experiment was conducted under greenhouse conditions. In addition to the current research findings, the present chapter will also highlight previous and current major research findings in boron nutrition and the role of B in nodule growth and symbiotic nitrogen fixation in soybean.

2. Materials and methods

A repeated greenhouse experiment was conducted. Cultivars of maturity group (MG) III Pella, Williams 82, Hutcheson, and Forest, were used. Seeds of soybean cultivar were germinated in flat trays in vermiculite, and then uniform size seedlings at V1 stage were transplanted into 9.45 L size pots. Soil in pots was a Dundee silt loam (fine-silty, mixed, active, thermic Typic Endoqualfs) with pH 6.3, 1.1% organic matter, a cation exchange capacity of 15 cmol/kg, and soil textural fractions of 26% sand, 56% silt, and 18% clay, average B concentration was 0.72 mg kg^{-1}. The soil contained an abundant native population of *B. japonicum*. Water stress was introduced as reported by Bellaloui [20,26]. Briefly, soil in pots were weighed and then saturated with deionized water and left to drain and weighed again to obtain the water field capacity using soil water sensors inserted in pots and measured by Soil Moisture Meter (WaterMark Company, Inc., Wisconsin, USA). Plants were divided into well watered (soil water potential between –15 to –20 kPa) (this was considered field capacity for the control plants), moderate water stress (soil water potential between -90 and -100 kPa), and severely water stressed (soil water potential between –150 to –200 kPa). Boron was foliar-applied as boric acid at a rate of 1.1 kg ha^{-1} once at flowering stage (R1-R2) and once at seed-fill stage (R5-R6). Combined treatments were well watered plants with no B (W-B), well watered plants with B (W+B); water stressed plants with no B (WS-B); water stressed plants with B (WS+B); severely water stressed plants with no B (SWS-B); severely water stressed plants with B (SWS+B). Samples were taken five days after the second B application for nitrate reductase assay to measure the rate nitrate reductase activity (NRA), nitrogenase, and leaf B. Mature seed were weighed at R8 (harvest maturity stage). Plants were considered fully matured when they reached R8 according to [27]. Greenhouse conditions were about 34°C ± 9°C during the day and about 28°C ± 8°C at night with a photosynthetic photon flux density (PPFD) of about 800 - 2300 μmol m^{-2} s^{-1}, as measured by Quantum Meter (Spectrum Technology, Inc., Illinois, USA). The big range of light intensity reflects a bright, sunny, or cloudy day. The source of lighting was a mixture of natural light, bulb light (60 W), cool white (250 W). To avoid differences in the day-length between the two experiments, the two experiments were conducted simultaneously at the same time and during the normal growing season (from April to September) for the Early Soybean Production System in the midsouth USA, and this is to be consistent with the normal photoperiod for soybean growth [10].

2.1. Nitrate reductase assay

The rate of nitrate reductase activity (NRA) was determined according to [28, 29]. Briefly, NRA was measured in the fully expanded leaves and nodules. Nodules were gently and carefully separated from roots and placed in NRA assay buffer solution. A fresh leaf sample of about 0.3 g was placed in 10 mL of potassium phosphate buffer at a concentration of 100 mM, pH 7.5, containing 1% (v/v) 1-propanol, in the flask. The incubation solution was vacuum-filtered for 1 min, and then flashed with nitrogen gas for 30 s, and then incubated at 30°C. A samples of 0.5 mL was taken at regular intervals (0, 60, 120, 180, and 300 min) for nitrite measurement. Samples were extracted with 5 mL of deionized water and reacted with 1.0 mL of 1% (w/v) sulfanilamide in 10% v/v HCl and 1.0 mL of *N*-naphthyl-(1)-ethylenediamine dihydrochloride (0.1%). Nitrite concentration in samples was measured by reading the absorbance at 540 nm after 30 minutes using a Beckman Coulter DU 800 spectro- photometer (Fullerton, CA). A standard curve was produced using KNO_2 as a source of NO_2 in the tested samples according to (Bellaloui et al., 2006). To measure the enzyme activity where there was no limiting concentration of NO_3 in the incubation culture solution (potential nitrate reductae activity, PNRA), exogenous NO_3 was added at a concentration of 10 mM as KNO_3.

2.2. Acetylene reduction assay

Destructive method for acetylene reduction assay was used. Three plants from each replicate were harvested five days after the second B application (at seed-fill stage). Nitrogenase activity was assayed using the acetylene reduction assay as described elsewhere [29-31]. Roots with nodules intact were excised and incubated in 60 mL plastic syringes. Roots from each replicate and from each treatment were placed in the syringes in the Mason jars and sealed. A 10% volume of air was then removed and replaced with an equal volume of acetylene. After 1 h of incubation at room temperature, duplicate 1.0 mL gas samples were removed and analyzed by gas chromatography for ethylene formation and carbon dioxide evolution. The gas chromatography (Agilent HP6960, Agilent Technologies, Wilmington, DE) was equipped with manual injector, injector loop, and sample splitter. A flame ionization detector (FID) and a thermal conductivity detector (TCD) were used. Using the sample loop and splitter, 0.25 mL of gas was directed into a 30 m length × 0.53 mm i.d. alumina megabore column (115-3532) connected to the FID, and 0.25 mL of sample was injected into a HP-PLOT D column (30 m length × 0.53 mm i.d. megabore with 40 μm film; 1905D Q04) connected to the TCD using helium as a carrier gas. Chromatographs were integrated using Chem Station software. Standard curves for ethylene and carbon dioxide were updated and produced for each day. Samples having <9% acetylene were not used in the analysis.. Nodules were carefully removed and counted, and then oven-dried at 60°C for 4 - 5 days.

2.3. Boron determination

The concentration of total B was measured in the fully expanded leaves after the second foliar B application, and in seeds at harvest maturity stage. Boron concentration was determined according to Azomethine—H method [19,32-34]. Briefly, 1 g of dry sample was placed in a porcelain crucible for ashing at 500°C for 8 hr. Samples then were extracted with 20 mL of 2

M HCl at 90°C for 10 min and azomethine-H solution containing 0.45% before the analysis (John et al., 1975) with a buffer solution contained 25% ammonium acetate, 1.5% EDTA, and 12.5% acetic acid. The concentration of B in the samples were determined spectrophotometrically by using color development after 45 minutes. Samples were read at 420 nm using a Beckman Coulter DU 800 spectrophotometer (Fullerton, California). Boron analysis in soil was conducted using Inductively Coupled Plasma spectrometry (ICP) using Thermo Elemental, Thermo Jarrell-Ash model 61E ICP, USA [10].

2.4. Analysis of $\delta^{15}N$ ($^{15}N/^{14}N$ ratio) and $\delta^{13}C$ ($^{13}C/^{12}C$ ratio) using natural abundance

Natural abundance of $\delta\ ^{15}N$ and ^{13}C isotopes was determined using about 0.9 mg of ground seeds. Isotopic analysis was conducted using a Thermo FinniGlyn Delta Plus Advantage Mass Spectrometer with a FinniGlyn ConFlo III, and Isomass Elemetal Analyzer (Bremen, Germany) according to [26, 35,36]. Isodat software version 2.38 was used to obtain Delta values [26]. The elemental combustion system was Costech ECS 4010 with an autosampler (Bremen, Germany).

2.5. Determination of seed sucrose

Seed sucrose concentration was measured in mature seeds. Sucrose concentration was measured according to [37,38] using an AD 7200 diode array feed analyzer (Perten, Springfield, IL). Briefly, about 25 g of seed were ground using a Laboratory Mill 3600 (Perten, Springfield, IL). Initial calibration equations were developed by the Department of Agronomy and Plant Genetics, University of Minnesota St Paul, MN using Thermo Galactic Grams PLS IQ software, developed by Perten company (Perten, Springfield, IL). Analyses of sugars were performed based on a seed dry matter basis [26, 37, 39].

2.6. Seed glucose determination

Glucose concentration in mature seeds was measured enzymatically using Glucose (HK) Assay Kit from Sigma, USA, Product Code GAHK-20, 2012) [40]. In this reaction, glucose is phosphorylated by adenosine triphosphate (ATP) in a reaction catalyzed by hexokinase. Glucose-6-phosphate (G6P) produced is then oxidized to form 6-phosphogluconate in the presence of oxidized nicotinamide adenine dinucleotide (NAD) in a reaction catalyzed by glucose-6-phosphate dehydrogenase (G6PDH). During this oxidation, an equimolar amount of NAD is reduced to NADH, and the increase in absorbance at 340 nm is directly proportional to glucose concentration in the sample. Mature seed samples were ground using a Laboratory Mill 3600 (Perten, Springfield, IL). A dry, ground sample of 0.1 mg was extracted with deionized water. The extraction procedure of glucose from seeds was conducted according to [26], and as instructed by Glucose (HK) Assay Kit from Sigma. The concentration of glucose was measured spectrophotometrically by reading the samples at 340nm using a Beckman Coulter DU 800 spectrophotometer (Fullerton, CA). The concentration of glucose was expressed as mg g dwt^{-1}.

2.7. Seed fructose determination

Fructose concentration is mature seeds was determined enzymatically according to Fructose Assay Kit from Sigma, USA, Product Code FA-20, 2012 [41]. In this reaction, fructose is phosphorylated by ATP in a reaction catalyzed by hexokinase, and the produced fructose 6-phosphate is then converted to G6P by phosphoglucose isomerase (PGI). The oxidation of G6P to 6-phosphogluconate takes place in the presence of NAD in the reaction catalyzed by glucose-6-phosphate dehydrogenase (G6PDH). An equimolar amount of NAD is reduced to NADH, and the consequent increase in absorbance at 340 nm is directly proportional to fructose concentration in a sample. A sample of 0.1 mg was extracted according to Fructose Assay Kit from Sigma, and is detailed in Bellaloui et al (2013) as instructed by Fructose Assay Kit from Sigma. The concentration of fructose in samples were measured spectrophotometrically by reading the samples at absorbance of 340nm using a Beckman Coulter DU 800 spectrophotometer (Fullerton, CA). The concentration of fructose was expressed as mg g dwt^{-1}.

2.8. Experimental design and statistical analysis

Treatments were arranged in a split plot design with irrigation as a main block and B treatment as sub-plot. Four replicates were used, and each replicate consisted of a pot containing three plants. Proc Mixed was used for data analysis of variance in SAS [42]. Means were separated by Fisher's least significant difference test at the 5% level of probability using Proc GLM analysis in SAS [42]. Since there were no interactions between the two experiments, the data were pooled and combined.

3. Results and discussion

Analysis of variance showed that boron application (T) and irrigation (IR) were significant for grain weight, nodule mass, nitrogen fixation (NF), nitrate reductase activity (NRA), N, B, and natural abundance of $\delta^{15}N$ and $\delta^{13}C$ (Table 1), and sugars (Table 2). Cultivar (CV) was significant for some parameters and not significant for others, indicating differences in cultivar responses to the B application and water stress. Both B application and IR significantly interacted (T×IR and E×T×IR interactions) for these parameters, indicating that B effects on these parameters depended on IR (watered or water stressed conditions). There were no significant interactions between Experiment (E) and T or IR, indicating that the effect of B application or IR had similar effect in each experiment (Table 1). Therefore, the data were pooled and combined [26].

3.1. Effect of B and water stress on grain weight, nodule mass and nodule number

Foliar B application to watered plants (W+B) resulted in significant increase ($P \leq 0.05$) in grain weight, nodule mass, and nodule number (Table 3). For example, in Pella cultivar the increase of grain weight, nodule mass, and nodule number was 18.7%, 38.5%, and 33.3 %, respectively. These parameters were different between cultivars and each cultivar responded differently to foliar B (Table 3). Previous research indicated that B plays a major role for plant growth and development [1, 2] and crop quality [4,19]. It was shown that foliar B improved seed set, seed yield, and seed quality of alfalfa [19] and sugar beet [4], and altered seed composition in soybean [10]. Previous research showed that B is an essential nutrient for the development of nitrogen-fixing root nodules in pea (*Pisum sativum*) [9]. A lower level of infection of the host plants with *Rhizobium* was noticed in plants grown in B-deficient medium compared to plants supplied with adequate B [9]. It was shown that that little or no ability to fix N_2 under B-deficient plants [11]. Recently, it was found that FB increased nodule weight under irrigated greenhouse conditions [43]. The current results showed that, even though B concentration in leaves was above the critical level (20 mg B kg^{-1}, critical level of B in leaves for normal plant growth) [44], FB resulted in a positive effect on seed and nodule weights, agreeing with previous research of those of [7, 43, 45]. Our results showed that foliar B increased grain weight, nodule mass and number due to the stimulatory effects of B on growth and development [9,11] and nodule improvement.

3.2. Effect of B and water stress on nitrogen fixation and nitrogen assimilation

Foliar application of B resulted in higher rates of nitrogen fixation (increase of nitrogenase), root respiration, and nitrogen assimilation (increase of nitrate reductase activity, NRA), (Table 4). Foliar B application to moderately water stressed plants (Table 4) increased grain weight, nodule number and mass, nitrogen fixation and assimilation. However, foliar B application to severely water stress plants did not result in an increase in these parameters because of the damaging effects of water stress to nitrogen metabolism enzymes, especially nitorgenase and nitrate reductase (Table 5). Researchers reported that B is an essential micronutrient for the development of nitrogen-fixing root nodules [11], and plants grown in B-deficient medium showed lower infection of the host plants with Rhizobium compared to plants supplied with adequate B [11]. It was reported that nodules showed little or no ability to fix N^2 in B-deficient plants, leading to N deficiency and necrosis of nodulated pea plants [9]. This indicated that nitrogen fixation in soybean was sensitive to B deficiency [9,12], and B deficiency can result in the reduction in early nodullin protein (ENOD2) in nodule parenchyma cells and malfunction of oxygen diffusion barrier [13]. It was hypothesized that B protects nitrogenase against oxygen damage by influencing membrane integrity and function [13] and may interact with membrane glycoproteins and glycolipids to maintain the proper conformation in nitrogen-fixing cells [5]. Although B has been shown to be essential for nodule growth and development, there is no convincing evidence that there is a direct effect of B on nitrogen metabolism [2,13,46,]. Nitrate assimilation, reflected by the key enzyme in nitrogen assimilation, was higher in W+B plants, indicating that B enhanced nitrate assimilation. The stimulatory effects of nitrogen assimilation by B was also reported by others (Bellaloui et al., 2011 AJPS. This is because nitrogen metabolism in legumes is both a result of both symbiotic N_2 fixation and mineral N assimilation processes. During this process, atmospheric N_2 is fixed by the enzyme nitrogenase in the bacteroids of nodules [47], and nitrate reduction (assimilation) is catalyzed by the enzyme nitrate reductase (NR). Both NR and nitrogenase enzymes coexist in nodules competing for reductant [48]. It appears that B may stimulate de novo synthesis and making nitrate (enzyme substrate) available for the enzyme nitrate reductase. Adding 0.5 mM B to the buffer solution increased NRA by 30% in WS+B compared with WS-B, and adding 10 mM NO_3 to the buffer solution increased NRA by 55% and 40% in leaves and nodules, respectively (data not shown). In SWS plants, adding B or NO_3 did not enhance NRA (data not shown). This indicated that both B and NO_3 stimulated NR enzyme somehow, maybe by facilitating nitrate availability in the cytoplasm to NR for reduction or enhancing nitrate translocation from the vacuoles to the cytoplasm, leading to higher NRA activity. This hypothesis was supported by the effect of B ion uptake [2,16] and the direct or indirect effects of B on the plasma membrane bound H+ ATPase (plasmalemma H+-ATPase activity) [7,17], cell wall structure and membrane integrity [2,7]. Our results are supported by [2] in that B may have an indirect influence on nitrate uptake and assimilation, and enhance NRA by inducing nitrate availability and increasing protein de novo synthesis as a result of nitrate absorption [49]. This observation is supported by [13] who found that adequate level of B increased NRA and decreased nitrate in xylem sap compared to deficiency level. The relationship between nitrogen fixation and nitrogen assimilation and how this relationship is influenced by foliar B and its impact on seed protein and oil and sugars is still not well established.

Source of variability	Seed weight	Nodule mass	Nodule number	ARA (NF)	Leaf NRA	NoduleNRA	B in leaves	N in leaves	$\delta^{15}N$	$\delta^{13}C$
Experiment (E)	NS	NS	NS	NS	NS	NS	NS	NS	NS	NS
Treatment (T)	**	**	***	**	***	**	***	*	*	**
Water stress (WS)	***	**	***	***	***	*	***	*	*	**
Cultivar (CV)	*	*	*	NS	NS	NS	NS	NS	NS	NS
ExT	NS	NS	NS	NS	NS	NS	NS	NS	NS	NS
ExWS	NS	NS	NS	NS	NS	NS	NS	NS	NS	NS
ExCV	NS	NS	NS	NS	NS	NS	NS	NS	NS	NS
TxWS	**	*	*	**	*	*	**	*	*	*
TxCV	*	*	*	*	*	*	*	**	*	*
CVxWS	**	**	**	*	**	*	*	*	*	*
ExTxWSxCV	**	**	*	**	**	**	*	*	*	**

* Significance at $P \leq 0.05$; ** Significance at $P \leq 0.01$; *** Significance at $P \leq 0.001$.

Table 1. Analysis of variance of the effects of foliar boron on seed weight (100 seed weight, g), nodule mass (mg $plant^{-1}$), nodule number $plant^{-1}$, [nitrogen fixation (acetylene reduction assay (ARA), μmol of C_2H_4 $plant^{-1}$ h^{-1})], leaf nitrate reductase activity (NRA, μmol NO_2 $g^{-1}h^{-1}$), and nodule NRA (μmol NO_2 $g^{-1}h^{-1}$), boron (B, mg kg^{-1})) and nitrogen (N, %) in leaves and seeds, and in $\delta^{15}N$ and in $\delta^{13}C$ isotope values in seeds in genotypes of maturity group III (Pella and William 82) and MG V (Hutcheson and Forest) under well watered and water stressed conditions (WS) with and without foliar boron (B) treatments (T) under greenhouse conditions [a].

Source of variability	Glucose	Fructose	Sucrose	Raffinose	Stachyose
Experiment (E)	NS	NS	NS	NS	NS
Treatment (T)	**	*	*	*	***
Irrigation (IR)	**	*	*	**	***
Cultivar (CV)	*	*	*	NS	NS
ExT	NS	NS	NS	NS	NS
ExIR	NS	NS	NS	NS	NS
ExCV	NS	NS	NS	NS	NS
TxIR	*	*	*	**	*
TxCV	**	*	*	*	*
CVxIR	*	**	*	*	*
ExTxIRxCV	*	*	*	**	*

* Significance at $P \leq 0.05$; ** Significance at $P \leq 0.01$; *** Significance at $P \leq 0.001$.

Table 2. Analysis of variance of the effects of foliar boron on sugars (mg g^{-1} dwt) in genotypes of maturity group III (Pella and William 82) and MG V (Hutcheson and Forest) under well watered and water stressed conditions (IR) with and without foliar boron (B) treatments (T) under greenhouse conditions [a].

				Watered soybean				
Variety	Boron	Grain weight (100 seed weight, g)	Nodule mass (mg dwt plant^{-1})	Nodule number plant^{-1}	ARA (NF) (μmol of C_2H_4 plant^{-1} h^{-1})	Root respiration (mmol of CO_2 evolved/g of root/h)	Leaf NRA (μmol NO_2 g^{-1} h^{-1})	Nodule NRA (μmol NO_2 $g^{-1}h^{-1}$)
Pella	W-B	16 a	65 a	33 a	11.6 a	7.5 a	5.6 ab	4.7 ab
W 82		15 a	68 a	28 b	10.5 a	7.8 a	4.9 b	5.2 a
Hutcheson		14 b	61 b	25 b	10.1 a	6.8 b	5.7 ab	3.8 b
Forrest		14 b	64 a	17 c	11.4 a	6.4 b	6.3 a	4.2 ab
Pella		19 a	90 a	44 a	15.4 a	9.7 b	7.5 ab	6.4 a
W 82	W+B	18 a	86 ab	38 abc	14.7 ab	10.5 ab	6.8 b	7.8 a
Hutcheson		16 b	84 b	36 c	13.2 b	10.6 ab	7.4 ab	5.2 b
Forrest		17 b	86 ab	38 abc	13.6 b	11.5 a	8.3 a	5.8 b

[a] Soybean plants were grown at field capacity at -15 to -20 kPa [10]. Soybeans were grown under greenhouse conditions similar to those in Bellaloui et al. (2011). Values within columns and within each B treatment sharing a letter are not significantly different (P>0.05) using Fishers' test. W 82=Williams 82.

Table 3. Effect of foliar boron on soybean seed weight (100 seed weight, g), nodule mass (mg plant^{-1}), nodule number plant^{-1}, ARA (μmol of C_2H_4 plant^{-1} h^{-1}), root respiration (mmol of CO_2 evolved/g of root/h), leaf NRA (μmol NO_2 $g^{-1}h^{-1}$), and nodule NRA (μmol NO_2 $g^{-1}h^{-1}$) in genotypes of maturity group III (Pella and William 82) and MG V (Hutcheson and Forest) under well watered conditions without boron (W-B) and with boron (W+B) under greenhouse conditions [a].

Watered stressed soybean								
Variety	**Boron**	**Grain weight (100 seed weight, g)**	**Nodule mass (mg dwt plant^{-1})**	**Nodule number plant^{-1}**	**ARA (NF)**	**Root respiration**	**Leaf NRA (µmol NO_2 $g^{-1}h^{-1}$)**	**Nodule NRA (µmol NO_2 $g^{-1}h^{-1}$)**
Pella	WS-B	9.4 a	45.3 b	22.4 a	7.6 a	4.3 b	3.6 a	2.4 b
W 82		10.3 a	51.6 a	24.1 a	8.7 a	4.9 b	2.7 b	3.6 a
Hutcheson		9.7 a	42.7 b	21.5 a	6.5 b	5.4 b	3.3 a	2.7 b
Forrest		10.1 a	46.3 b	14.5 b	5.7 c	6.3 a	4.7 a	2.5 b
Pella		11.1 a	56.7 b	28.6 a	10.5 a	5.4 c	5.4 b	4.7 ab
W 82	WS+B	12.5 a	66.4 a	31.4 a	9.7 a	7.6 b	7.5 a	5.3 a
Hutcheson		12.6 a	57.3 b	30.7 a	10.4a	8.7 a	5.8 b	4.7 ab
Forrest		11.5 a	63.2 a	19.6 b	7.4 b	9.8 a	6.3 a	3.9 b

[a] Soybean plants were grown under water stress (WS) (-90 to -100 kPa soil water potential). Soybeans were grown under greenhouse conditions similar to those previously reported [10]. Values within columns and within each B treatment sharing a letter are not significantly different (P>0.05) using Fishers' test. W 82–Williams 82.

Table 4. Effect of foliar boron on soybean seed weight (g), nodule mass (mg plant^{-1}), nodule number plant^{-1}, ARA ((µmol of C_2H_4 plant^{-1} h^{-1}), root respiration (mmol of CO_2 evolved/g of root/h), leaf NRA (µmol NO_2 $g^{-1}h^{-1}$), and nodule NRA (µmol NO_2 $g^{-1}h^{-1}$) in genotypes of maturity group III (Pella and William 82) and MG V (Hutcheson and Forest) under water stress conditions without boron (WS-B) and with boron (WS+B) under greenhouse conditions [a].

Severe watered stressed soybean								
Variety	**Boron**	**Grain weight (100 seed weight, g)**	**Nodule mass (mg dwt plant^{-1})**	**Nodule number plant^{-1}**	**ARA (NF)**	**Root respiration**	**Leaf NRA (µmol NO_2 $g^{-1}h^{-1}$)**	**Nodule NRA (µmol NO_2 $g^{-1}h^{-1}$)**
Pella	SWS-B	4.4 a	21.5 b	20.7 a	4.6 a	2.1 c	2.1 a	0.7 a
W 82		5.5 a	28.5 a	21.6 a	5.3 a	3.3 b	1.5 b	0.9 a
Hutcheson		5.3 a	27.5 a	19.6 a	4.7 a	3.6 b	2.4 a	0.8 a
Forrest		4.7 b	22.6 b	21.5 a	3.2 b	4.7 a	1.7 b	0.8 a
Pella		4.3 c	25.4 a	22.6 b	5.1 a	2.6 b	2.4 a	0.8 a
W 82	SWS+B	6.0 a	30.6 a	26.5 a	4.8 a	2.4 b	0.9 b	0.6 a
Hutcheson		5.5 b	26.5 a	29.7 a	3.4 b	3.8 a	2.1 a	0.8 a
Forrest		4.1 c	21.6 b	19.6 b	3.2 b	4.9 a	1.2 b	0.7 a

[a] Soybean plants were grown under severe water stress (soil water potential between –150 to –200 kPa). Soybeans were grown under greenhouse conditions similar to those previously reported [10]. Values within columns and within each B treatment sharing a letter are not significantly different (P>0.05) using Fishers' test. W 82=Williams 82.

Table 5. Effect of foliar boron on soybean seed weight (g), nodule mass (mg plant^{-1}), nodule number plant^{-1}, nitrogen fixation (ARA, µmol of C_2H_4 plant^{-1} h^{-1}), root respiration (mmol of CO_2 evolved/g of root/h), leaf NRA (µmol NO_2 $g^{-1}h^{-1}$), and nodule NRA (µmol NO_2 $g^{-1}h^{-1}$) in genotypes of maturity group III (Pella and William 82) and MG V (Hutcheson and Forest) under severe water stress conditions without boron (SWS-B) and with boron (SWS+B) under greenhouse conditions [a].

Foliar boron application increased B in leaves and seed in watered plants (Figure 1). No significant B concentration differences were observed between leaves and seeds B in each watered treatment in each cultivar, indicating that B movement from leaves to seeds was not limited. In severely water-stressed plants, application of foliar B did not significantly increase B in leaves, and B movement from leaves to seed was limited, indicated by the large accumulation of B in leaves and small accumulation of B in seeds. Similar trend of N in leaves and seed was noticed (Figure 2), indicating a close relationship between B and N.

3.3. Effects of B and water stress on seed sugars

Since B plays an important role in carbohydrate mobility and since carbohydrates are a source of reducing power in nitrogen assimilation, sugar profiling was also investigated. Our research demonstrated that foliar B application resulted in higher sucrose, glucose, and fructose under irrigated conditions, but under severe water stress these mono and disaccharides sugars decreased, but stachyose and raffinose increased (Table 6,7,8).

This indicated that there was a redistribution of sugars under severe water stress, and this shift may provide plants with an adaptive mechanism to tolerate the stress.

Foliar boron resulted in higher seed sucrose, glucose, and fructose concentrations in W+B plants, indicating B involvement in sugar metabolism and synthesis. The involvement of B in sugar synthesis and distribution is not understood, but B involvement in sugar movement and metabolism was previously reported [2,3,50]. The decrease of sucrose, glucose, and fructose in SWS-B and SWS+B plants compared with W+B and W-B indicated that B enhanced accumulation of sucrose, glucose, and fructose concentrations may be due to B role in sugar movement within the plants, and that severe water stress limited sugars movement due reduction of B uptake and stomatal conductance. The increase of seed stachyose concentration in seeds of SWS-B and SWS+B plants indicated that severe water stress affects the distribution of sugar fractions, in our case the increase of stachyose and raffinose and decrease of sucrose, glucose, and fructose concentrations. The increase of stachyose under water stress may indicate the role of stachyose in plant tolerance to biotic and abiotic stress [14,15]. It was also reported that raffinose and galactinol levels may play an important role in plant tolerance to biotic and abiotic stress [14,15], and the accumulation of galactinol and raffinose may protect the plant from drought [51], and the activity of sucrose synthase, the main enzyme involved in sucrose hydrolysis in nodules, was significantly inhibited under drought conditions [52,53]. The biological functions of raffinose and stachyose are not clear [54], but previous research reported that oligosaccharides (sucrose, raffinose, and stachyose) are related to seed quality [55] and the acquisition of desiccation tolerance during seed development and maturation.

Soybean seed sugars are important to soybean seed industry because they determine the quality of seeds beside protein and oil. This is because soybean seed with high raffinose and stachyose concentrations are undesirable and have negative effects on the nutritive value of soymeal and seed consumed by human. Stachyose and raffinose are indigestible by humans and animals, especially monogastric animal such as chicken and pigs, causing flatulence or diarrhea [56]. On the other hand, low raffinose and stachyose levels in soybean seed are desirable [57]. High level of seed sucrose, glucose, and fructose are desirable because it improves taste and flavor of tofu, soymilk, and natto [58]. Currently, soybean cultivars with improved sugar profiles have been released to the market [58], and breeding for desirable sugars in soybean or agricultural practices to improve seed sugars are needed.

			Watered soybean			
Variety	Boron	Glucose (mg g^{-1})	Fructose (mg g^{-1})	Sucrose (mg g^{-1})	Raffinose (mg g^{-1})	Stachyose (mg g^{-1})
Pella	W-B	1.5 b	0.73 a	35.4 b	5.4 a	32.6 b
W 82		2.1 a	0.65 b	42.4 a	4.8 b	43.6 a
Hutcheson		1.5 b	0.64 b	32.7 b	5.3 ab	42.8 a
Forrest		2.2 a	0.61 b	30.8 b	5.7 a	41.6 a
Pella		2.3 b	0.93 b	57.4 b	5.4 ab	35.4 b
W 82	W+B	2.8 a	1.22 a	63.2 a	5.1 b	44.6 a
Hutcheson		2.1 b	0.89 b	53.8 b	5.7 a	42.6 a
Forrest		2.9 a	0.94 b	59.7 b	5.2 b	46.5 a

[a] Soybean plants were grown at field capacity at -15 to -20 kPa according to Bellaloui et al., (2011). Soybeans were grown under greenhouse conditions similar to those previously reported [10]. Values within columns and within in each B treatment sharing a letter are not significantly different (P>0.05) using Fishers' test. W 82=Williams 82.

Table 6. Effect of foliar boron on soybean seed sugars in genotypes of maturity group III (Pella and William 82) and MG V (Hutcheson and Forest) under well watered conditions without boron (W-B) and with boron (W+B) under greenhouse conditions [a].

Variety	Boron	Glucose (mg g^{-1})	Fructose (mg g^{-1})	Sucrose (mg g^{-1})	Raffinose (mg g^{-1})	Stachyose (mg g^{-1})
Pella	WS-B	0.8 b	0.74 a	19.3 a	6.5 b	45.5 b
W 82		1.1 b	0.65 b	16.4 b	7.3 a	52.5 a
Hutcheson		1.5 a	0.42 c	21.3 a	7.4 a	47.3 b
Forrest		0.9 b	0.69 b	15.4 b	6.4 b	50.3 a
Pella		1.5 b	0.76 a	27.5 ab	7.1 a	41.6 b
W 82	WS+B	1.5 b	0.75 a	25.4 b	6.2 b	48.7 a
Hutcheson		1.9 a	0.64 b	31.2 a	7.4 a	50.3 a
Forrest		1.6 b	0.72 a	30.5 a	6.3 b	48.5 a

[a] Soybean plants were grown under water stress (WS) (-90 to -100 kPa soil water potential). Soybeans were grown under greenhouse conditions similar to those previously reported [10]. Values within columns and within each B treatment sharing a letter are not significantly different (P>0.05) using Fishers' test. W 82=Williams 82.

Table 7. Effect of foliar boron on soybean seed sugars in genotypes of maturity group III (Pella and William 82) and MG V (Hutcheson and Forest) under water stress conditions without boron (WS-B) and with boron (WS+B) under greenhouse conditions [a].

Variety	Boron	Glucose (mg g^{-1})	Fructose (mg g^{-1})	Sucrose (mg g^{-1})	Raffinose (mg g^{-1})	Stachyose (mg g^{-1})
Pella	SWS-B	0.65 a	0.54 a	17.5 a	8.5 b	73.5 a
W 82		0.53 b	0.43 b	14.3 b	9.5 a	65.4 b
Hutcheson		0.51 b	0.47 b	15.3 ab	8.3 b	73.7 a
Forrest		0.48 c	0.52 a	11.6 c	9.5 a	65.7 b
Pella		0.59 a	0.53 a	19.5 a	8.5 b	62.1 b
W 82	SWS+B	0.48 c	0.47 b	13.2 b	9.5 a	73.6 a
Hutcheson		0.52 b	0.51 a	17.6 a	8.7 b	68.5 ab
Forrest		0.51 b	0.48 b	14.2 b	7.2 c	68.5 ab

[a] Soybean plants were grown under severe water stress (soil water potential between −150 to −200 kPa). Soybeans were grown under greenhouse conditions similar to those previously reported [10]. Values within columns and within each B treatment sharing a letter are not significantly different (P>0.05) using Fishers' test. W 82=Williams 82.

Table 8. Effect of foliar boron on soybean seed sugars in genotypes of maturity group III (Pella and William 82) and MG V (Hutcheson and Forest) under severe water stress conditions without boron (SWS-B) and with boron (SWS+B) under greenhouse conditions [a].

3.4. Effect of B and water stress on $\delta^{15}N$ ($^{15}N/^{14}N$ ratio) and $\delta^{13}C$ ($^{13}C/^{12}C$ ratio)

Foliar B did not result in changes in $^{15}N/^{14}N$ or $^{13}C/^{12}C$ ratios, but significant differences in these ratios were observed between irrigated and non-irrigated soybean with or without foliar B (Figure 3). The alteration of $^{15}N/^{14}N$ by increasing ^{15}N (derived from soil nitrogen that is used for nitrate assimilation) and decreasing ^{14}N (derived from atmospheric nitrogen that is used for nitrogen fixation) indicated that the source of nitrogen use changed, and plants favored N from soil over atmospheric nitrogen, indicating that nitrogenase is more sensitive than nitrate reductase under water stress. The mechanisms of this shift are not understood, but one possible explanation is that the shift in $^{15}N/^{14}N$ may reflect a possible mechanism to compensate for the inhibition of nitrogen fixation under water stress conditions. Previous research indicated that $\delta^{15}N$ values in the xylem and plant tissues were associated with acquired N, and changed with N metabolism [59]. The increase in $\delta^{13}C$ or higher $^{13}C/^{12}C$ ratio (less negative) in seed under severe water stress conditions indicated that the source of carbon fixation used was shifted. Previous research reported that that the $\delta^{13}C$ value in plant tissues can be affected by water supply [60], plant physiology [61], and mycorrhizal infection [62]. The level of $\delta^{13}C$ was dependent on the environmental factors and their association with plant gas exchange, stomatal conductance, and CO_2 fixation [63]. It was found that drought stress leads to stomatal closure and ^{13}C fixation increase, resulting in less discrimination against $\delta^{13}C$ [64,65]. Previous research indicated that the the shift in $^{13}C/^{12}C$ ratio was a result of a shift in carbon fixation metabolism from ribulose bisphosphate (RuBP) carboxylase pathway to phosphoenolpyruvate carboxylase (PEP). This shift resulted in $\delta^{13}C$ enrichment [60]. It was found that in C3 species, to which soybean belongs, carbon isotope composition changes among and between

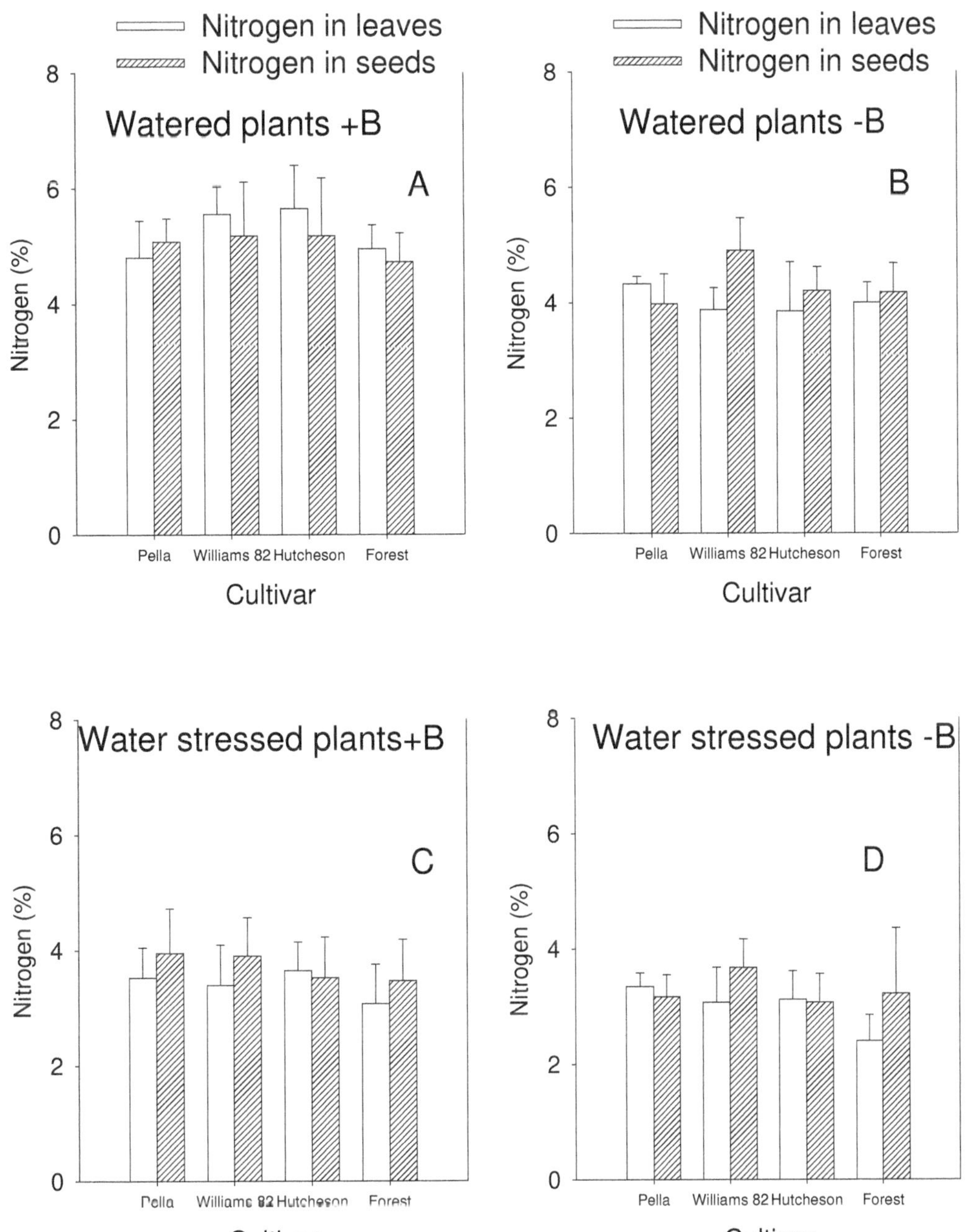

Figure 1. Effects of foliar boron application (1.1 kg B ha^{-1}) on boron concentration in leaves and seed in soybean genotypes in watered (A,B) and severe water stressed (C,D) soybean genotypes. Soybean plants were grown under severe water stress (soil water potential between −150 to −200 kPa). Soybeans were grown under greenhouse conditions similar to those previously reported [10].

genotypes correlated with water use efficiency, and the stable isotope ^{13}C would be discriminated against during photosynthesis, leading to a smaller ^{13}C to ^{12}C ratio [66]. The enrichment of ^{13}C may be due closure of stomatal conductance under severe water stress, leading to ^{13}C fixation increase and less ^{13}C discrimination [67,68]. Our current results are in agreement with previous reports that environmental stresses, including drought, can alter $\delta^{13}C$ due to the drought effects on the balance between stomatal conductance and carboxylation [67,68,69].

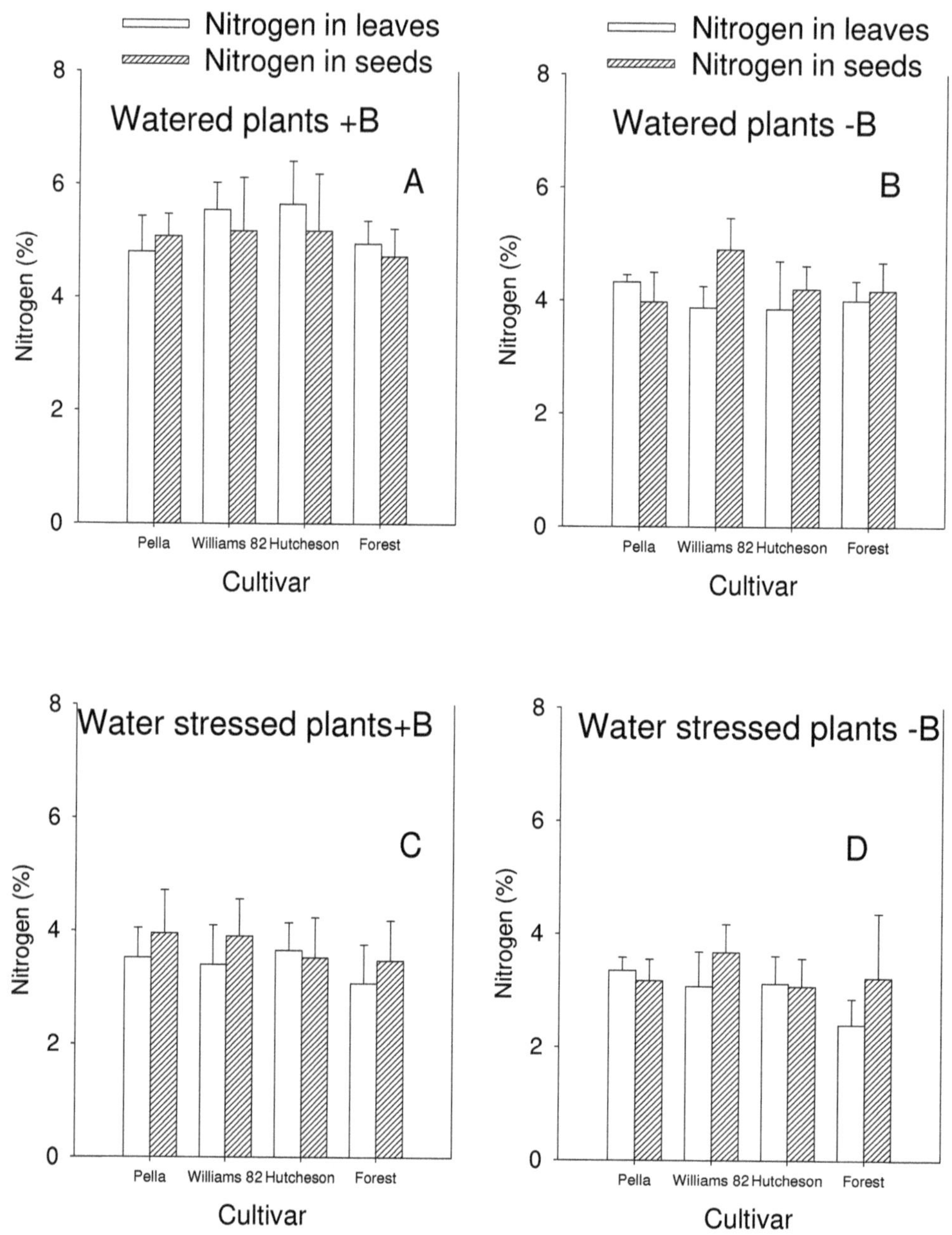

Figure 2. Effects of foliar boron application (1.1 kg B ha^{-1}) on nitrogen percentage in leaves and seed in soybean genotypes in watered (A,B) and severe water stressed (C,D) soybean genotypes. Soybean plants were grown under severe water stress (soil water potential between –150 to –200 kPa). Soybeans were grown under greenhouse conditions similar to those previously reported [10].

During carbon fixation by photosynthesis, the naturally occurring stable isotope ^{13}C is discriminated against, and plants would have a smaller ^{13}C to ^{12}C ratio than ^{13}C to ^{12}C ratio in fixed CO_2 of the air, suggesting a possible use of this technique to select for water use efficiency (Farquhar et al., 1989). Our results demonstrated that $\delta\,^{15}N$ and $\delta^{13}C$ values changed and enrichment occurred under water stress conditions, suggesting that both nitrogen and carbon metabolism pathways were affected during water stress, impacting seed production and seed quality.

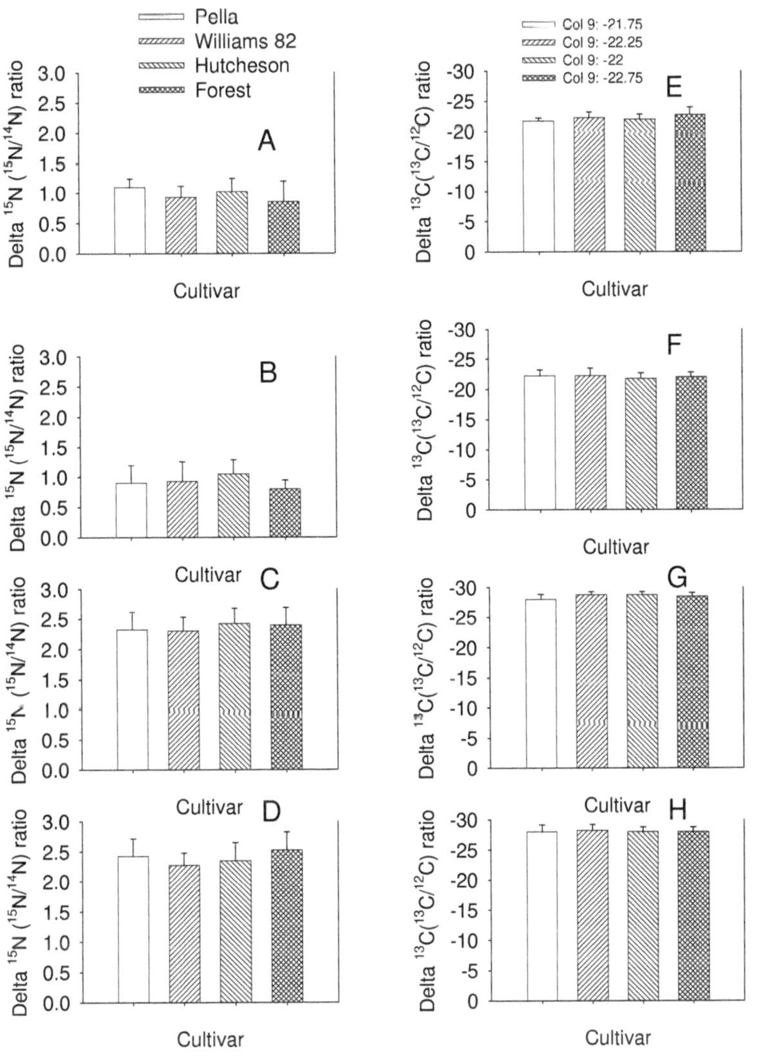

Figure 3. Effects of foliar boron application (1.1 kg B ha-1) on seed $\delta^{15}N$ ($^{15}N/^{14}N$ Ratio) (A-C) and $\delta^{13}C$ ($^{13}C/^{12}C$ Ratio) (D-H) in soybean genotypes in watered plants with B (A) and without B (B); in water stressed plants with B (C) and without B (D); in soybean genotypes in watered plants with B (E) and without B (F); in water stressed plants with B (G) and without B (H); Soybean plants were grown under severe water stress conditions (soil water potential between –150 to –200 kPa). Soybeans were grown under greenhouse conditions similar to those previously reported [10]

4. Conclusions

Foliar boron application resulted in nodule growth by increasing the number and mass of nodules under well watered or moderate water stress conditions. Also, foliar boron resulted in higher nitrogen fixation and nitrogen assimilation under well watered or moderate water stress conditions. Foliar B application under severe water stress did not enhance nodule number or mass, nitrogen fixation and nitrogen assimilation. This is because severe water stress altered nitrogen and carbon fixation as indicated by changes in the values of ^{15}N and ^{13}C natural isotopes. Nitrogen fixation is more sensitive than nitrogen assimilation under severe water stress conditions. Foliar B enhanced seed sugars under well watered conditions, but sever water stress resulted in redistribution of sugar fractions by increasing monosaccharides such as glucose and fructose, decreasing sucrose as a result, but increasing both raffinose and stachyose due to their possible roles in drought and the acquisition of desiccation tolerance during seed development and maturation. Increasing sugars by foliar B is desirable as high glucose, fructose, and sucrose contribute to soybean seed quality by improving the taste and flavor of soymeal based products such as tofu, soymilk, and natto.

Although our research showed that B has beneficial effects on nodules and nitrogen fixation and seed quality, further research is needed to test these findings under field and drought conditions in multi-year and multi-location experiments so that recommendations can be made.

Acknowledgements

We thank Sandra Mosley for lab assistance and greenhouse management. Also, we thank Leslie Price for his technical assistance on nitrogen and carbon isotopes measurements. The U.S. Department of Agriculture (USDA) prohibits discrimination in all its programs and activities on the basis of race, color, national origin, age, disability, and where applicable, sex, marital status, familial information, political beliefs, reprisal, or because all or part of an individual's income is derived from any public assistance program. (Not all prohibited bases apply to all programs.) Persons with disabilities who require alternative means for communication of program information (Braille, large print, audiotape, etc.) should contact USDA's TARGET Center at (202) 720-2600 (voice and TDD). To file a complaint of discrimination, write to USDA, Director, Office of Civil Rights, 1400 Independence Avenue, S.W., Washington, D.C. 20250-9410, or call (800) 795-3272 (voice) or (202) 720-6382 (TDD). USDA is an equal opportunity provider and employer.

Author details

Nacer Bellaloui[1], Alemu Mengistu[2], My Abdelmajid Kassem[3], Craig A. Abel[4] and L.H.S. Zobiole[5]

1 USDA-ARS, Crop Genetics Research Unit, Stoneville, USA

2 USDA-ARS, Crop Genetics Research Unit, Jackson, TN, USA

3 Plant Genomics and Biotechnology Lab, Department of Biological Sciences, Fayetteville State University, Fayetteville, NC, USA

4 Corn Insects and Crop Genetics Research, Iowa State University, Ames, Iowa, USA

5 Crop Protection, R&D Dow AgroSciences – Brazil, Cascavel, Paraná, Brazil

References

[1] Pilbeam DJ, Kirkby EA. The Physiological Role of Boron in Plants. Journal of Plant Nutrition 1983; 6, 563–582.

[2] Marschner H. Mineral Nutrition of Higher Plants. San Diego, CA (Academic Press) 1995; p379-396.

[3] Brown PH, Bellaloui N, Dandekar AM, Hu H. Transgenically Enhanced Sorbitol Synthesis Facilitates Phloem Boron Transport and Increases Tolerance of Tobacco to Boron Deficiency. Plant Physiology 1999; 119, 17-20.

[4] Dordas C, Apostolides G, Goundra O. Boron Application Affects Seed Yield and Seed Quality of Sugar Beets. Journal of Agricultural Sciences 2007; 145, 377–384.

[5] Brown PH, Bellaloui N, Wimmer MA, Bassil ES, Ruiz J, Hu H, Pfeffer H, Dannel F, Romheld V. Boron in Plant Biology. Plant Biology 2002; 4, 205–223.

[6] Hu H, Brown PH. Localization of Boron in Cell Walls of Squash and Tobacco and its Association with Pectin: Evidence for a Structural Role of Boron in the Cell Wall. Plant Physiology 1994; 105, 681–689.

[7] Schon MK, Blevins DG. Foliar Boron Applications Increase the Final Number of Branches and Pods on Branches of Field-Grown Soybeans. Plant Physiology 1990; 92, 602-607.

[8] Bellaloui N, Brown PH, Dandekar AM. Manipulation of *in vivo* Sorbitol Production Alters Boron Uptake and Transport in Tobacco. Plant Physiology 1999; 119, 735-742.

[9] Bolaños L, Brewin NJ, Bonilla I. Effects of Boron on Rhizobium-Legume Cell--Surface Interactions and Node Development. Plant Physiology 1996; 110, 1249–1256.

[10] Bellaloui N, Abbas HK, Gillen AM, Abel CA. Effect of Glyphosate-Boron Application on Seed Composition and Nitrogen Metabolism in Glyphosate-Resistant Soybean. Journal of Agriculture and Food Chemistry 2009a; 57, 9050–9056.

[11] Bolaños L, Esteban E, de Lorenzo C, Fernández-Pascual M, deFelipe MR, Garate A, Bonilla I. Essentiality of Boron for Symbiotic Dinitrogen Fixation in Pea (*Pisum sativum*) Rhizobium Nodules. Plant Physiology 1994; 104, 85–90.

[12] Carpena RO, Esteban E, Sarro MJ, Peñalosa J, Gárate A, Lucena JJ, Zornoza P. Boron and Calcium Distribution in Nitrogen-Fixing Pea Plants. Plant Sciences 2000; 151, 163–170.

[13] Bonilla I, Mergold-Villasenor C, Campos ME, Sanchez N, Perez H, Lopez I, Castrejon L, Sanchez F, Cassab, GI. The Aberrant Cell Walls of Boron-Deficient Bean Root Nodules have no Covalently Bound Hydroxyproline-/Proline-Rich Protein. Plant Physiology 1997; 115, 1329–1340.

[14] Bellaloui N, Mengistu A, Fisher DK, Abel CA. Soybean Seed Composition as Affected by Drought and *Phomopsis* in Phomopsis Susceptible and Resistant Genotypes. Journal of Crop Improvement 2012a; 26, 428-453.

[15] Bellaloui N, Mengistu A, Zobiole LHS, Shier WT. Resistance to Toxin-Mediated Fungal Infection: Role of Lignins, Isoflavones, other Seed Phenolics, Sugars, and Boron in the Mechanism of Resistance to Charcoal Rot Disease in Soybean. Toxin Reviews 2012b; 31, 16-26.

[16] Goldbach H. Influence of Boron Nutrition on Net Uptake and Efflux of ^{32}P and ^{14}C-Glucose in *Helianthus annuus* Roots and Cell Cultures of *Daucus carota*. Journal of Plant Physiology 1985; 118, 431–438.

[17] Camacho-Cristóbal JJ, González-Fontes A. Boron Deficiency Decreases Plasmalemma H+-Atpase Expression and Nitrate Uptake, and Promotes Ammonium Assimilation into Asparagine in Tobacco Roots. Planta 2007; 226, 443-451.

[18] Camacho-Cristóbal JJ, Rexach J, González-Fontes A. Boron in Plants: Deficiency and Toxicity. Journal Integrative Plant Biology 2008; 50, 1247-1255.

[19] Dordas C. Foliar Boron Application Improves Seed Set, Seed Yield, and Seed Quality of Alfalfa. Agronomy Journal 2006; 98, 907–913.

[20] Sinclair TR, Purcell LC, King CA, Sneller CH, Chen P, Vadez V. Drought Tolerance and Yield Increase of Soybean Resulting from Improved Symbiotic N_2 Fixation. Field Crops Research 2007; 101, 68–71.

[21] Malavolta E, Vitti GC, Oliveira AS. Avaliação do Estado Nutricional das Plantas: Princípioseaplicações. Piracicaba: Potafos; 1997.

[22] Brown P.H, Hu H. Phloem Mobility of Boron is Species Dependent: Evidence for Phloem Mobility in Sorbitol-rich Species. Annals of Botany 1996; 77, 497-505.

[23] Planchet E, Rannou O, Ricoult C, Boutet-Mercey S, Maia-Grondard A, Limami AM. Nitrogen Metabolism Responses to Water Deficit Act Through both Abscisic Acid (ABA)-Dependent and Independent Pathways in *Medicago truncatula* During Post-Germination. Journal of Experimental Botany 2011; 62, 2, 605–615.

[24] Sinclair TR, Serraj R. Legume Nitrogen Fixation and Drought. Nature 1995; 378, 344.

[25] Serraj R, Sinclair TR, Purcell LC. Symbiotic N_2 Fixation Response to Drought. Journal of Experimental Botany 1999; 50, 143–155.

[26] Bellaloui N, Hu Y, Mengistu A, Kassem MA, Abel, CA. Effects of Foliar Boron Application on Seed Composition, Cell Wall Boron, and Seed δ^{15}N and δ ^{13}C isotopes in Water-stressed Soybean Plants. Frontiers in Plant Nutrition 2013; 4, 1-12.

[27] Fehr WR, Caviness CE. Stages of Soybean Development. Iowa Agriculture and Home Economics Experiment Station. Iowa State University, Ames, Iowa; 1977.

[28] Klepper L, Hageman RH. The Occurrence of Nitrate Reductase in Apple Leaves. Plant Physiology 1969; 44, 1, 110-114.

[29] Bellaloui N, Reddy KN, Zablotowicz RM, Mengistu A. Simulated Glyphosate Drift Influences Nitrate Assimilation and Nitrogen Fixation in Non-Glyphosate- Resistant Soybean, Journal of Agriculture and Food Chemistry 2006; 54, 9, 3357-3364.

[30] Hardy RWF, Holsten D, Jackson EK, Burns RC. The Acetylene-Ethylene Assay for Nitrogen Fixation: Laboratory and Field Evaluation. Plant Physiology 1968; 43, 1185-1205.

[31] Zablotowicz RM, Focht DD, Cannell GH. Nodulation and N Fixation of Field Grown California Cowpeas as Influenced by Irrigated and Drought Conditions. Agronomy Journal 1981; 73, 9-12.

[32] Lohse G. Microanalytical Azomethine-H Method for Boron Determination in Plant Tissue. Communications in Soil Science and Plant Analysis 1982; 13, 2.

[33] Bellaloui N, Reddy KN, Gillen AM, Abel CA. Nitrogen Metabolism and Seed Composition as Influenced by Foliar Boron Application in Soybean. Plant and Soil 2010a; 336, 143–155.

[34] Bellaloui N, Smith JR, Gillen AM, Ray JD. Effect of Maturity on Seed Sugar in the Early Soybean Production System as Measured on Near-Isogenic Soybean Lines. Crop Science 2010b; 49, 608–620.

[35] Bellaloui N, Mengistu A. Seed Composition Is Influenced by Irrigation Regimes and Cultivar Differences in Soybean. Irrigation Science 2008; 26, 261-268.

[36] Bellaloui N, Mengistu A, Paris RL. Soybean Seed Composition in Cultivars Differing in Resistance to Charcoal Rot (*Macrophomina phaseolina*). Journal of Agricultural Sciences 2008; 146, 667-675.

[37] Wilcox JR, Shibles RM. Interrelationships Among Seed Quality Attributes in Soybean. Crop Science 2001; 41, 11–14.

[38] Bellaloui N, Mengistu A, Fisher DK, Zobiole L HS, Abbas HK. Irrigation Management: Effects of Soybean Diseases on Seed Composition in Genotypes Differing in their Disease Resistance under Irrigated and Non-irrigated Conditions. In: Irrigation Management, Technologies, Ali MH (ed.), Chapter 8, Nova Publishers, 2012; p193-228.

[39] Boydak E, Alpaslan M, Hayta M, Gercek S, Simsek M. Seed Composition of Soybeans Grown in the Harran Region of Turkey as Affected by Row Spacing and Irrigation. Journal of Agricultural and Food Chemistry 2002; 50, 718-720.

[40] Sigma, USA. Glucose (HK) Assay Kit Product Code GAHK-20. ProducInformation, 2012.http://www.sigmaaldrich.com/etc/medialib/docs/Sigma/Bulletin/gahk20bul.Par.0001.File.tmp/gahk20bul.pdf (accessed July 29 2013).

[41] Sigma, USA. Fructose Assay Kit, Product Code FA-20. ProductIninformation, 2012. http://www.sigmaaldrich.com/etc/medialib/docs/Sigma/Bulletin/fa20bul.Par.0001.File.tmp/fa20bul.pdf (accessed July 29 2013).

[42] SAS. SAS 9.1 TS LeVel 1M3, Windows Version 5.1.2600, SAS Institute: Cary; 2001.

[43] Yamagishi M, Yamamoto Y. Effects of Boron on Nodule Development and Symbiotic Nitrogen Fixation in Soybean Plants. Soil Science and Plant Nutrition 1994; 40, 265-274.

[44] Mills HA, Jr. Jones JB. Plant Analysis Handbook II. Micromacro Publisher: Athens; 1996.

[45] Reinbott TM, Blevins DG. Response of Soybean to Foliar-Applied Boron and Magnesium and Soil-Applied Boron. Journal of Plant Nutrition 1995; 18, 179-200.

[46] Shelp BJ. Physiology and Biochemistry of Boron in Plants. In: Gupta UC (ed.) Boron and its Role in Crop Production, USA, Boca Raton, FL. CRC Press; 1993; p 53–85.

[47] Kanayama Y, Kimura K, Nakamura Y, Ike T. "Purification and Characterization of Nitrate Reductase from Nodule Cytosol of Soybean Plants. Physiologia Plantarum 1999; 105, 3, 396-401.

[48] Caba JM, Lluch C, Ligero F. Distribution of Nitrate Reductase Activity in *Vicia faba*: Effect of Nitrate and Plant Genotype. Physiologia Plantarum 1995; 93, 667-672.

[49] Ruiz JM, Baghour M, Bretones G, Belakbir A, Romero L. Nitrogen Metabolism in Tobacco Plants (*Nicotiana Tabacum* L.): Role of Boron as a Possible Regulatory Factor. International Journal of Plant Sciences 1998; 159, 121–126.

[50] Perica S, Bellaloui N, Greve C, Hu H, Brown PH. Boron Transport and Sugar Metabolism in Olive. Journal of the American Society for Horticultural Science 2001; 126, 291-296.

[51] Taji T, Ohsumi C, Iuchi S, Seki M, Kasuga M, Kobayashi M, Yamaguchi-Shinozaki K, Shinozaki K. Important. Roles of Droughtand Cold inducible Genes for Galactinol Synthase in Stress Tolerance in *Arabidopsis thaliana.* Plant Journal 2002; 29, 417–426.

[52] Streeter JG. Effects of Drought on Nitrogen Fixation in Soybean Root Nodules. Plant, Cell & Environment 2003; 26, 1199–1204.

[53] González EM, Gordon AJ, James CL, Arrese-Igor C. The Role of Sucrose Synthase in the Response of Soybean Nodules to Drought. Journal of Experimental Botany 1995; 46, 1515-1523.

[54] Ren C, Bilyeu KD, Beuselinck PR. Composition, Vigor, and Proteome of Mature Soybean Seeds Developed Under High Temperature. Crop Science 2009; 49, 1010–1022.

[55] Wilson RF. Seed Composition. In Boerma H, Specht JE (eds), Soybeans: Improvement, Production, and Uses, Madison, WI. American Society of Agronomy, Crop Science Society of America, and Soil Science Society of America: 2004; p621–668.

[56] Liu K. Soybeans Chemistry, Technology, and Utilization. New York, NY 1997: Chapman & Hall.

[57] Obendorf RL, Horbowicz M, Dickerman AM, Brenac P, Smith ME. Soluble Oligosaccharides and Galactosyl Cyclitols in Maturing Soybean Seeds in Planta and *In Vitro*. Crop Science 1998; 38, 78–84.

[58] Hou A, Chen P, Alloatti J, Li D, Mozzoni L, Zhang B, Shi A. Genetic Variability of Seed Sugar Content in Worldwide Soybean Germplasm Collections. Crop Science 2009; 49, 903-912.

[59] Yoneyama T, Handley LL, Schrimgeour CM, Fisher DB, Raven JA. Variations of the Natural Abundances of Nitrogen and Carbon Isotopes in *Triticum aestivum* with Special Reference to Phloem and Xylem Exudates. New Phytologist 1997; 137, 205-213.

[60] O'Leary M.H. Environmental Effects on Carbon Isotope Fractionation in Terrestrial Plants. In: Wada E, Yoneyama T, Minagawa M, Ando T, Fry BD (eds), Kyoto, Kyoto University Press, 1995; p517-530.

[61] Kumarashinghe KS, Kirda C, Arag M, Zapata F, Danso KSA. ^{13}C isotope Discrimination Correlated with Biological Nitrogen Fixation in Soybean (*Glycine max* L. Merrill). Plant and Soil 1992; 139, 145–147.

[62] Högberg P. ^{15}N Natural Abundance as a Possible Marker of the Ectomycorrhizal Habit of Trees in Mixed African Wood Land. New Phytology 1990; 115, 483-486.

[63] Livingston NJ, Guy RD, Sun ZJ, Ethier GJ. The Effects of Nitrogen Stress on the Stable Carbon Isotope Composition, Productivity and Water Use Efficiency of White Spruce (*Piceaglauca* (Moench) Voss) Seedlings. Plant Cell & Environment. 1999; 22, 281–289.

[64] Sun ZJ, Livingston, NJ, Guy RD, Ethier GJ. Stable Carbon Isotopes as Indicators of Increased Water Use Efficiency and Productivity in White Spruce (*Picea glauca* (Moench) Voss) Seedlings. Plant Cell & Environment. 1996; 19, 887–894.

[65] Matsushima M. (2007). Nitrogen and Water Availabilities and Competitiveness of Bluejoint: Spruce Growth and Foliar Carbon-13 and Nitrogen-15 Abundance. *S.S.S.A. J.* 71, 2007; p1547-1556.

[66] Farquhar GD, Richards RA. Isotopic Composition of Plant Carbon Correlates with Water-Use Efficiency of Wheat Genotypes. Australian Journal of Plant Physiology 1984; 11, 6, 539-552.

[67] Fotelli MN, Rennenberg H, Holst T, Mayer H, Geßler A. Carbon Isotope Composition of Various Tissues of Beech (*Fagus sylvatica*) Regeneration Is Indicative of Recent Environmental Conditions within the Forest Understory. New Phytologist 2003; 159, 1, 229-244.

[68] Kume A, Satomura T, Tsuboi N, Chiwa M, Hanba YT, Nakane K, Horikoshi T, Saku-

gawa H. Effects of Understory Vegetation on the Ecophysiological Characteristics of an Overstory Pine, Pinus Densiflora. Forest Ecology and Management 2003; 176, 195-203.

[69] Robinson D, Handley LL, Scrimgeour CM, Gordon DC, Forster BP, Ellis RP. Using Stable Isotope Natural Abundances (δ^{15}N and δ^{13}C) to Integrate the Stress Responses of Wild Barley (*Hordeum spontaneum* C. Koch.) Genotypes. Journal of Experimental Botany 2000; 51, 342, 41-50.

4

Nitrogen Fixation Outside and Inside Plant Tissues

C.P. Chanway, R. Anand and H. Yang

1. Introduction

Nitrogen is one of the most important elements in biological systems, comprising the main building blocks of nucleic acids, enzymes and proteins among its multiple functions. In nature, it exists primarily in the gaseous form and constitutes approximately 78% of the atmosphere. Despite its abundance, nitrogen (N) is one of the most growth-limiting nutrients in terrestrial and aquatic ecosystems (Dalton & Krammer, 2006) because its gaseous form is inert and unusable by most living organisms except for nitrogen fixing microorgansisms. For it to become biologically available, atmospheric nitrogen must be transformed or "fixed" from its inert gaseous form (N_2) to ammonia (NH_3), which can then be assimilated into a variety of important biochemicals. This transformation, which requires a large amount of energy to break apart the triple-bonded N atoms that comprise gaseous N_2, is called 'nitrogen fixation' (NF). Nitrogen is fixed naturally through energy-releasing abiotic processes such as lightening, forest fires and volcanic activity. These processes produce oxides of N in the atmosphere that subsequently dissolve in rain and descend to the ground as NH_3 molecules. Approximately 12% of annual global NF is fixed in this way (Bezdicek & Kennedy, 1998). Fertilizer production using high temperatures and pressures in the Haber-Bosch process occurs widely and accounts for approximately 20% of annual global NF (Bezdicek & Kennedy, 1998). However, the process is fossil-fuel intensive and consumes 3-5% of the world's natural gas annually (Myrold & Bottomley, 2007). Alternatively, NF occurs through the normal metabolic activity of many prokaryotic microorganisms, known as diazotrophs, through a process commonly referred to as biological nitrogen fixation (BNF). This essentially "free" process is responsible for the addition of almost all biologically available N that enters terrestrial ecosystems, some 140 million metric tons per year (Bezdicek & Kennedy, 1998; Galloway et al., 2008). Biological nitrogen fixation is an ATP-demanding process that is catalyzed by the enzyme complex known as nitrogenase, which is found in many members of the Bacteria and Archaea (Galloway et al., 2008). Terrestrial BNF occurs primarily in the soil, by either free-living diazotrophs or those associated to varying degrees with plants (see below).

2. Free-living bacteria

Free-living diazotrophic bacteria are those that do not associate with plants (*cf.* rhizosphere bacteria below) and are found in soils that are free from the direct influence of plant roots. These microorganisms are ubiquitous in terrestrial and aquatic environments and are physiologically very diverse (Reed et al., 2011). Since most soils are C and N limited, the amount of N_2 they fix in soil is restricted by access to energy sources, *i.e.,* substrates to generate adenosine triphosphate (ATP) and micronutrients required for the synthesis and functioning of nitrogenase (Reed et al., 2011). BNF by free-living diazotrophs is also limited by the severe oxygen sensitivity of nitrogenase (Postgate, 1998), which is a problem that has been at least partially overcome in different ways by diazotrophs participating in nitrogen-fixing symbioses (see below). Antagonistic microbial interactions such as parasitism and competition for nutrients (Cacciari et al.,1986; Bashan & Holguin, 1997; Bashan et al., 2004) further reduce the amount of N_2 they fix. While the general belief is that free-living diazotrophs do not contribute large quantities of fixed N to most terrestrial ecosystems, perhaps 3-5 kg/ha/yr (Postgate, 1998; Newton, 2007), their cumulative N contributions are thought to be important in some tropical and temperate forest ecosystems (Cleveland et al., 1999; Gehring et al., 2005; Reed et al., 2007; 2008).

3. Rhizosphere bacteria

The rhizosphere is defined as soil that surrounds plant roots and is under their direct metabolic influence (Curl & Truelove, 1987). Proximity to plant roots is important for soil organisms as actively growing plant roots deposit approximately 20% of annual photosynthate in the rhizosphere (Nguyen, 2003), but depending on the type of plant and its growth stage, more than 50% of newly-fixed carbon may be deposited in the rhizosphere at any given time. The soluble carbon compounds that plants deposit through their root systems are known as root exudates. These comprise a wide range of carbon compounds (amino acids, peptides, proteins, enzymes, "growth factors", vitamins and phytohormones) (Grayston et al., 1997; Jones et al., 2004; Shi et al., 2012) and are released continuously during the growing season through a process known as root exudation (Jones et al., 2004). This process has been shown to significantly stimulate growth and population sizes of most soil microorganisms, but effects are particularly noticeable in soil bacteria and fungi. The degree of stimulation is significant: in comparison with bulk soil not under the influence of plant roots, the rhizosphere typically supports 5-100 x larger bacterial and fungal populations than non-rhizosphere or "bulk" soil (Warembourg, 1997; Dobelaere et al., 2003). Due to their ability to fix N_2, diazotrophs can have a competitive advantage over non-N_2 fixing bacteria in the rhizosphere and prevail in it particularly when soil N is limited (Döbereiner & Pedrosa, 1987). In addition to stimulating their own growth, rhizosphere diazotrophs representing several genera (*e.g.*, *Acetobacter, Azoarcus, Azospirillum, Azotobacter, Beijerinckia, Burkholderia, Enterobacter, Herbaspirillum, Klebsiella, Paenibacillus* and *Pseudomonas*) have been shown to enhance the growth of the plants that generate a suitable rhizosphere. These include agriculturally-important species such as rice, wheat, barley, potato and several vegetable crops (Dobelaere et al., 2003). While it is logical to assume diazotrophic rhizosphere bacteria enhance plant growth through BNF, these microorganisms are often capable of producing plant growth-enhancing phytohormones and pathogen-suppressing antibiotics (Chanway, 2002). Many are also able to enhance the availability of N, P and S in the rhizosphere enzymatically, rendering it difficult to conclude with certainty the mechanism by which plant growth is stimulated.

4. Phyllosphere

The leaf surface, or phyllosphere is another microsite known to be colonized by a wide range of microorganisms, including diazotrophic bacteria (Lindow & Brandl, 2003). While comparatively little work has been done on phyllosphere-colonizing diazotrophs, it is likely that their contribution to plant N nutrition is modest, owing to the energy constraints and problems associated with oxygen toxicity of nitrogenase such N_2 fixing microorganisms would experience. Nevertheless, phyllosphere-colonizing diazotrophs should be evaluated further for possible contributions to plant nutrition.

5. Cyanobacterial associations

Cyanobacteria are prokaryotes belonging to the domain *Bacteria* that are capable of fixing carbon, through oxygenic photosynthesis, as well as nitrogen through BNF. As a result, cyanobacteria are nutritionally independent to a large degree (Meeks, 1988) and occupy diverse habitats, ranging from freshwater and oceanic ecosystems, temperate soils to extreme environments such as hot springs and deserts (Herrero et al., 2001). To counter the adverse effects of oxygen evolution from photosynthesis on BNF, diazotrophic cyanobacteria have developed strategies such as temporal separation of oxygenic photosynthesis and micro-oxic NF (Cerveny et al., 2013) and formation of thick-walled, non-photosynthetic heterocysts with microaerobic interiors that are conducive for NF (Wolk et al., 1994).

Many diazotrophic cyanobacterial species enter into symbioses with eukaryotes including phytoplankton, fungi and terrestrial plants. Cyanobacterial symbionts (cyanobionts) in these associations may contribute a significant portion of N required for growth of both organisms through BNF in N-limited aquatic and terrestrial environments (Schell & Alexander, 1973; Hobara et al., 2006). In the ocean, they are frequently found in association with diatoms (Ferrario et al., 1995) and brown algae (Carpenter, 1972), while in fresh water, the cyanobiont *Anabaena azollae* forms a symbiotic association with water ferns belonging to the genus

Azolla (Talley et al., 1977). The *Azolla-Anabaena* symbiosis often occurs in rice culture, where it results in enhanced rice performance due to a N fertilizer effect from BNF (Yanni, 1992). Diazotrophic cyanobacteria have also been found in the coralloid roots of cycads (Gehringer et al., 2010), auricles of liverworts (Adams & Duggan, 2008), slime cavities of hornworts (Adams & Duggan, 2008), stem glands of *Gunnera* (Bergman et al., 1992), hyphae of lichens (Jayasinghearachchi & Seneviratne, 2004) and on the leaf surfaces of mosses (Solheim & Zielke, 2002), where they fix and transfer N to the non-N_2 fixing partner. To facilitate locating a suitable symbiotic partner, some cyanobacteria belonging to the order *Nostocales* can differentiate into motile segments termed hormogonia that travel chemotactically toward a potential partner (Campbell & Meeks, 1989). If the partner is suitable, hormogonia differentiate back into vegetative cells and enter into an active diazotrophic symbiosis where they fix and provide N for both partners.

6. Legume x *Rhizobium* symbiosis

The legume x *Rhizobium* symbioses is a well-known mutualism involving plants from the angiosperm family *Leguminosae* (synonym *Fabaceae*) and bacteria belonging to the family *Rhizobiaceae* (Postgate 1998). In this symbiosis, diazotrophic soil bacteria belonging to the genus *Rhizobium* (or closely related genera) seek out and infect roots of suitable legume plant hosts using a complex chemical signaling system. Bacteria then colonize certain root cortex cells and initiate formation of a new plant organ, the root nodule. Bacteria proliferate within root nodule cells and then differentiate into a nitrogen fixing form called a bacteroid, to fix N_2. The plant vascular system is continuous with that of the root nodule, which enables newly fixed N to be rapidly translocated to other parts of the plant most in need of N. Energy for BNF is provided to bacteroids from plant photosynthesis and the oxygen concentration in root nodule cells is tightly regulated by an iron containing protein very similar in composition to hemoglobin called leghemoglobin. Bacteroids in this coadapted symbiosis are capable of high rates of BNF, *i.e.*, up to 600 kg/ha/yr, particularly when compared to BNF in the rhizosphere (15-25 kg/ha/yr) or by free-living diazotrophs (3-5 kg/ha/yr).

The legume x *Rhizobium* symbiosis has been studied widely from ecological, agronomic and molecular biological perspectives to not only enhance the nitrogen-fixing efficacy of existing symbioses but to determine if similar associations might also be developed with non-legume crops (Oldroyd et al., 2011, Udvardi & Poole, 2013). Notwithstanding impressive gains in our understanding of this symbiosis, its inherent complexity currently precludes our ability to extend effective nodulation or nitrogen fixing capacity to non-legume species (Beatty & Good, 2011). However, systemic infection of non-legumes with endophytic diazotrophic bacteria has been observed in several plant species and in some cases, appears to satisfy the majority of the "host" plant's N requirement with amounts of BNF similar to that of legume root nodules (see *Sugarcane and Other Crops* and *Nitrogen Fixation in Gymnosperms* below).

7. Actinorhizal symbiosis

The actinorhizal symbiosis refers to a root nodule-forming, nitrogen-fixing symbiotic relationship that is functionally analogous to the legume x *Rhizobium* symbiosis but distinct in most of the details. It is restricted to members of a small group of woody, non-legume pioneer species known as Actinorhizal plants and diazotrophs belonging to a single genus, *Frankia*, in a phylum of mostly filamentous soil bacteria, the Actinobacteria. Actinorhizal plants comprise approximately 200 plant species in 24 genera belonging to 8 plant families: *Casuarinaceae, Betulaceae, Myricaceae, Elaeagnaceae, Coriariaceae, Rhamnaceae, Datiscaceae* and *Rosaceae* (Huss-Danell, 1997). All but some members of *Datiscaceae* are shrubs or trees, and all are relatively shade-intolerant pioneering species that are able to colonize N-poor sites due to their ability to enter into the actinorhizal symbiosis (Crocker & Major, 1955; Chapin et al., 1994). *Frankia* spp. are notoriously difficult to grow in culture hence our understanding of these microorganisms and the actinorhizal symbioses lags far behind that of the well-characterized legume x *Rhizobium* associations. Owing to their recalcitrance to growth in culture, taxonomy of the genus *Frankia* is poorly developed, with only one species designation since a member of this genus was first reproducibly isolated over 30 years ago (Callaham et al., 1978). Numerous strains of *Frankia* have since been isolated from inside surface-sterilized root nodules and are designated as *Frankia* spp. followed by a strain number or name.

In contrast to the easily cultured, gram negative rod shaped cells that typify *Rhizobium* (and relatives) species, *Frankia* spp. strains possess gram-variable cells of three distinct types: filamentous vegetative hyphae, reproductive spores and N_2-fixing vesicles (Benson & Silvester, 1993). The infection of plant roots and subsequent formation of root nodules bear some similarity to these processes in *Rhizobium*-infected legumes, but the origin and composition of *Frankia*-induced root nodules differs significantly (Wall & Berry, 2008). Unlike the *de novo* structure of a legume root nodule, actinorhizal root nodules are comprised of numerous tightly or loosely packed lobes, each of which originates from the pericycle indicating that it is a modified lateral root. Unlike *Rhizobium* bacteroids, the filamentous *Frankia* hyphae differentiate into an active nitrogen fixing form by developing numerous vesicles inside the root nodule or they may differentiate into spores and contribute nothing to the N economy of the plant. The tendency to sporulate in the nodule appears to be a strain specific characteristic that is influenced by the identity of plant host species. The oxygen concentration in root nodules is regulated primarily by varying the number and thickness of nitrogen fixing vesicle wall layers, however, the elegant heme protein leghemoglobin, which facilitates oxygen diffusion and supply in legume root nodules, has also been detected in nodules of some actinorhizal plants. Physiological and structural differences between actinorhizal and legume root nodules suggest that actinorhizal symbioses are less evolved and less efficacious in NF compared to legume x *Rhizobium* symbioses, but an effective actinorhizal symbiosis can fix several hundred kilograms of N per hectare annually, an amount which is similar to an effective legume symbiosis.

8. Bacterial endophytes

The presence of microorganisms in plant tissues might reasonably be considered an indication that a disease state is imminent, however this is not necessarily the case. Several decades ago, Trevet and Hollis (1948) reported the occurrence of bacteria within tissues of healthy potato plants and several studies have since demonstrated that internal tissues of healthy plants are colonized by bacteria. The term 'bacterial endophytes' has been used to describe bacteria that reside within living plant tissues without causing disease (Wilson, 1995; van der Lelie et al., 2009), however it does not differentiate whether such bacteria are (i) truly harmless, (ii) latent pathogens (Sinclair & Cerkauskas, 1996) or (iii) able to elicit production of symbiotic structures such as root nodules on the host. We use the term 'endophyte' in this paper to describe bacteria that 'can be detected at a particular moment within the tissue of apparently healthy plant hosts' (Schultz & Boyle, 2005) without inducing disease or organogenesis (Iniguez et al., 2005). In contrast to free-living, rhizosphere or phyllosphere microorganisms, bacterial endophytes are better protected from abiotic stresses such as extreme variations in temperature, pH, nutrient and water availability as well as biotic stresses such as competition (Loper et al., 1985; Cocking, 2003; Rosenblueth & Martinez-Romero, 2006). In addition, bacterial endophytes colonize niches that are more conducive to forming mutualistic relationships with plants through NF, for example, as suggested in sugarcane and other crops (see below) (Richardson et al., 2009).

8.1. Bacterial endophytes of sugarcane and other agricultural crops

In the 1980's, Brazilian researchers were perplexed by the consistently high yields of field-grown sugarcane, an N-demanding crop, without exogenous N fertilizer application and looked for a microbiological explanation for this apparently anomalous observation. After it was determined that rhizospheric NF did not occur at sufficient rates to facilitate high sugarcane yields, Cavalcante & Döbereiner (1988) looked for microorganisms within sugarcane tissues that might be involved and isolated a diazotrophic bacterium, *Gluconoacetobacter diazotrophicus*, previously known as *Acetobacter diazotrophicus*. Its ability to establish high endophytic populations and to fix N_2 in high sucrose concentrations (Boddey et al., 1991) at low pH (Boddey et al., 1991; Stephan et al., 1991), conditions which typify sugarcane tissues, led to the suggestion that this diazotrophic bacterium could satisfy almost all of the sugarcane N requirements. Several other N-contributing diazotrophic endophytes were subsequently found to associate with sugarcane including two *Herbaspirillum* species (Cavalcante & Döbereiner, 1988; Baldani et al., 1992; 2002), *Azoarcus* spp. (Reinhold et al., 1993) and *Azospirillum brasilense* (de Bellone and Bellone, 2006). However, more recent studies indicate that the primary source of diazotrophy may involve a consortium of bacteria, possibly including uncultured strains, that live on or inside plant tissues (Burbano et al., 2011; Taulé et al., 2012). Since the discovery of diazotrophic endophytes in sugarcane (*Saccharum officinarum* L.) (Ruschel et al., 1975), several other agriculturally important crop species including rice (*Oryza sativa*) (Shrestha & Ladha, 1996), maize (*Zea mays* L.) (Montañez et al., 2009) and kallar grass

(*Leptochloa fusca* L.) (Malik et al., 1997) have been postulated to receive significant amounts of fixed N_2 in this way.

Despite sugarcane's apparent potential to derive much of its N from BNF, it has not been proven that *G. diazotrophicus* or any endophytic diazotroph is the primary causal agent of N accumulation by this plant in the field. Non-culturable endophytes with plant growth-promoting potential have been detected in sugarcane tissues (Hallmann et al., 1997; Mendes et al., 2007). In addition, all known culturable bacterial endophytes including *G. diazotrophicus* colonize the rhizosphere as well as internal plant tissues, rendering it difficult if not impossible to distinguish endophytic from non-endophytic BNF. There is also no clear evidence that N is transferred directly from endophytic diazotrophs to the host plant, unlike other symbiotic N_2 fixing systems where specific nitrogenous compounds such as ureides in soybean (Herridge, 1982) and citrulline in alder (Leaf et al., 1958) are synthesized to transport fixed N from the site of NF to other host tissues. It is generally accepted that the N transfer to the host from rhizospheric NF results from the release of mineralized N from dead bacterial cells (Dobbelaere et al., 2003; Momose et al., 2009; Mia & Shamsuddin, 2010). Whether or not a similar mechanism operates with endophytic diazotrophs remains unanswered.

If endophytic diazotrophs are ultimately proven to be the primary cause of BNF and growth promotion of their host plants, such a plant x microbe association would represent another type of mutualistic symbiosis where the plant provides photosynthate and a competition-free, microaerobic environment for effective N_2 fixation (Hallman et al., 1997; Reinhold-Hurek & Hurek, 1998a,b; Santi et al., 2013) for microorganisms in return for plant growth-promoting amounts of N from BNF. In contrast to the legume x *Rhizobium* and actinorhizal symbioses, no symbiosis-specific structures (*e.g.*, nodules or chambers housing bacteria) have been detected in infected host, though Anand & Chanway (2013 a)observed occasional plant cells filled with diazotrophic *Paenibacillus polymyxa* in inoculated lodgepole pine tissues (Fig. 1b).

8.2. Nitrogen fixation in gymnosperms: Effects on seedling growth and N content

Lodgepole pine (*Pinus contorta* var. *latifolia* (Dougl.) Engelm.), a commercially important gymnosperm species indigenous to western North America, is capable of growing in very rocky substrates and is notable for its ability to thrive on nutrient poor, N limited soils (Weetman et al., 1988; Chapman & Paul, 2012). Based on earlier work with lodgepole pine suggesting that rhizospheric BNF contributed only small amounts of N to seedlings (Chanway & Holl, 1991) as well as reports that BNF in sugarcane was endophytic, we searched for endophytic diazotrophs in lodgepole pine as a possible explanation for the ability of this species to grow on N-deficient substrates. We successfully isolated several *Paenibacillus* strains that possessed significant acetylene reduction activity from extracts of surface-sterilized lodgepole pine seedling and tree tissues (Bal et al., 2012) and tested them for BNF with lodgepole pine and western red cedar seedlings in a ^{15}N soil dilution assay (Bal & Chanway, 2012a,b). When pine was reintroduced to one of the strains, *Paenibacillus polymyxa* strain P2b-2R, and grown in a very N-limited soil, seedlings were found to derive more than half (66%) of their foliar N from BNF, but their growth was *inhibited* compared to non-inoculated controls 9 months after planting (Bal & Chanway, 2012a). Similar effects, *i.e.*, BNF and seedling growth inhibition, were observed with western red cedar (Bal & Chanway, 2012b).

Based on these results, we hypothesized that soil N depletion would eventually restrict the growth rate of control seedlings to a point where they would be outperformed by N_2-fixing seedlings, and set up longer term growth experiments to evaluate this possibility. After a 13-month growth period, pine seedlings treated with P2b-2R were observed to derive most of their foliar N (79%) from BNF (Anand et al., 2013), which was confirmation of the ability of lodgepole pine to fix N after colonization by *P. polymyxa*. When compared with previous results, we found that inoculated seedlings grown for 7, 9 and 13 months derived 30%, 66% (Bal & Chanway, 2012a) and 79% (Anand et al., 2013), respectively, of their foliar N from the atmosphere. The progressive increase in the proportion of N derived from BNF *i.e.*, %Ndfa, with seedling age suggests that BNF is an important component of N nutrition of pine in N-limited soil. Urquiaga et al. (1992) also observed an increasing reliance on BNF with seedling age, and concomitant decreasing soil N, in sugarcane cv. Krakatau. The %Ndfa rose from 6% to 55% during the interval 100 - 250 days after emergence. In addition to high %Ndfa, 13-month old seedlings treated with P2b-2R also accumulated significantly more biomass (78%) than controls, and had overcome the growth inhibition observed in seedlings of younger ages. While *P. polymyxa* possesses several characteristics that can result in plant growth promotion (Chanway, 2002), the enhanced performance of 13-month old seedlings was likely caused by an increase in the amount of N derived from BNF because the proportion of foliar N from BNF (68%-79%) (Anand et al., 2013) in older seedlings was only marginally higher than that of nine-month old seedlings (64%-66%) (Bal & Chanway, 2012a), but the concentration of foliar N (2.35%) (Anand et al., 2013) was 5-fold greater than nine-month old seedlings. The difference in total foliar N content, *i.e.*, foliar N concentration x foliar biomass, was even greater and represented an elevation of foliar N concentration from a level considered to be very severely N deficient in all nine-month old seedlings (Bal & Chanway, 2012a) and the 13-month old controls (Anand et al., 2013) to one that is adequate for healthy lodgepole pine (Ballard & Carter, 1986). The comparatively early onset of BNF and delayed seedling growth response suggest that development of a fully effective N_2-fixing bacterial population, able to enhance foliar N concentration in pine, is not a rapid process. Establishment of fully effective BNF in pine does not appear to depend on the population size of endophytic diazotrophs because P2b-2R colonization of root, stem and needle tissues in older seedlings did not differ significantly from younger seedlings (Anand et al., 2013), though it is possible that external root colonization was quantitatively related to BNF. However, if BNF is endophytic and requires physiological modifications of bacteria, *e.g.*, differentiation of *Paenibacillus* into a N_2-fixing form similar to *Rhizobium* bacteroids (Postgate, 1998), or plants, *e.g.*, the establishment of specialized sites of BNF within plant tissues or cells, to be fully effective, some time may be required for them to complete. Our observations of intracellular colonization by P2b-2R*gfp* (Figure 1; Anand & Chanway, 2013a) are consistent with this idea but further research is required to evaluate the physiology of *P. polymyxa* in association with pine as well as the relationship of bacterial population size to BNF.

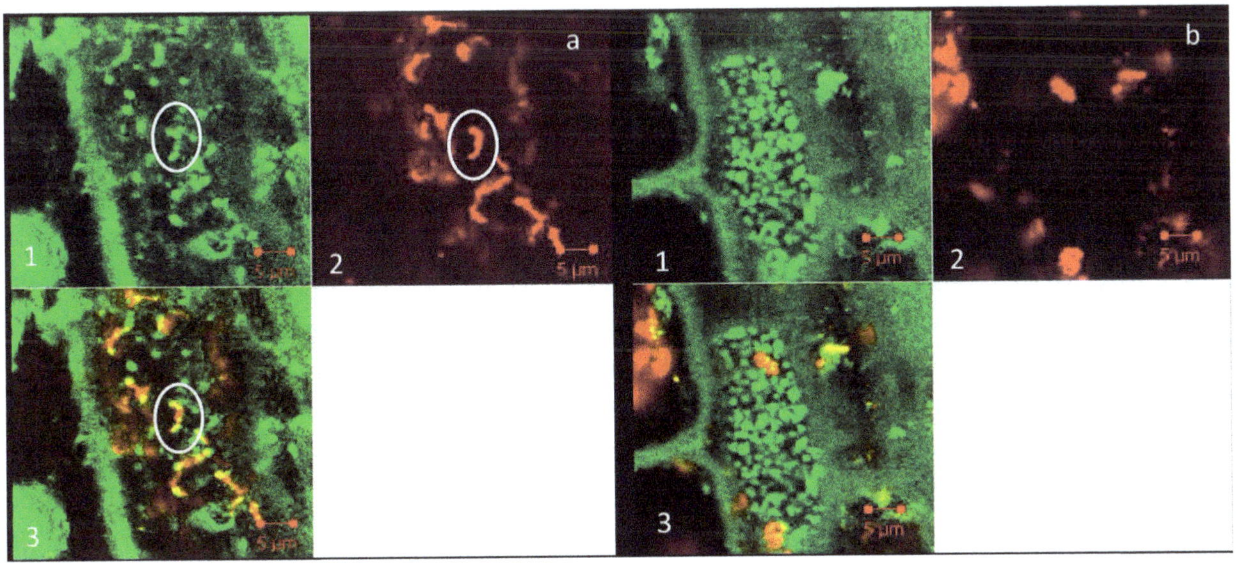

Figure 1. Colonization of cortex cells from a pine stem by green fluorescent protein (GFP) labeled *Paenibacillus polymyxa* strain P2b-2R*gfp* using confocal laser scanning microscropy. Stem sections were viewed using green channel light (Figure 1a & b, panel 1) to visualize bright green GFP-labeled bacteria or red channel light (Figure 1a & b, panel 2) to visualize chloroplasts within pine cells. Using green channel light, GFP-labeled bacteria appear as green dots in a single pine cell (Figure 1a & b, panel 1). Under red light, chloroplasts appear as larger red organelles in the same single pine cells (Figure 1a & b, panel 2). Cells viewed using green and red channel light simultaneously (Figure 1a & b, panel 3) show GFP-labeled bacteria and chloroplasts within a pine cell. The white circle in Figure 1a, panel 3 encloses (i) several GFP-labeled bacterial cells, (ii) a single chloroplast in Figure 1a, panel 2 and GFP-labeled bacteria associated very closely with the chloroplast in Figure 1a, panel 3. Complete methodology is described in Anand and Chanway (2013a).

8.3. Nitrogen fixation in gymnosperms: Colonization of seedlings

The ability of *P. polymyxa* to colonize gymnosperm seedlings internally and externally has been evaluated in several greenhouse experiments (Bent & Chanway, 2002; Bal & Chanway, 2012a,b; Anand et al., 2013), but the ability of a green fluorescent protein-tagged derivative of P2b-2R to colonize internal pine tissues (Anand & Chanway, 2013a) provided convincing evidence that P2b-2R is endophytic, notwithstanding possible differences in behavior of GFP-tagged and wild type cells (van der Lelie et al., 2009). The population densities of P2b-2R inside pine tissues were observed to be comparable to densities reported for endophytic diazotrophs of crop plants such as rice (*Oryza sativa* L.), 10^4 - 10^5 cfu/g tissue (Elbeltagy et al., 2001), surgarcane, 10^5 - 10^7 (Sevilla et al., 2001) and grape (*Vitis vinifera* L.), 10^5 -10^9 (Compant et al., 2005) as well as hybrid spruce (*Picea glauca* x *P. engelmannii*), 10^3 – 10^5 (Shishido et al., 1999) and other tree species, 10^1 – 10^7 (Izumi, 2011). Recovery of P2b-2R from stem and needle tissues suggests bacteria migrate from the roots and soil to aerial plant organs: the observation that root and stem populations decreased while needle populations increased (Anand et al., 2013) supports this idea. Compant et al. (2005) observed a similar trend in grape plantlets inoculated with *Burkholderia sp.* PsJN and suggested that the stem acts as a transportation corridor for bacteria to reach leaves, which they considered a sink for endophytes. However, it is also possible that primordial pine stem and needle tissues were colonized at germination by P2b-2R residing on the seed coat and in the spermosphere, resulting in growth of bacterial populations *in situ* after shoots expanded, but the mechanism and patterns by which P2b-2R colonizes aerial pine tissues cannot be affirmed without further study.

We also observed intact cells from different pine stem sections that were colonized internally by GFP-labeled bacteria (Anand & Chanway, 2013a). In some cases, GFP-labeled bacteria were observed in pine cells in close proximity to chloroplasts (Fig. 1a), which raises the possibility that bacteria colonized microsites near these energy-generating organelles. GFP labelling were also observed tightly packed within other pine cells (Fig. 1b). Whether bacteria in either of these pine cells (Fig. 1) were fixing N cannot be determined, but these unique endophytic colonization patterns warrant further study. Endophytic bacteria have also been observed inside cells of grape (Compant et al., 2005), grasses (Hurek et al., 1994), sugarcane (James & Olivares, 1998) and poplar (*Populus deltoides* x *P. nigra*) (van der Lelie et al., 2009). Intracellular colonization of the pine stem cortex was rare and we were unable to inspect needle tissues for P2b-2R*gfp* so it is unclear if colonized pine cells are of biological importance. GFP-labeled bacteria may simply have detected micro-colonies of bacteria feeding on dead or dying plant cells in the stem cortex. Alternatively, it is tempting to hypothesize a possible role for intracellular bacteria in BNF considering the foliar ^{15}N dilution observed in colonized seedlings (Bal and Chanway, 2012a,b; Anand et al., 2013; Anand & Chanway, 2013b). Could these cells be specialized seats of N_2-fixation, perhaps similar to *Rhizobium* bacteroids inside cells of legume root nodules? You et al. (1991) reported root cortex cells of rice, a non-leguminous plant, contained diazotrophic bacteria and some species of algae have been found to harbor endophytic bacteria near their chloroplasts (Colombo, 1978; Preisig & Hibberd, 1984). While this is a very interesting possibility, it is premature to focus on intracellular BNF in pine without more convincing supporting data.

Results with lodgepole pine strongly suggest that seedlings can fix N_2 when colonized by *P. polymyxa* and demonstrate that N_2-fixing seedlings develop healthy foliar N levels and enhanced growth in very N-limited soil. Western red cedar may also benefit from treatment with this bacterium (Bal & Chanway, 2012b; Anand & Chanway, 2013b), but amount of BNF and plant growth promotion were reduced compared to pine. Because *P. polymyxa* P2b-2R originated from pine seedling tissue, it is tempting to speculate that some degree of plant x microbe specificity may exist in this type of association. Research with pine and cedar raises several other questions regarding the ecological importance of endophytic diazotrophs in gymnosperms. For example, is *P. polymyxa* unique in its ability to enter into this type of interaction with gymnosperms or is this a comparatively common trait in conifer forests? If *P. polymyxa* is unique, how widespread is its distribution? How long must pine and diazotroph coexist before effective N fixation occurs and is BNF restricted to the seedling stage? What are the effects of soil N on colonization and BNF by *P. polymyxa*? It is possible that endophytic diazotrophs play a key role in natural regeneration of gymnosperms at N-poor sites but whether this is true and their importance in forest ecosystems can only be ascertained through future studies.

9. Conclusions and future research

Biological nitrogen fixation is a free and environmentally benign process through which biologically useful N can be generated for plant growth. Because nitrogen fixation is most effective when bacteria associate with plants to some degree, future research should focus on the various known plant associative and symbiotic nitrogen fixing systems. However, owing to the complexity of root-nodule based symbioses, colonization of plants by endophytic diazotrophs seems to hold the greatest potential for expansion of BNF to plant species that do not normally fix N_2. Initial steps might include mass screening of known endophytic diazotrophs with commercially important plant species while continuing studies of effective nitrogen fixing associations involving plants that are colonized by endophytic diazotrophs. Such studies will help to elucidate the molecular, physiological and ecological details of this potentially useful plant x microbe interaction.

Acknowledgements

Funding for this study was provided by an NSERC Discovery Grant to CPC.

Author details

C.P. Chanway[1], R. Anand[2] and H. Yang[1]

1 Dept of Forest and Conservation Sciences, University of British Columbia, Vancouver, Canada

2 Children's Hospital, Vancouver, Canada

References

[1] Adams, D.G. & Duggan, P.S. (2008). Cyanobacteria-bryophyte symbioses. *Journal of Experimental Botany*, Vol.59, pp.1047-1058

[2] Anand, R & Chanway, C.P. (2013a). Detection of GFP-labeled *Paenibacillus polymyxa* in auto-fluorescing pine seedling tissues. *Biology & Fertility of Soils*, Vol.49, pp.111-118 doi: 10.1007/s0037401207279

[3] Anand, R. & Chanway, C.P. (2013b). N2-fixation and growth promotion in cedar colonized by an endophytic strain of Paenibacillus polymyxa. Biology & Fertility of Soils, Vol.49, pp.235-239 doi: 10.1007/s0037401207359

[4] Anand, R.; Grayston, S. & Chanway C.P (2013). N_2-fixation and seedling growth promotion of lodgepole pine by endophytic *Paenibacillus polymyxa. Microbial Ecology,* Vol.66, pp.369-374 doi: 10.1007/s0024801301961

[5] Bal A.S.; Anand, R.; Berge, O. & Chanway, C.P. (2012). Isolation and identification of diazotrophic bacteria from internal tissues of *Pinus contorta* and *Thuja plicata.* Can J For Res., Vol.42, pp.807–813

[6] Bal, A.S. & Chanway, C.P. (2012a). Evidence of nitrogen fixation in lodgepole pine inoculated with diazotrophic *Paenibacillus polymyxa. Botany,* Vol.90, pp.891–896

[7] Bal, A. & Chanway, C.P. (2012b). ^{15}N foliar dilution of western red cedar in response to seed inoculation with diazotrophic *Paenibacillus polymyxa. Biology & Fertility of Soils,* Vol.48, pp.967–971 doi: 10.1007/s0037401206999

[8] Baldani, J.I.; Reis, V.M.; Baldani, V.L.D. & Döbereiner, J. (2002). A brief story of nitrogen fixation in sugarcane—reasons for success in Brazil. *Func Plant Biol.,* Vol.29, pp. 417–423

[9] Baldani, V.L.D.; Baldani, J.L.; Olivares, F. & Dobereiner, J. (1992). Identification and ecology of *Herbaspirillum seropedicae* and the closely related *Pseudomonas rubrisubalbicans. Symbiosis,* Vol.13, pp.65–73

[10] Ballard, T.M. & Carter, R.E. (1986). Evaluating forest stand nutrient status. B.C. *Min. For., Victoria. Land Manage.,* Rep.29

[11] Bashan, Y. & Holguin, G. (1997). *Azospirillum* plant relationships : environmental and physiological advances (1990-1996). *Can. J. Microbiol.,* Vol.43, pp.103-121

[12] Bashan, Y.; Holguin, G. & de-Bashan, L.E. (2004). *Azospirillum*-plant relationships : physiological, molecular, agricultural, and environmental advances (1997-2003). *Can. J. Microbiol.,* Vol. 50, pp.521-577

[13] Beatty, P.H. & Good A.G. (2011). Future prospects for cereals that fix nitrogen. *Science,* Vol.333, pp.416-417

[14] Benson, D.R. & Silvester, W.B. (1993). Biology of *Frankia* strains, actinomycete symbionts of actinorhizal plants. *Microbiol Rev.,* Vol.57, pp.293–319

[15] Bent, E. & Chanway, C.P. (2002). Potential misidentification of a spore-forming *Paenibacillus polymyxa* isolate as an endophyte by using culture-based methods. *Applied and Environmental Microbiology,* Vol.68, pp.4650-4652

[16] Bergman, B.; Johansson, C. & Söderback, E. (1992). The Nostoc-Gunnera symbiosis. *New Phytol.,* Vol.122, pp.379–400

[17] Bezdicek, D.F. & Kennedy A.C. (1998) *In* Lynch JM, Hobbie JE (eds), Microorganisms in Action. Blackwell, Oxford, UK

[18] Boddey, R.M.; Urquiaga, S.; Reis, V. & Dobereiner, J. (1991). Biological nitrogen fixation associated with sugar cane. *Plant and Soil,* Vol.137, pp.111-117

[19] Burbano, C.S.; Liu, Y.; Rösner, K.L.; Reis, V.M.; Caballero-Mellado, J.; Reinhold-Hurek B. & Hurek, T. (2011). Predominant *nif*H transcript phylotypes related to *Rhizobium rosettiformans* in field-grown sugarcane plants and in Norway spruce. *Environ Microbiol Rpt.,* Vol.3, pp.383–389

[20] Cacciari, I.; Del Gallo, M.; Ippoliti, S.; Lippi, D.; Pietrosanti, T. & Pietrosanti, W. (1986). Growth and survival of *Azospirillum brasilense* and *Arthrobacter giacomelloi* in binary continuous culture. *Plant Soil.,* Vol. 90, pp. 107-116

[21] Callaham, D.; Del Tredici, P. & Torrey, J.G. (1978). Isolation and cultivation in vitro of the actinomycete causing root nodulation in *Comptonia. Science* Vol.199, pp.899–902

[22] Campbell, E.L. & Meeks, J.C. (1989). Characteristics of hormogonia formation by symbiotic *Nostoc* spp. in response to the presence of *Anthoceros punctatus* or its extracellular products. *Applied and Environmental Microbiology,* Vol.55, pp.125-131

[23] Carpenter, E.J. (1972). Nitrogen fixation by a blue-green epiphyte on pelagic *Sargassum. Science,* Vol.178, pp.1207-1208

[24] Cavalcante, V.A. & Döbereiner, J. (1988). A new acid-tolerant nitrogen-fixing bacterium associated with sugarcane. *Plant Soil.,* Vol.108, pp.23-31

[25] Cerveny, J.; Sinetova, M.A.; Valledor, L.; Sherman, L.A. & Nedbal, L. (2013). Ultradian metabolic rhythm in the diazotrophic cyanobacterium *Cyanothece* sp. ATCC 51142. *Proc Natl Acad Sci U.S.A.,* Vol.110, pp.13210-13215

[26] Chanway, C.P. & Holl, F.B. (1991). Biomass increase and associative nitrogen fixation of mycorrhizal *Pinus contorta* seedlings inoculated with a plant growth promoting *Bacillus* strain. *Can J Bot.,* Vol.69, pp.507–511

[27] Chanway, C.P. (2002). Plant growth promotion by *Bacillus* and relatives, pp. 219-235. *In* Berkeley R, Heyndrickx M, Logan N, De Vos, P (eds), Applications and systematics of *Bacillus* and relatives. Blackwell, Oxford, UK

[28] Chapin, F.S.; Walker, L.R.; Fastie, C.L. & Sharman, L.C. (1994). Mechanisms of primary succession following deglaciation at glacier bay, Alaska. *Ecological Monographs,* Vol.64, pp.149-175

[29] Chapman, W.K. & Paul, L. (2012). Evidence that northern pioneering pines with tuberculate mycorrhizae are unaffected by varying soil nitrogen levels. *Microb Ecol.,* Vol.64, pp.964–972

[30] Cleveland, C.; Townsend, A.; Schimel, D.; Fisher, H.; Howarth, R.; Hedin, L.; Perakis,

S.; Latty, E.; Von Fischer, J.; Elseroad, A. & Wasson, M. (1999). Global patterns of terrestrial biological nitrogen (N_2) fixation in natural ecosystems. *Global Biogeochem. Cycles.*, Vol.13, pp.623-645

[31] Cocking, E. (2003). Endophytic colonization of plant roots by nitrogen-fixing bacteria. *Plant and Soil,* Vol.252, pp.169-175

[32] Colombo, P.M. (1978). Occurrence of endophytic bacteria in siphonous algae. *Phycologia,* Vol.17, pp.148–151. doi: 10.1093/biomet/78.3.691

[33] Compant, S.; Reiter, B.; Sessitsch, A.; Nowak, J.; Clement, C. & Ait Barka, E. (2005). Endophytic colonization of *Vitis vinifera* L. by plant growth-promoting bacterium *Burkholderia* sp. strain PsJN. *Appl. Environ. Microbiol.*, Vol.71, pp.1685-1693

[34] Crocker, R.L. & Major, J. (1955). Soil development in relation to vegetation and surface age at glacier bay, Alaska. *The Journal of Ecology,* Vol.43, pp.427-448

[35] Curl, E.A. & Truelove, B. (1986). The Rhizosphere. Springer, New York

[36] Dalton, D.A. & Kramer, S. (2006). Nitrogen-fixing bacteria in non-legumes. In Springer Netherlands, Dordrecht pp. 105-130

[37] de Bellone S.C. & Bellone, C.H. (2006). Presence of endophytic diazotrophs in sugarcane juice. *World Journal of Microbiology and Biotechnology,* Vol.22, pp.1065-1068

[38] Dobbelaere, S.; Vanderleyden, J. & Okon, Y. (2003). Plant growth-promoting effects of diazotrophs in the rhizosphere. *Critical Reviews in Plant Sciences,* Vol.22, pp.107-149

[39] Döbereiner, J. &Pedrosa, F.O. (1987). Nitrogen-fixing bacteria in non-leguminous crop plants. Science Tech, Inc., Madison, Wisconsin

[40] Elbeltagy, A.; Nishioka, K.; Sato, T.; Suzuki, H.; Ye, B.; Hamada, T.; Isawa, T.; Mitsui, H. & Minamisawa, K. (2001). Endophytic colonization and in planta nitrogen fixation by a *Herbaspirillum* sp. isolated from wild rice species. *Appl Environ Microbiol.,* Vol.67, pp.5285–5293

[41] Galloway, J.N.; Townsend, A.R.; Erisman, J.W.; Bekunda, M.; Cai, Z.; Freney, J.R.; Martinelli, L.A.; Seitzinger, S.P. & Sutton, M.A. (2008). Transformation of the nitrogen cycle: recent trends, questions, and potential solutions. *Science,* Vol.320, pp. 889-892

[42] Gehring, C.; Vlek, P.L.G.; de Souza, L.A.G. & Denich, M. (2005). Biological nitrogen fixation in secondary regrowth and mature rainforest of central Amazonia. *Agric Ecosyst Environ.*, 111: 237-252

[43] Gehringer, M.M.; Pengelly, J.J.; Cuddy, W.S.; Fieker, C.; Forster, P.I. & Neilan, B.A. (2010). Host selection of symbiotic cyanobacteria in 31 species of the Australian cycad genus: *Macrozamia* (*Zamiaceae*). *Mol Plant Microbe Interact.,* Vol.23, pp.811-22

[44] Grayston, S.J.; Vaughan, D. & Jones, D. (1997). Rhizosphere carbon flow in trees, in comparison with annual plants: the importance of root exudation and its impact on microbial activity and nutrient availability. *Applied Soil Ecology,* Vol.5, pp.29–56

[45] Hallmann, J.; Quadt-Hallmann, A.; Mahaffee, W.F. & Kloepper, J.W., 1997. Bacterial endophytes in agricultural crops. *Can. J. Microbiol.*, 43: 895-914

[46] Hallmann, J.; Quadt-Hallmann, A.; Mahaffee, W.F. & Kloepper, J.W. (1997). Bacterial endophytes in agricultural crops. *Canadian Journal of Microbiology,* Vol.43, pp.895–914

[47] Herrero, A.; Muro-Pastor, A. & Flores, E. (2001). Nitrogen control in cyanobacteria. *J.Bacteriol.*, Vol.183, pp.411-425

[48] Herridge, D.F. (1982). Relative abundance of ureides and nitrate in plant tissues of soybean as a quantitative assay of nitrogen fixation. *Plant Physiol.,* Vol.70, pp.1–6

[49] Hobara S.; McCalley C.; Koba K.; Giblin A.E.; Weiss, M.S.; Gettel, G.M. & Shaver, G.R. (2006). Nitrogen fixation in surface soils and vegetation in an Arctic tundra watershed: A key source of atmospheric nitrogen. *Arct Antarct Alp Res.,* Vol.38, pp.363–372

[50] Hurek, T.B.; Reinhold-Hurek, B.; Montagu, M.B. & Kellenberger, E. (1994). Root colonization and systematic spreading of Azoarcus sp strain BH72 in grasses. *J Bacteriol.,* Vol.176, pp.1913-1923

[51] Huss-Danell, K. (1997). Actinorhizal symbioses and their N_2 fixation. *New Phytologist,* Vol.136, pp.375-405

[52] Iniquez, A.L.; Dong, Y.; Carter, H.D.; Ahmer, B.M.; Stone, J.M. & Triplett, E.W. (2005). Regulation of enteric endophytic bacterial colonization by plant defenses. *Mol Plant Microbe Interact.,* Vol.18, pp.169-178

[53] Izumi, H. (2011). Diversity of endophytic bacteria in forest trees. In: Pirttilä AM, Frank AC (eds) Endophytes of forest trees. Biology and Applications Series: Forestry Sciences Vol. 80. Springer, Heidelberg, pp 95–105

[54] James, E.K. & Olivares, F.L.(1998). Infection and colonization of sugar cane and other gramineaceous plants by endophytic diazotrophs. *Critical Reviews in Plant Sciences,* Vol.17 pp.77–119

[55] Jayasinghearachchi, H.S. & Seneviratne, G. (2004). Can mushrooms fix atmospheric nitrogen? *J. Biosci.,* Vol.29, pp.293–296

[56] Jones, D.L.; Hodge, A. & Kuzyakov, Y. (2004). Plant and mycorrhizal regulation of rhizodeposition. *New Phytologist,* Vol.163, pp.459 –480

[57] Leaf, G.; Gardner, I.C. & Bond, G. (1958). Observations on the composition and metabolism of the nitrogen-fixing root nodules of *Alnus. Journal of Experimental Botany,* Vol.9, pp.320-31

[58] Lindow, S.E. & Brandl, M.T. (2003). Microbiology of the phyllosphere. *Appl. Environ.*, Vol.69, pp.1875-1883

[59] Loper, J.E.; Haack, C. & Schroth, M.N. (1985). Population dynamics of soil Pseudomonads in the rhizosphere of potato (*Solanum tuberosum* L.). *Applied and Environmental Microbiology*, Vol.49, pp.416-422

[60] Ferrario, M.E.; Villafane, V.; Helbling, W. & Holm-Hansen, O. (1995). The occurrence of the symbiont Richelia in Rhizosolenia and Hemiaulus in the North Pacific. *Revista Brasileira de Biologia*, Vol. 55, pp.439–443

[61] Malik, K.A.; Bilal, R.; Mehnaz, S.; Rasul, G.; Mirza, M.S. & Ali, S. (1997). Association of nitrogen-fixing, plant-growth-promoting rhizobacteria (PGPR) with kallar grass and rice. *Plant and Soil*, Vol.194, pp.37–44

[62] Meeks, J.C. (1998). Symbiosis between nitrogen-fixing cyanobacteria and plants. *Bioscience*, Vol.48, pp.266-276

[63] Mendes, R.; Pizzirani-Kleiner, A.A.; Araujo, W.L.; Raaijmakerset, J.M. (2007). Diversity of cultivated endophytic bacteria from sugarcane: genetic and biochemical characterization of *Burkholderia cepacia* complex isolates. *Appl Environ Microbiol.*, Vol.73, pp. 7259–7267. doi: 10.1128/AEM.01222-07

[64] Mia , M.A.B. & Shamsuddin, Z.H. (2010). Nitrogen fixation and transportation by rhizobacteria: a scenario of rice and banana. *International Journal of Botany*, Vol.6, pp. 235-242

[65] Momose, A.; Ohtake, N.; Sueyoshi, K..; Sato, T.; Nakanishi, Y.; Akao, S. & Ohyama, T. (2009). Nitrogen fixation and translocation in young sugarcan (*Saccharum officinarum* L.) plants associated with endophytic nitrogen-fixing bacteria. *Microbes Environ.*, Vol. 24, pp.224-230

[66] Montañez, A.; Abreu, C.; Gill, P.R.; Hardarson, G. & Sicardi, M. (2009). Biological nitrogen fixation in maize (*Zea mays* L.) by ^{15}N isotope-dilution and identification of associated culturable diazotrophs. *Biol Fertil Soils* Vol.45, pp.253–263. doi:10.1007/s00374-008-0322-2

[67] Myrold, D.D. & Bottomley, P.J. (2007). Biological N inputs. In Soil microbiology, ecology and biochemistry. Elsevier, Burlington, MA, pp. 365-388

[68] Newton, W.E. 2007. Physiology, biochemistry, and molecular biology of nitrogen fixation. In Biology of the Nitrogen Cycle. Elsevier, Amsterdam pp. 109-129

[69] Nguyen, C. (2003). Rhizodeposition of organic C by plants: mechanisms and controls. *Agronomie* Vol.23, pp.375 –396

[70] Oldroyd, G.E.D.; Murray, J.D.; Poole, P.S. & Downie, J.A. (2011). The rules of engagement in the legume-rhizobial symbiosis. *Annu. Rev. Genet.*, Vol.45, pp.119–44

[71] Postgate, J.R. (1998). Nitrogen fixation. Cambridge University Press, Cambridge, UK

[72] Preisig, H.R. & Hibberd, D.J. (1984). Virus-like particles and endophytic bacteria in Paraphysomonas and Chromophysomonas (Chrysophyceae). *Nord J Bot.*, Vol.4, pp. 279–285

[73] Reed, H.E. & Martiny, J.B.H. (2007). Testing the functional significance of microbial composition in natural communities. *FEMS Microbiol. Ecol.*, Vol.62, pp.161-170

[74] Reed, S.C.; Cleveland, C.C. & Townsend., A.R. (2008). Tree species control rates of free-living nitrogen fixation in a tropical rain forest. *Ecology* Vol.89, pp.2924-2934

[75] Reed., S.C.; Cleveland, C.C. & Townsend, A.R. (2011). Functional ecology of free-living nitrogen fixation: a contemporary perspective. *Annu. Rev. Ecol. Evol. Syst.*, Vol.42, pp.489–512

[76] Reinhold, B.; Hurek, T.; Gillis, M.; Hoste, B.; Vancanneyt, M.; Kersters, K. & De Ley, J. (1993). *Azoarcus* gen.nov. nitrogen-fixing proteobacteria associated with roots of kallar grass (*Leptochloa fusca* L. Kunth), and description of two species, *Azoarcus indigens* sp. nov. and *Azoarcus communis* sp. nov. *International Journal of Systematic Bacteriology*, Vol.43, pp.574–584

[77] Reinhold-Hurek, B. & Hurek, T. (1998a). Interactions of gramineous plants with *Azoarcus* spp. and other diazotrophs: identification, localization, and perspectives to study their function. *Critical Reviews in Plant Sciences*, Vol.17, pp.29–54

[78] Reinhold-Hurek, B. & Hurek, T. (1998b). Life in grasses: diazotrophic endophytes. *Trends in Microbiology*, Vol.6, pp.139–144

[79] Richardson, A.E.; Barea, J.M.; NcNeill, A.M. & Prigent-Combaret, C. (2009). Acquisition of phosphorus and nitrogen in the rhizosphere and plant growth promotion by microorganisms. *Plant Soil.*, Vol.321, pp.305-339

[80] Rosenblueth, M. & Martinez-Romero, E. (2006). Bacterial endophytes and their interactions with hosts. *MPMI*, Vol.19, pp. 827-837

[81] Ruschel, A.; Henis, Y. & Salati, E. (1975). Nitrogen ^{15}N tracing of N_2-fixation with soil grown sugarcane seedlings. *Soil Biol. Biochem.*, Vol.5, pp.83–89

[82] Santi, C.; Bogusz, D. & Franche, C. (2013). Biological nitrogen fixation in non-legume plants. *Annals of Botany*, doi:10.1093/aob/mct048

[83] Schell, D.M. & Alexander, V. (1973). Nitrogen fixation in Arctic coastal tundra in relation to vegetation and micro-Relief. *Arct.*, Vol.26, pp.130–137

[84] Schultz, B. & Boyle, C. (2005). The endophyte continuum. *Mycological Research*, Vol. 109, pp.661-686

[85] Sevilla, M.; Burris, R. H.; Gunapala, N. & Kennedy, C. (2001). Comparison of benefit to sugarcane plant growth and ^{15}N incorporation following inoculation of sterile

plants with *Acetobacter diazotrophicus* wild-type and Nif mutant strains. *Mol. Plant-Microbe Interact.*, Vol.14, pp.358–366

[86] Shi, S.; O'Callaghan, M.; Jones, E.E.; Richardson, A.E.; Walter, C.; Stewart, A. & Condron, L. (2012). Investigation of organic anions in tree root exudates and rhizosphere microbial communities using in situ and destructive sampling techniques. *Plant Soil.*, Vol.359, pp.149-163

[87] Shishido, M. & Chanway, C.P. (1999). Spruce growth response specificity after treatment with plant growth-promoting pseudomonads. *Can. J. Bot.*, Vol.77, pp.22-31

[88] Shrestha, R.K. & Ladha, J.K. (1996). Genotypic variation in promotion of rice nitrogen fixation as determined by nitrogen ^{15}N dilution. *Soil Sci. Soc. Am. J.*, Vol.60, pp. 1815-1821

[89] Sinclair, J.B. & Cerkauskas, R.F. (1996). Latent infection vs. endophytic colonization by fungi. In: Redlin SC, Carris LM (eds) Endophytic fungi in grasses and woody plants. APS, St Paul, MN, pp 3-30

[90] Solheim, B. & Zielke, M. (2002). Associations between cyanobacteria and mosses. In: Rai AN, Bergman B, Rasmussen (eds) Cyanobacteria in symbiosis. Dordrecht: Kluwer Academic Publishers, 37–152

[91] Stephan, M.P.; Oliveira, M.; Teixeira, K.R.S.; Martinez-Drets, G. & Döbereiner, J. (1991). Physiology and dinitrogen fixation of *Acetobacter diazotrophicus*. *FEMS Microbiol. Lett.*, Vol.77, pp.67–72

[92] Talley, S.N.; Talley, B.J. & Rains, D.W. (1977). Nitrogen fixation by azolla in rice fields. In Genetic Engineering for Nitrogen Fixation (ed) Hollander, A. pp 259-282. New York: Plenum Press

[93] Taulé, C.; Mareque, C.; Barlocco, C.; Hackembruch, F.; Reis, V.M.; Sicardi, M. & Battistoni, F. (2012). The contribution of nitrogen fixation to sugarcane (*Saccharum officinarum* L.), and the identification and characterization of part of the associated diazotrophic bacterial community. *Plant Soil.*, Vol.356, pp.35–49

[94] Trevet, I.W. & Hollis, J.P. (1948). Bacteria in storage organs of healthy plants. *Phytopathology*, Vol.38, pp.960-967

[95] Udvarte, M. & Poole, P.S. (2013). Transport and metabolism in legume-Rhizobia symbioses. *Annu. Rev. Plant Biol.*, Vol.64, pp.781–805

[96] Urquiaga S.; Cruz K.H.S. & Boddey R.M. (1992). Contribution of nitrogen fixation to sugar cane: nitrogen-15 and nitrogen-balance estimates. *Soil Science Society of America Journal*, Vol.56, pp.105-114

[97] van der Lelie, D.; Taghavi, S.; Monchy, S. et al. (2009). Poplar and its bacterial endophytes: coexistence and harmony. *Critical reviews in Plant Sciences*, Vol.28, pp.346-358

[98] Wall, L.G. & Berry, A.M. (2008). "Early interactions, infection and nodulation in actinorhizal symbiosis" in Nitrogen-Fixing Actinorhizal Symbioses, K. Pawlowski and W. E. Newton (Eds), pp. 147–166, Springer, Dordrecht, The Netherlands

[99] Warembourg, F.R. (1997). The 'rhizosphere effect': a plant strategy for plants to exploit and colonize nutrient-limited habitats. *Bocconea,* Vol.7, pp.187-194

[100] Weetman, G.F.; Fournier, R.M. & Schnorbus, E. (1988). Lodgepole pine fertilization screening trials: four-year growth response following initial predictions. *Soil Sci Soc Am J.,* Vol.52, pp.833–839

[101] Wilson, D. (1995). Endophyte: the evolution of a term, and clarification of its use and definition. *Oikos,* Vol.73, pp.274-276

[102] Wolk, C.P.; Ernst, A. & Elhai, J. (1994). Heterocyst metabolism and development. *The Molecular Biology of Cyanobacteria,* Vol.1, pp.769-823

[103] Yanni, Y.G. (1992). The effect of cyanobacteria and azolla on the performance of rice under different levels of fertilizer nitrogen. *World Journal of Microbiology and Biotechnology,* Vol.8, pp.132-136

[104] You, C.B.; Song, W.; Wang, H.X.; Li, J.P.; Lin, M. & Hai, W.L. (1991). Association of *Alcaligenes faecalis* with wetland rice. *Plant and Soil,* Vol.137, pp.81-85

5

Symbiotic of Nitrogen Fixation Between Acid Aluminium Tolerant *Bradyrhizobium japonicum* and Soybean

Nisa Rachmania Mubarik and Tedja-Imas Sunatmo

1. Introduction

Indonesia is a tropical country in Southeast Asia region, located between the Asia and Australia continents. In most parts of Indonesia, climate variation and high of rainfall causes intensive leaching, soil becomes low content of alkaline and the pH tend to acidic. Indonesia has acid dry land area approximately 102.8 million hectares, but only 55.8 million hectares are suitable for agricultural [1]. The arid lands in Indonesia which are generally formed from mineral soil are acidic (pH 4.6 to 5.5) and poor of nutrients. One effort to increase the soil fertility and plant productivity on acid dry land with planting legumes, such as soybean. Inoculation of root nodule bacteria on soybean plant could enhance soybean quality and its productivity [2 & 3]. Some varieties of acid tolerant soybean, such as Tanggamus, Sibayak, Seulawah, Ratai, and Nanti are issued by the Research Institute for Legumes plants and Tuber Crops Indonesia could grow at acidic soil with pH 4.5-5.0 and produced soybean up to 2000 Kg/hectares on the right growing conditions [4]. Soybeans generally grow in soil at pH 5.5-6.0 while the optimum pH is 6.8. Below pH 4.7 soybean production will decline. It is related to the chemical properties of acid soil, that is high levels of aluminium, high P fixation, iron and manganese concentration increases to the toxic level, sensitive to erosion, and poor biotic status under a low pH conditions [5]. Soybean production could be increase by symbiosis with root nodule bacteria. The effectiveness of symbiotic bacteria in legume root nodules is strongly influenced by the soil conditions. Keyser and Munns [5] suggested that aluminum (Al) with a high concentration (50 μM) is one of the stress factors that can inhibit the growth and prolong the lag phase of root nodulating bacteria. Richardson *et al.* [6] also stated that the Al concentration of 7.5 μM at pH 4.8 can inhibit the expression of nod genes that play a role in nodulation. Furthermore, Johnson and Wood [7] stated that the Al^{3+} cation can bind to PO_4^{3-} of DNA thereby inhibiting

DNA replication and transcription. Therefore, strains of acid and high Al-tolerant root nodulating bacteria which have symbiotic effectiveness with soybean are needed to explore.

2. Acid aluminium tolerant *Bradyrhizobium japonicum*

Some strains of root nodulating bacteria tolerant to acid soil conditions have been reported [8]. The bacteria has ability to fix atmospheric nitrogen (N_2) and and convert into ammonium (NH_3) [9]. *Bradyrhizobium japonicum* is one of root nodule bacteria that can contribute on soybean growth by providing fixed nitrogen in nodules of soybean plants [2]. *Bradyrhizobium* is included to the family Rhizobiaceae. This family consists of four genera, namely *Agrobacterium, Bradyrhizobium, Phyllobacterium,* and *Rhizobium*. The characteristics of *Bradyrhizobium* are rod-shaped, nonspore-forming cells, motile with one polar or subpolar flagelum, aerobic, Gram-negative, cell-sized of 0.5-0.9 μm and 1.2-3.0 μm, the optimum growth temperature is 25-30°C at pH 6-7 [10]. *Bradyrhizobium* is known as slow growing bacteria with a generation time ranged 7-20 hours. The bacterial growth on yeast mannitol agar (YMA) needs 5-7 days incubation on room temperature.

Bradyrhizobium japonicum has sticky consistency and slimy (mucoid) when grown on media containing carbohydrates. The mucus is an extracellular polysaccharide that serves to maintain bacterial survival in environmental conditions with the concentration of acid and aluminum (Al) is high. Strains of *B. japonicum* have more slimy colony and generally more tolerant on acid-Al stress conditions compared to the dry type colony [8 & 11]. There are not all bacteria categorized as tolerant of acid (pH 4.0-4.5) are also a high Al tolerant. Some strains of *B. japonicum* were tolerant on an acid condition, even at the pH level 4.0-4.5. Twenty five strains of *B. japonicum* has been selected for acid tolerance (pH 4.5) consist of Al 50 μM, Mn 200 μM, Ca 50 μM, and low P 5 μM [12]. One of the *B. japonicum* (BJ) isolate namely BJ 11 wt (wild-type) has the highest tolerance on acid and had a good ability to grow on pH 4.5 media (Figure 1).

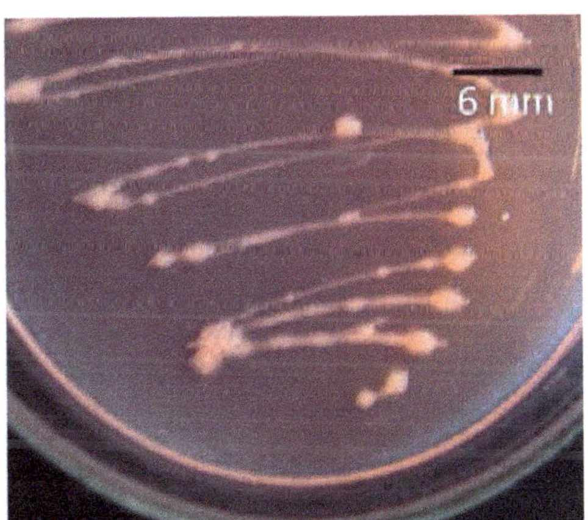

Figure 1. The growth of *Bradyrhizobium japonicum* BJ11 (wt) on pH 4.5 yeast-extract mannitol agar containing 0.0025% congo red at 10 days incubation in room temperature

Root nodulating bacteria can be distinguished from other bacteria by growing it on media yeast extract mannitol agar (YMA) consist of 10 g/L mannitol, 0.5 g/L K_2HPO_4, 0.2 g/L $MgSO_4.7H_2O$, 0.2 g/L NaCl, and 0.5 g/L yeast extract and containing 0.0025% congo red. Root nodulating bacteria can not absorb congo red or less, and the colony is colorless or pale white [2]. Bradyrhizobia growing on agar media are classified into three types based on the appearance of colonies, such as: small dry (SD), large mucoid (LM), large watery (LW), and dimorphism [13]. Colony of SD type is round, convex, translucent, and diameter of <1 mm. The LM type is circular, convex, slimy, relatively translucent, and diameter> 1 mm. The LW type is irregular shapes, flat, watery, translucent, and diameter> 1 mm. Dimorphism type strain is called to strain with a mixture of SD and LM type. Colony type can be used to predict tolerant or sensitive strain to acid-Al condition. A small dry colony type strain is more sensitive to acid-Al compare to large one and wet type colony [8]. BJ 11 is the slow-growing colony, circular shape, convex elevation, slimy, translucent, and diameter of colony > 1 mm, it is categorized large mucoid. Other root nodulating bacteria has fast growing and acid producing is classified as genus *Rhizobium*, whereas the slow-growing and alkaline producing reaction belong to the genus *Bradyrhizobium*. The growth reaction on YMA medium which is acidic or alkaline is determined by adding 0.0025% bromothymol blue. Colony of root nodulating bacteria that produce acid reaction is yellow [15].

3. Symbiotic effectiveness

Effective strains of *Bradyrhizobium japonicum* produce an effective root nodules on their host. Usually one strain of root nodulating bacteria is used as an inoculum for one variety of soybean plant. Selection should be done from large number of tested strains by using a suitable host-plant on soil and climatic conditions of the host habitat [2].

Symbiotic effectiveness is the relative ability of an association between legume and root nodulating bacteria. Effective nodule consist of leghemoglobin, that is an iron-containing red protein binding with O_2 that controls the partial pressure of O_2 (pO_2) in the nodule [15]. When pO_2 was below or above normal condition (0.21 atm), it could decrease the activity of N_2 fixation. Leghemoglobin is induced by the interaction between *Bradyrhizobium* with soybean.

Effective nodule tends to be large size, reddish, and able to fix nitrogen gas from air. In addition, the effective root nodules have a limited number and distribution, usually found on the main root and secondary first root [14]. Ineffective nodules tend to be small, numerous, greenish white (pale), unable to fix nitrogen from air and spread the root system [14].

Symbiotic effectiveness of acid-tolerant soybean with acid-Al tolerant *B. japonicum* could be done by using Leonard bottle modified that consists of two volumes of 700 ml bottles of ketchup. One bottle is cut at the base and used for growth media that contains sand and charcoal. Other bottle cut at the neck and is used as a reservoir for the nutrient solution [16]. The lower bottle is filled with 300 ml of N-free nutrient solution of pH 4.5 [17] and 100 ml of N-free nutrient solution poured into growing medium in the mixture form of sand and coconut shell charcoal about 480 gram. Before used, sand is sieved and washed with clean water several times until clean and dry. Each bottle is covered with cement paper and sterilized by autoclaving at 121°C and 1 atmosphere for 2 hours. Two days sprouts of soybean in Leonard jar. Each sprout was inoculated with 10^8 cells ml^{-1} of *B. japonicum*. N-free nutrient solution and nutrient solution contained 5 mM KNO_3 used as control treatments, respectively. Symbiotic effectiveness value (SE) is measured based on percentage of dry weight of plants inoculated with tested strain toward dry weight of plants treated with KNO_3 or reference strain. *Bradyrhizobium japonicum* USDA 110 is used as a reference strain and completely genomic sequenced [18].

The effectiveness of symbiosis can be observed in several ways viz. the determination of plant dry weight, total N content, and nitrogenase activity [2]. Dry weight of the plant is still considered relevant for evaluating the effectiveness of symbiotic root nodulating bacteria with soybean plants, because plant dry weight significantly correlated with total N content [14]. Plant dry weight is usually correlated with the dry weight of root nodules. Upper plant dry weight is used as a parameter to evaluate the binding of N, because as much as 70% of the fixing N is transported to the upper plant [14].

The symbiotic interaction between soybean and root nodulating bacteria played an important role in increasing the plant growth of soybean plant. Effectiveness of a root nodulating bacteria in fixing nitrogen is affected by the compatibility between bacteria and the soybean plant. Mubarik *et al.* [19] described that inoculation of BJ 11 (wt) root nodulating bacteria could increase the height of soybean plant and shoot dry weight until 37 days after planting (DAP) (Figure 2). The nodule dry weight was positively correlated with the ability of plants to fix N and shoot dry weight. The value of symbiotic effectiveness, shoot dry weight, and N uptake of BJ 11 was higher than USDA 110 as reference strain (Table 1).

Figure 2. Soybean plant (37 days after planting) grow on a Leonard bottle using N-free nutrient solution pH 4.5 + Al 50 µM: (1) without inoculation BJ 11 and (2) inoculated with BJ 11 [16]

Treatment	Number of nodule (nodule plant $^{-1}$)	Dry-weight of nodule (g plant $^{-1}$)	Height of plant (cm)	Nitro-genase activity (μmol C_2H_4 plant $^{-1}$ hour $^{-1}$)	Dry-weight of shoot (g plant $^{-1}$)	N uptake (mg N plant $^{-1}$)	SEN(%)	SER (%)
BJ 11 (wt)	17	0.0397	71.4	12.54	0.8447	16.88	155.37	144.55
USDA 110	12	0.0241	63.0	12.21	0.6164	13.63	114.92	100.00
Control N	0	0	46.0	0	0.5509	13.58	100.00	96.26
Control NO	0	0	37.3	0	0.4561	6.91	83.88	77.55

0 =no detection, N:without BJ inoculation consist of 5 mM KNO_3, N0: without BJ inoculation and without 5mM KNO_3, Symbiotic Effectiveness (SE) against N/R.

Table 1. Effect of inoculation of *B. japonicum* on soybean cultivar Slamet at 37 DAP using N-free solution at pH 4,5 + Al 50 μM (Mubarik *et al.* 2012)

Bradyrhizobium japonicum is able to form nodules and fix nitrogen. Nodule formation on the roots of leguminous plants generally through the following stages: (i) the introduction of a suitable partner on the part of the plant and bacterial attachment to root hairs, (ii) the hair root invasion by bacteria through thread-forming infection threads, (iii) the bacteria moves to the main root through infection threads, (iv) the formation of bacteria in plant cells called bacteroids, and (v) plant and bacterial cell division that is constantly and will produce the mature root nodules [15]. Stages of nodulation (nodule formation) is controlled by the *nod* genes.

The source of energy for nitrogen fixation in bacteroids depends on host photosynthate which is transported through the membrane simbiosome in the form intermediate product of the tricarboxylic acid cycle (Krebs cycle) such as succinic acid, fumaric and malic acid which is a electron donor to produce ATP and reduce N_2. Pyruvic acid is the reductant that involved directly as an electron donor in the nitrogenase system [15]. The N_2 binding reaction that occurs in bacteroids as follows:

$$N_2 + 8e + 8H^+ + 16\ MgATP \xrightarrow{\text{nitrogenase}} 2NH_3 + H_2 + 16\ MgADP + 16\ Pi$$

Complex of nitrogenase reduces the triple bond of N_2 into ammonia molecules. Nitrogenase enzyme activity can be measured by the acetylene reduction technique. Acetylene (C_2H_2) can be used as an alternative substrat to N_2. Reduction of N_2 and acetylene by nitrogenase as follows:

$$N_2 + 12\ ATP + 6e^- + 6H^+ \rightarrow 2\ NH_3 + 12\ ADP + 12\ Pi$$

$$C_2H_2 + 4\ ATP + 2e^- + 2H^+ \rightarrow C_2H_4 + 4ADP + 4Pi$$

The comparison between the substrate N_2 reduction by C_2H_2 is 3:1, and according to calculation [20] the total amount of N fixed by plants (μg) = μmol C_2H_4 x 28.

While the C_2H_2 reduction can provide a useful tool for detecting N_2-fixing activity in both legumes and non-legumes plants, the method is unsuitable for measuring N_2 fixation at field scales. There are some suitability of methods for quantifying N_2 fixation for crop legumes, such as measurement of N difference, relative ureide method, ^{15}N natural abundance, and ^{15}N isotope dilution [21]. But none of the methods for assessing N_2 fixation is perfect. Some additional informations are needed to support the N_2 fixation data, such as assessment of nodulation, growth analysis, rooting patterns of N_2 fixing and companion non N_2-fixing plants, determination of mineral N soil, and soil analysis [21].

4. Greenhouse experiments of symbiotic between acid aluminium tolerant *B. japonicum* and soybean on acid soils

Situmorang *et al.* [22] prepared media for soybean cultivication by using mixed composition of 1200 g acid soil (pH 4.5) and 800 g peat in a polybag. Peat is used as an additional organic matter to the soil. Acid soils and peat are prepared by drying and filtering using 2 mm pore of diameter. The media is sterilized by autoclave at 121°C and 2 atm for one hour. The media is inoculated with 20% (v/w) of 10^8 cells/ml bacterial culture. Positive control media is added with 5 mM KNO_3. Plant harvest are divided into two groups, at the 50 DAP to crop nodules and 75-108 DAP to crop pods of legume. Three isolates are used viz. BJ 11 (wt), and its mutant BJ 11(5) and BJ 11(19). Wahyudi *et al.* [23] has been constructed several strains of acid-aluminium tolerant *B. japonicum* with increased symbiotic effectiveness through transposon TN5 mutagenesis, such as BJ 11(5), BJ 11(19), BJ 11 (20), and KDR 15 (37). The mutants could grow better on acid pH (4.0-4.5) and when each mutant inoculated to soybean plants will influenced better of symbiotic effectiveness, plant height, shoot and root weight, number of flowers, pods, dry weight of 100 seeds, and plant N-content [22].

Inoculation of BJ 11(19) isolate increased number of seeds and pods higher than the other treatments [22]. Acid tolerant soybean such as Slamet generally has weight 12.5 g of 100 seeds [24]. BJ 11 (19) showed the highest 13.5 g of 100 seeds. Pods that were already formed then were filled with photosynthate to form seeds. Numbers of seeds are effected by the number and size of pods. Higher number of pods also produce higher numbers of seeds [25].

Further experiments are done in acid soil plots (pH 4.5). Totally 12 plot experiments, each plot measured 1 m x 2 m x 0.2 m filled with 45 kg of acid soils (pH 4.5) and 10% (w / w) peat or rice husk as innoculant carrier. Each plot planting with soybean sprouts each with a spacing of 20 x 40 cm^2. Amount of inoculant (about 1.0 x 10^8 cells/ml) in peat-carrier is applied to each plot. Every hole on plot planted with 5 seedling soybeans and to be reduced to 3 plants at 30 days after planting. Each plot is separated by a distance of 1 m from other plot. Results of plot experiment showed that the effectiveness of symbiotic BJ 11 (19) with soybean is significantly had higher value on the plant height, dry weight of upper crop, root nodules, nodule number, nitrogenase activity, and weight of 100 seeds. Treatment of compost before planted soybean in acid soils could produce better crops and increase producing of soybean seeds compare to without compost (Figure 3). The compost consists of plant residues and soil microbes that can improve acid soil structure becomes more fertile and porous.

5. Viability test of acid-aluminium *B. japonicum* inoculant using peat as carrier

Viability of *B. japonicum* shoud be tested before used as an inoculant on fields experiments. Handayani *et al.* [26] conducted to test the viability of strains of acid-aluminium tolerant after a period of storages (1, 2, and 3 months) both at room temperature (± 25 °C) and 10 °C. The inoculant of *B. japonicum* BJ 11 (wt), and its mutants viz. BJ 11 (5) and BJ 11 (19), were tested by using sterilized peat as carrier (Figure 4). Peat is an decaying-organic material containing humic acid and organic-C and N which suitable for microbial growth. The result of viability test showed that there were an interaction between strain types, temperature, and a period of storage. The Inoculant of BJ 11 (19) which was stored at temperature 10 °C for 2 months showed the highest viability at 2,5 x 10^8 cell/g inoculants (Table 2).

(1) (2) (3) (4) (5) (6)

Figure 3. Growth of acid tolerant soybean variety Slamet 38 DAP on plot experiments: (1) control without inoculation, (2) control without inoculation + compost, (3) inoculation with BJ 11 (wt), (4) inoculation with BJ 11 (wt) + compost, (5) inoculation with BJ 11 (19), (6) inoculation with BJ 11 (wt) + compost.

Strain	Temperature	Storage periode (months)		
		1	2	3
BJ 11 (5)	Room	9.8 x 10^7 cdef	2.8 x 10^7 f	1.3 x 10^8 abcdef
	10 °C	1.4 x 10^8 abcdef	1.2 x 10^8 abcdef	7.6 x 10^7 def
BJ 11 (19)	Room	2.4 x 10^8 ab	2.0 x 10^8 abc	1.1 x 10^8 bcdef
	10 °C	1.8 x 10^8 abcd	2.5 x 10^8 a	1.8 x 10^8 abcd
BJ11 (wt)	Room	1.6 x 10^8 abcde	1.1 x 10^8 bcdef	4.2 x 10^7 ef
	10 °C	1.3 x 10^8 abcdef	1.1 x 10^8 bcdef	1.9 x 10^8 abcd

Numbers on the same column followed by the same letter were not significantly different based on Duncan Multiple Range Test (α = 0.05)

Table 2. Viability of three acid aluminium tolerant *B. japonicum* strains (cell. g^{-1}) stored at room temperature (± 25 °C) and 10 °C at 3 months storage [26]

Figure 4. The formula of inoculant acid-aluminium tolerant *B. japonicum* containing 10^9 cells g^{-1} using peat as carrier. Each pack contains 0.5 kg of inoculant for 10 kg of soybean seeds

6. Field trial of application of acid-aluminium tolerant *B. japonicum* on soybean

There are three locations for field trials to apply of the formula acid-aluminium tolerant *B. japonicum* on soybean viz. Jasinga (West Java), Sukadana (Province Lampung), and Tambang

Ulang (Province South Kalimantan). Planting sites prepared a total area 1 hectare. Before planting on the field, the chemical contents of the soil and total plate count of soil bacteria are analyzed (Table 3 & 4). There are not found indigenous *B. japonicum* on all of field trial locations before symbiotic effectiveness treatments.

Parameter	Soil contents		
	Jasinga	Sukadana	Tambang Ulang
C (%)	1.18	0.73	2.54
N (%)	0.16	0.11	0.20
P (%)	0.0342	0.0142	0.0749
Mg (%)	0.02	0.01	0.02
K (%)	0	0	0.01
Ca (%)	0.23	0.14	0.67
C/N (%)	7.38	6.64	12.7
Al-dd	2.56	0.88	0.15
Capacity of cation exchange	7.55	2.45	7.14
pH : aquadest	4.61	5.56	6.18
pH :KCl	4.72	4.76	4.81

Table 3. Chemical properties of soil at the field trial locations

Field location	Numbers of cell (cfu/ml)	Numbers of *B. japonicum* (cfu/ml)
Jasinga-West Java	4.0×10^5	0
Sukadana -Lampung	5.9×10^5	0
Tambang Ulang -South Kalimantan	7.4×10^5	0

Table 4. Total plate count of bacteria and total of indigenous *B. japonicum* isolated from soil on planting sites

The field trial was conducted to examine the efficiency of BJ 11 (wt) and BJ 11 (19) on the growth, nodulation and yield of soybean variety Tanggamus and Anjasmoro. Tanggamus is one of leading variety which can adapt to dry acid soil, Anjasmoro generally showed good adaptation on paddy fields.

The seeds were coated with the inoculum formula before sowing. Seeds were sown by hand in each hole and planted 3 seeds per hole at a depth of 3 cm, distance of hole 20 cm x 40 cm. Fertilizer was placed at other hole besides of seeds hole. Watering was carried regularly if no rain. Removal of weeds or grasses are done as far as possible.

Soybean seed are sown by hand in a hole at soil. There were three seeds per polybag. Soybean seeds were selected based on the same size and healthy (able to shoot). Some treatments were conducted to soybean seed as follows: 1. inoculated by *B. japonicum* galur BJ 11, 2. inoculated by BJ 11 and application with 100 % N fertilizer; 3. inoculated by *B. japonicum* galur BJ 11 and application with 50 % N fertilizer + 50% compost; 4. Control treatment: without inoculant, without inoculant + 100% N fertilizer, without inoculant + 50 % N fertilizer + 50% compost.

Each treatments were done at 150-200 m^2 and replicated two times per treatment. Mineral fertilization 100% N treatment consisted of 100 Kg ha^{-1} urea + 200 Kg ha^{-1} TSP (trisodium phosphate) + 100 Kg ha^{-1} KCl. For 50% N consisted of a half dose of urea + 200 Kg ha^{-1} TSP + 100 Kg ha^{-1} KCl + compost 1000 Kg ha^{-1}. Compost was spread out at land surface one week before seeds planting. The compost only consisted of decaying plants and decomposed by microbes. There are not found rhizobia in compost, and consist of phosphate solubilizing bacteria as much as 320 cell.ml^{-1}. Urea used twice at one planting period viz a half dose at seeds planting and the rest at 30 days after planting (DAP) [27].

Growth parameters such as plant height at 30 days after planting (DAP), number of pods at 90 DAP, total number of seeds, total of seed weight, and weight of 100 seeds numbers of pods compare to control were determined. Growth parameters were measured from 10 plants per treatments. Data were analyzed using completely randomized design and the means at p<0.05 level of significance.

The results of field experiments showed that there were a significant effect of *B. japonicum* inoculation for soybean variety Tanggamus and Anjasmoro which grown at Jasinga –West Java, Tanah Laut-South Kalimantan and Sukadana-Lampung compared to control, without inoculants and fertilizer (Table 5, 6 & 7). Inoculation BJ 11 formula showed a better response on soybean growth than control, treatment without fertilizer and inoculant. Plants inoculated with BJ 11 (wt) and its mutant BJ 11 (19) showed higher plant height, number of pods, and seeds, weight of 100 seeds compare to control. To improve field-scale of soybean production in acid soils still need N- fertilizer, but the application of inoculant *B. japonicum* can reduce a half of N fertilizer.

Anjasmoro

Treatment	Plant height at 45 DAP (cm)	Number of branch	Number of flower	Plant height at 90 DAP (cm)	Number of branch at 90 DAP	Number of pods	Number of seed	Total of seed weight (g)	Weight of 100 seeds (g)
BJ 11 (19) + 1 N	30.6 d	1 c	28.2 cd	34.2 c	1.5 c	10.8 c	17.5 d	2.83 cd	13.42 b
BJ 11 (19) + 1/2 N + C	35.8 b	0.7 cd	31.2 bc	41 b	2.6 a	16.2 b	29.4 bc	4.33 bc	12.97 b
BJ 11 (19)	31.9 cd	0.3 de	19.9 e	37.5 bc	1.4 c	10 c	18.3 d	2.27 de	12.60 b
BJ 11 (WT) + 1 N	40.1 a	1.7 b	35.4 b	46.5 a	2.3 ab	19.7 b	35 b	4.55 b	13.21 b
BJ 11 (WT) + 1/2 N + C	42 a	2.3 a	45.2 a	51.9 a	2.5 a	31.3 a	51.1 a	6.29 a	12.55 b
BJ 11 (WT)	33.6 bc	0.4 de	21.8 de	34.c1	0.1 d	5.5 c	7.1 e	1.02 e	15.06 a
1 N	41.3 a	1.6 b	36.4 b	38.1 bc	1.6 bc	16.6 b	25,5 bcd	3.43 bcd	12.75 b
1/2 N + C	42.2 a	2.1 ab	29.9 bc	38.1 bc	1.7 bc	17.1 b	29.8 bc	4.23 bc	13.17 b
Control	26.4 e	O e	20.5 e	34.7 c	1.8 bc	10.9 c	21.8 cd	2.95 cd	13.26 b

Anjasmoro

Treatment	**Plant height at 45 DAP (cm)**	**Number of branch**	**Number of flower**	**Plant height at 90 DAP (cm)**	**Number of branch at 90 DAP**	**Number of pods**	**Number of seed**	**Total of seed weight (g)**	**Weight of 100 seeds (g)**

BJ 11 = BJ 11 inoculant formula; N = 100 Kg.Ha^{-1} urea + 200 Kg. Ha^{-1} TSP and 100 Kg.Ha^{-1} KCl); ½ N = 50 Kg.Ha^{-1} urea + 200 Kg. Ha^{-1} TSP and 100 Kg.Ha^{-1} KCl; C = compost. Control = without fertilizer (NPK) and inoculants. Numbers on the same column followed by the same letter were not significantly different based on Duncan Multiple Range Test (α = 0.05).

Tanggamus

Treatment	Plant height at 45 DAP (cm)	Number of branch	Number of flower	Plant height at 90 DAP (cm)	Number of branch at 90 DAP	Number of pods	Number of seed	Total of seed weight (g)	Weight of 100 seeds (g)
BJ 11 (19) + 1 N	30.6 d	1 c	28.1 cd	32.6 cd	1.6 cd	14.5 bc	23.9 cd	2.37 cd	7.95 b
BJ 11 (19) + 1/2 N + C	35.8 b	0.7 cd	31.2 bc	33.2 cd	1 d	12.1 cd	17.8 d	1.38 de	7.11 b
BJ 11 (19)	31.9 cd	0.3 de	19.8 e	36.1 bc	1.6 cd	21.8 a	29.2 bc	2.68 bcd	8.32 b
BJ 11 (WT) + 1 N	40.1 a	1.7 b	35.4 b	43 a	2.1 bc	23.5 a	34.9 abc	3.72 ab	10.74 a
BJ 11 (WT) + 1/2 N + C	42.1 a	2.4 a	45.1 a	42 ab	2.8 ab	24. 8 a	38.6 ab	3.42 abc	9 ab
BJ 11 (WT)	33.6 bc	0.4 de	21.8 de	41.7 ab	3.1 a	25.9 a	42.6 a	4.02 a	8.95 ab
1 N	41.3 a	1.6 b	36.3 b	31.7 cd	1.3 cd	19.5 ab	25.1 cd	2.32 cd	8.66 ab
1/2 N + C	42.2 a	2.1 ab	29.9 bc	29.4 cd	0.9 d	10.6 cd	15.3 de	1.39 de	9.42 ab
Control	26.4 e	O e	20.5 e	27.4 d	0.1 e	5.5 d	5 e	0.42 e	8.44 ab

BJ 11 = BJ 11 inoculant formula; N = 100 Kg.Ha^{-1} urea + 200 Kg. Ha^{-1} TSP and 100 Kg.Ha^{-1} KCl); ½ N = 50 Kg.Ha^{-1} urea + 200 Kg. Ha^{-1} TSP and 100 Kg.Ha^{-1} KCl; C = compost. Control = without fertilizer (NPK) and inoculants. Numbers on the same column followed by the same letter were not significantly different based on Duncan Multiple Range Test (α = 0.05).

Table 5. Growth of Anjasmoro and Tanggamus cultivar soybean plants on treatment with acid-aluminium tolerant *B. japonicum* formula on acid soil at Jasinga- West Java

Anjasmoro

Treatment	**Plant height at 30 DAP (cm)**	**Plant height at 90 DAP (cm)**	**Numbe r of leaf at 90 DAP**	**Number of branch at 30 DAP**	**Number of branch at 90 DAP**	**Number of pod at 90 DAP**	**Number of seed**	**Total of seed weight (g)**	**Weight of 100 seeds (g)**
BJ 11 (19) + 1 N	38.3 de	66.8 a	33.1 a	3 ab	3.3 ab	54.8 a	109.2 a	17.3 ab	14.7 abc
BJ 11 (19) + 1/2 N + C	38.3 de	61.5 b	28 bc	2.7 abc	3.1 ab	60.2 a	71.5 c	10.4 cd	13.5 bc
BJ 11 (19)	41.6 bc	52.8 c	26.4 cd	2.2 c	3 ab	41.2 b	123.1 a	18.7 a	13.4 bc
BJ 11 (WT) + 1 N	53 a	68.9 a	22.6 de	2.5 abc	2.3 d	34.7 bc	81 bc	14.1 bc	15 abc

Anjasmoro										
Treatment	**Plant height at 30 DAP (cm)**	**Plant height at 90 DAP (cm)**	**Number of leaf at 90 DAP**	**Number of branch at 30 DAP**	**Number of branch at 90 DAP**	**Number of pod at 90 DAP**	**Number of seed**	**Total of seed weight (g)**	**Weight of 100 seeds (g)**
BJ 11 (WT) + 1/2 N + C	44. 3 b	71.2 a	29 abc	2 c	3 ab	58.2 a	75.9 c	12.3 cd	15.3 ab
BJ 11 (WT)	40.9 cd	69.9 a	32 ab	3.3 a	3.6 a	67.7 a	101.1 ab	17.5 ab	13.3 c
1 N	53.4 a	50.8 cd	21.6 ef	3 ab	2.9 bc	27.9 cd	46.3 d	6 e	16.1 a
1/2 N + C	35.8 e	47.6 d	21.2 ef	2.3 bc	3.1 ab	27.2 cd	63.1 cd	9.2 de	13.3 c
Control	41 cd	45.8 d	17.3 f	3.2 a	2.4 cd	15.6 d	40.6 d	5 e	14.6 abc

BJ 11 = BJ 11 inoculant formula; N = 100 $Kg.Ha^{-1}$ urea + 200 Kg. Ha^{-1} TSP and 100 $Kg.Ha^{-1}$ KCl); ½ N = 50 $Kg.Ha^{-1}$ urea + 200 Kg. Ha^{-1} TSP and 100 $Kg.Ha^{-1}$ KCl; C = compost. Control = without fertilizer (NPK) and inoculants. Numbers on the same column followed by the same letter were not significantly different based on Duncan Multiple Range Test (α = 0.05).

Tanggamus									
Treatment	Plant height at 30 DAP (cm)	Plant height at 90 DAP (cm)	Number of leaf at 90 DAP	Number of branch at 30 DAP	Number of branch at 90 DAP	Number of pod at 90 DAP	Number of seed	Total of seed weight (g)	Weight of 100 seeds (g)
BJ 11 (19) + 1 N	28.6 d	62.5 a	32.9 a	1.2 ab	3 ab	37.9 abc	49.3 bcd	4.3 bc	11.0 ab
BJ 11 (19) + 1/2 N + C	30.4 cd	59.4 ab	24.8 b	1.5 a	2.7 bc	33.6 bcd	56.2 bc	4.9 bc	10.4 b
BJ 11 (19)	51.1 c	50.5 c	18.5 c	1.5 a	2.2 c	21.8 e	47.1 cd	4.8 bc	10.3 b
BJ 11 (WT) + 1 N	33.8 ab	57.3 b	31.8 a	1.7 a	3.2 ab	41.1 ab	74.9 a	6.9 a	11.4 ab
BJ 11 (WT) + 1/2 N + C	35 a	59 ab	32.6 a	1.7 a	2.7 bc	45.1 a	64.9 ab	6.8 a	10.9 ab
BJ 11 (WT)	33.9 ab	55.4 b	25 b	0.6 bc	2.6 bc	31.6 cd	64.4 ab	5.8 ab	12.1 a
1 N	34.9 a	45.8 c	23.3 bc	1.9 a	2.5 bc	21.7 e	36.7 d	3.5 c	10.3 b
1/2 N + C	32.4 bc	45.7 c	21 1 bc	1.7 a	2.7 bc	19.8 e	38.8 d	4.1 c	11.4 ab
Control	28.8 d	48.9 c	22.7 bc	0.5 c	3.5 a	26.3 de	50.3 bcd	4.4 bc	10.6 b

BJ 11 = BJ 11 inoculant formula; N = 100 $Kg.Ha^{-1}$ urea + 200 Kg. Ha^{-1} TSP and 100 $Kg.Ha^{-1}$ KCl); ½ N = 50 $Kg.Ha^{-1}$ urea + 200 Kg. Ha^{-1} TSP and 100 $Kg.Ha^{-1}$ KCl; C = compost. Control = without fertilizer (NPK) and inoculants. Numbers on the same column followed by the same letter were not significantly different based on Duncan Multiple Range Test (α = 0.05).

Table 6. Growth of Anjasmoro and Tanggamus cultivar soybean plants on treatment with acid-aluminium tolerant *B. japonicum* formula on acid soil at Tambang Ulang-South Kalimantan

Anjasmoro

Treatment	Plant height at 30 DAP (cm)	Plant height at 45 DAP (cm)	Number of branch at 30 DAP	Number of branch at 90 DAP	Number of pod at 90 DAP	Number of seed	Total of seed weight (g)	Weight of 100 seed (g)
BJ 11 (19) + 1 N	40.9 b	61.5 ab	1.5 ab	2.4 a	54.9 a	60.6 ab	7.6 bc	12.1 ab
BJ 11 (19) + 1/2 N + C	46.7 a	66.2 a	1.9 a	2.9 a	60.2 a	64.6 ab	10.3 a	13.6 ab
BJ 11 (19)	22.4 d	67.8 a	0 c	0 b	41.2 b	13.4 e	1 e	10.5 ab
BJ 11 (WT) + 1 N	46.7 a	50.2 bc	1.4 ab	2.4 a	34.8 bc	73.9 a	9.5 ab	16.1 a
BJ 11 (WT) + 1/2 N + C	40.6 b	44.4 c	1 b	2.4 a	58.2 a	66.2 ab	7.2 bc	11.2 ab
BJ 11 (WT)	25.3 d	64.7 a	0.2 c	0.2 b	67.7 a	35.2 d	3.1 de	8.8 b
1 N	43.2 ab	68.4 a	1.5 ab	3.0 a	27.9 cd	53.8 bc	7 c	11.7 ab
1/2 N + C	40.9 b	64.3 a	1.6 ab	2.1 a	27.2 cd	57.4 bc	6.4 c	10.1 ab
Control	33.6 c	55.3 abc	1 b	0.9 b	15.6 d	42.9 cd	4.1 d	9.6 b

BJ 11 = BJ 11 inoculant formula; N = 100 $Kg.Ha^{-1}$ urea + 200 Kg. Ha^{-1} TSP and 100 $Kg.Ha^{-1}$ KCl); ½ N = 50 $Kg.Ha^{-1}$ urea + 200 Kg. Ha^{-1} TSP and 100 $Kg.Ha^{-1}$ KCl; C = compost. Control = without fertilizer (NPK) and inoculants. Numbers on the same column followed by the same letter were not significantly different based on Duncan Multiple Range Test (α = 0.05).

Tanggamus

Treatment	Plant height at 30 DAP (cm)	Plant height at 45 DAP (cm)	Number of branch at 30 DAP	Number of branch at 90 DAP	Number of pod at 90 DAP	Number of seed	Total of seed weight (g)	Weight of 100 seed (g)
BJ 11 (19) + 1 N	32.4 bc	61 a	3.5 ab	4.3 ab	62.3 a	131.6 a	10.7 b	7.4 a
BJ 11 (19) + 1/2 N + C	35.4 b	60.3 a	3.7 ab	4.5 a	61.4 a	127 a	10.2 b	7.3 a
BJ 11 (19)	25.1 d	34.5 f	0.5 d	1.2 c	18.4 b	33.1 c	2.2 c	6.8 a
BJ 11 (WT) + 1 N	31.1 c	51 bcd	3.1 b	4.7 a	51.6 a	128.5 a	9.7 b	7.8 a
BJ 11 (WT) + 1/2 N + C	36.1 a	57 ab	3.9 ab	4.3 ab	58.3 a	130.1 a	10.1 b	7.4 a
BJ 11 (WT)	26.3 d	42.1 e	1.2 cd	1.8 c	21.3 b	43.3 bc	3 c	7.1 a
1 N	33 abc	53.6 bc	4.4 a	4.8 a	69.1 a	145.3 a	16.1	7.3 a
1/2 N + C	32.3 bc	59.4 a	3.1 b	4.3 ab	60.4 a	129 a	10.6 b	7.9 a
Control	29.7 c	47.2 d	1.6 c	3.1 b	35.2 b	73.1 b	5.5 bc	7.8 a

BJ 11 = BJ 11 inoculant formula; N = 100 $Kg.Ha^{-1}$ urea + 200 Kg. Ha^{-1} TSP and 100 $Kg.Ha^{-1}$ KCl); ½ N = 50 $Kg.Ha^{-1}$ urea + 200 Kg. Ha^{-1} TSP and 100 $Kg.Ha^{-1}$ KCl; C = compost. Control = without fertilizer (NPK) and inoculants. Numbers on the same column followed by the same letter were not significantly different based on Duncan Multiple Range Test (α = 0.05).

Table 7. Growth of Anjasmoro and Tanggamus variety soybean plants on treatment with acid-aluminium tolerant *B. japonicum* formula on acid soil at Sukadana- Lampung

7. Conclusion

Effectiveness symbiotic between soybean and acid-toleran aluminium root nodule bacteria, such as *Bradyrhizobium japonicum* BJ 11 played an important role on increasing the plant growth on acid soil (pH> 4.5). The bacteria provided fixed nitrogen to soybean plant and then support growth and development of plants. Soybean plants inoculated with *B. japonicum* strain BJ 11 (wild-type) and its mutant BJ 11 (19) showed better growth than control without inoculation in greenhouse and field trial experiments. B. japonicum inoculant on peat as carrier showed high viability and stability during storages.

Acknowledgements

This project was supported by Integrated Excelence Research, The Ministry of Research and Technology, Republic of Indonesia in 1996 to TI Sunatmo and Incentive Programs for Applied Research, The Ministry of Research and Technology, Republic of Indonesia in 2007-2009 to NR Mubarik. We thank our colleaguaes and students for support in the project.

Author details

Nisa Rachmania Mubarik and Tedja-Imas Sunatmo

*Address all correspondence to: nrachmania@ipb.ac.id

Department of Biology, Faculty of Mathematics and Natural Sciences, Bogor Agricultural University, Jalan Agatis, IPB Dermaga, Bogor, Indonesia

References

[1] Mulyani A. Perkembangan potensi lahan kering masam (*The development potential of acid dryland*) Sinar Tani 2006; Edition 24-30 Mei 2006.

[2] Somasegaran P, Hoben HJ. Methods in Legume-Rhizobium Technology. Hawaii: University of Hawaii; 1994.

[3] Abbasi MK, Majeed A, Sadiq A, Khan SR. Application of *Bradyrhizobium japonicum* and phosphorus fertilization improved growth, yields and nodulation of soybean in the sub-humid hilly region of Azad Jammu and Khasmir, Pakistan. Plant Prod Sci 2008; 11: 368-376.

[4] Suhartina. Perkembangan dan Deskripsi Varietas Unggul Kedelai 1998-2002. (*Development and Description of Soybean Varieties 1998-2002*). Malang: Crops Research Institute for Legumes and Tubers; 2003.

[5] Keyser HH, Munns DN. Tolerance of rhizobia to acidicity, alumunium and phosphate. Soil Sci Soc Am J. 1979; 43: 59-523.

[6] Richardson AE, Richard JS, Michael AD, Barry GR. Expression of nodulation genes in *Rhizobium leguminosarium* biovar *trifolii* is affected by low pH and by Ca and Al ions. Appl Environ Microbiol. 1988; 10: 2541-2548.

[7] Johnson AC, Wood M. DNA, a possible site of action of alumunium in *Rhizobium* spp. Appl Environ Microbiol. 1991; 56: 3629-3633.

[8] Ayanaba A, Asanuma S, Munns DN. An agar plate method for rapid screening of Rhizobium tolerance to acid alumunium stress. Soil Sci Am J. 1983; 47:256-258.

[9] Atlas RM, Barta R. Microbial Ecology, Fundamental and Application. 4th Ed.. Menlo Park: Addison Wesley Longman, Inc; 1998.

[10] Holt JG, Krieg NR, Sneath PHA, Staley JT, Williams ST. Bergey's Manual of Determinative Bacteriology. 9th Ed.. Baltimore: Williams & Wilkins; 1994.

[11] Cunningham SD, Munns DN. Correlation between extracellular polysaccharide production and acid tolerance in *Rhizobium*. Soil Sci Soc Am J.1984; 48: 1273-1276.

[12] Endarini T, Wahyudi AT, Tedja-Imas. Seleksi galur-galur *Bradyrhizobium japonicum* indigenus toleran medium asam-Al (*Selection of Bradyrhizobium japonicum indigenous strains on acid-Al tolerant medium*). Hayati 1995; 2: 15-81.

[13] Fuhrmann J. Symbiotic effectiveness of indigenous soybean bradyrhizobia as related to serological, morphological, rhizobitoxine and hydrogenase phenotypes. Appl Environ Microbiol. 1990; 56: 224-229.

[14] Vincent JM. A Manual for the Practical Study of the Root Nodule Bacteria IBP Hand Book 15. Oxford: Blackwell Sci.; 1970.

[15] Madigan MT, Martinko JM, Parker J. Brock Biology of Microorganisms. 9th Ed. Upper Saddle River: Prentince Hall; 2000.

[16] Habibah H. Efektivitas inokulan *Bradyrhizobium japonicum* toleran asam- aluminium terhadap pertumbuhan kedelai kultivar Slamet. (*Effectiveness of* Bradyrhizobium japonicum *acid aluminum tolerant inoculant on growth of soybean cultivar Slamet*) Master thesis. Bogor: Graduate School of Bogor Agricultural University; 2008.

[17] Alva AK, Edwards DG, Caroll BJ, Asher CJ, Greehoff PM. Nodulation and early growth of soybean mutants with increased nodulation capacity under acid soil infertility factors. Agron J. 1988; 80: 836-84.

[18] Kaneko T, Nakamura Y, Sato S, Minamisawa K, Uchiumi T, Sasamoto S, Watanabe A, Idesawa K, Iriguchi M, Kawashima K, Kohara M, Matsumoto M, Shimpo S, Tsur-

uoka H, Wada T,Yamada M, Tabata S. Complete genomic sequence of nitrogen-fixing symbiotic bacterium Bradyrhizobium japonicum USDA110. DNA Res. 2002;9(6): 189-197.

[19] Mubarik NR, Habibah H, Wahyudi AT. Greenhouse experiments of symbiotic effectiveness of acid-aluminium tolerance *Bradyrhizobium japonicum* on soybean plant. In Nejadkoorki F, editor. International Conference on Applied Life Sciences. Rijeka,Croatia:InTech.pp.337-342;2012.DOI:10.5772/52498. Online: http://www.intecho- pen./ books/ international conference-on-applied-life-sciences.

[20] Masterson CL, Murphy PM. The acetylene reduction technology. In: Subba Rao NS, editor. Recent Advance in Biological Nitrogen Fixation. New Delhi: Oxford and IBH Publ. pp. 8-33; 1979.

[21] Unkovich M, Herridge D, Peoples M, Cadisch G, Boddey B, Giller K, Alves B, Chalk P. Measuring Plant-associated Nitrogen Fixation in Agricultural Systems. Canberra: ACIAR. 2008.

[22] Situmorang ARF, Mubarik NR, Triadiati. The use of acid-alumunium tolerant *Bradyrhizobium japonicum* inoculant for soybean grown on acid soils. Hayati J Biosci 2009;16:157-160.

[23] Wahyudi AT, Suwanto A, Tedja-Imas, Tjahyoleksono A. Screening of acid- alumunium tolerant *Bradyrhizobium japonicum* strain analysis of marker genes and competition in planta. Aspac J Mol Biol Biotechnol 1998; 6:13-20.

[24] Sunarto. 1995. Pemuliaan kedelai untuk toleran terhadap tanah masam dan keracunan Al. (*The breeding soybeans for tolerance to acid soils and Al toxicity*). J Indust Pangan 4:98-99.

[25] Harun MU, Ammar M. Respon kedelai (*Glycine max*) terhadap *Bradyrhizobium japonicum* strain Hup+ pada tanah masam. (*Response of soybean (Glycine max) against* Bradyrhizobium japonicum *strain Hup+ on acid soils*). J Pertanian Indonesia 2001; 3:111-115.

[26] Handayani L, Mubarik NR, Wahyudi AT. Uji viabilitas inokulan *Bradyrhizobium japonicum* toleran asam aluminium pada bahan pembawa gambut. Biosainstifika 2009; 2: 75-82.

[27] Mubarik NR, Imas T, Wahyudi AT, Triadiati, Suharyanto, Widiastuti H. The use of acid-alumunium tolerant *Bradyrhizobium japonicum* formula for soybean grown on field acid soil.On line: World Academy of Science, Engineering and Technology 2011; 53: 879-882 (eISSN 2010-3778). Printed: World Academy of Science, Engineering and Technology 2011; 77: 1280- 1283 (pISSN 2010- 376X).

6

Nitrogen Fixing Cyanobacteria: Future Prospect

Ahmed A. Issa, Mohamed Hemida Abd-Alla and
Takuji Ohyama

1. Introduction

Cyanobacteria are often called "blue-green algae", this name is convenient for talking about organisms in water that make their own food, but does not reflect any relationship between the cyanobacteria and other organisms called algae. Cyanobacteria are relatives to bacteria, not eukaryotes, and it is only the chloroplast in eukaryotic algae to which cyanobacteria are related. Some cyanobacteria are aquatic and photosynthetic, that is, they live in water, and can manufacture their own food. They are quite small and usually unicellular, though they often grow in colonies large enough to see. In fact, it may surprise you then to know that the cyanobacteria are still around; they are one of the largest and most important groups of bacteria on earth (Berry *et al,* 2008). The great contribution of cyanobacteria is the origin of plants chloroplast with which plants make food for themselves is actually a cyanobacterium living within the plant's cells. Sometime in the late Proterozoic or in the early Cambrian, cyanobacteria began to take up residence within certain eukaryote cells, making food for the eukaryote host in return for a home. This event is known as endosymbiosis, and is also the origin of eukaryotic mitochondrion (Issa *et al.,* 2002). Majority of cyanobacteria are aerobic photoautotrophs, their life processes require only water, carbon dioxide, inorganic substances and light. Photosynthesis is their principal mode of energy metabolism. In the natural environment, however, it is known that some species are able to survive long periods in complete darkness. Furthermore, certain cyanobacteria show a distinct ability for heterotrophic nutrition (Fay, 1965). Cyanobacteria might be the first plants to colonies bare areas of rock and soil. Adaptations, such as ultraviolet absorbing sheath pigments, increase their fitness in the relatively exposed land environment. Many species are capable of living in soil and other terrestrial habitats, where they are important in the functional processes of ecosystems and cycling of nutrient elements (Whitton, 1992). The prominent habitats of cyanobacteria are limnic and marine environments. They flourish in water that is salty, brackish or fresh, in cold and hot springs, and in environments where no other microalgae can exist. Most marine forms (Humm and Wicks, 1980) grow along the shore as benthic vegetation in the zone between high and low tide marks. Cyanobacteria comprise a large component of marine plankton with global distribution (Gallon *et al.*, 1996). A number of freshwater species are also able to withstand relatively high concentrations of sodium chloride. It appears that many cyanobacteria isolated from coastal environments tolerate saline environments (i.e. are halotolerant) rather than require salinity (i.e. are halophilic). As frequent colonisers of euryhaline (very saline) environments, cyanobacteria are found in salt works and salt marshes, and are capable of growth at combined salt concentrations as high as 2-3 (%) (Reed *et al.*, 1984). Freshwater localities with diverse trophic states are prominent habitats for cyanobacteria. Numerous species characteristically inhabit, and can occasionally dominate, both near-surface epilimnic and deep, euphotic, hypolimnic waters of lakes (Whitton, 1973). Others colonise surfaces by attaching to rocks or sediments, sometimes forming mats that may tear loose and float to the surface. Cyanobacteria have an impressive ability to colonise infertile substrates such as volcanic ash, desert sand and rocks (Dor and Danin, 1996). They are extraordinary excavators, boring hollows into limestone and special types of sandstone (Weber *et al.*, 1996). Another remarkable feature is their ability to survive extremely high and low temperatures. Cyanobacteria are inhabitants of hot springs (Castenholz, 1973), mountain streams (Kann, 1988), Arctic and Antarctic lakes (Skulberg, 1996) and snow and ice (Kol, 1968; Laamanen, 1996). The cyanobacteria also include species that run through the entire range of water types, from polysaprobic zones to katharobic waters (Van Landingham, 1982).

Once known as blue-green algae, cyanobacteria are the most diverse photosynthetic bacteria. The gram negative bacteria have chlorophyll *a* and photosystems I and II that allow them to perform oxygenic photosynthesis. Unlike most bacteria, cyanobacteria lack α-ketoglutarate dehydrogenase and therefore do not use the citric acid cycle for carbohydrate metabolism, but the pentose phosphate pathway. With such great diversity there has been some controversy on how to classify cyanobacteria. *Bergey's Manual* has divided the organism into five subsections. The classical taxonomy of cyanobacteria divides these organisms into five 'subsections' or orders, three for non-heterocystous types and two for heterocystous types (Castenholz, 2001; Castenholz and Waterbury, 1989). The non-heterocystous cyanobacteria comprise Subsection I (Chroococcales), which are unicellular cyanobacteria that reproduce by binary fission; Subsection II (Pleurocapsales) are unicellular cyanobacteria that produce daughter cells smaller than the parent; and Subsection III (Oscillatoriales) consists of cyanobacteria that produce filaments of cells known as trichomes. All three subsections have N_2-fixing representatives (Bergmanetal., 1997). The classification of N_2-fixing cyanobacteria based on behaviour are shown in table (1) and photographed in figure (1). Heterocyst formation is an important aspect to nitrogen fixation. The filamentous cells differentiate into heterocysts when the cells are deprived of dissolved inorganic nitrogen. A heterocyst consists of a thick cell wall and only contains photosystem I for ATP production. Photosystem II is degraded to prevent O_2 production. O_2 inhibits nitrogenase, the enzyme responsible for N_2-fixation. We will discuss the contribution of cyanobacterial nitrogen fixer organisms' in ecosystem and future prospects.

A. Cyanobacteria that can fix N_2 aerobically
A1.Cyanobacteria that separate N_2 fixation from oxygenic photosynthesis in space. Includes heterocystous genera, for example, *Anabaena*.
A2.Cyanobacteria that separate N_2fixation from oxygenic photosynthesis in time. Includes non-heterocystous genera, such as *Gloeothece*, *Cyanothece* and *Lyngbya*
A3.Cyanobacteria that separate N_2 fixation from oxygenic photosynthesis both in space and in time. Includes non-heterocystous genera, such as *Trichodesmium* and *Katagnymene*
B. Cyanobacteria that can fix only N_2 either anaerobically or microaerobically
Many non-heterocystous cyanobacteria,forexample, *Plectonemaboryanum*.

Table 1. A classification of N_2-fixing Cyanobacteria based on behavior

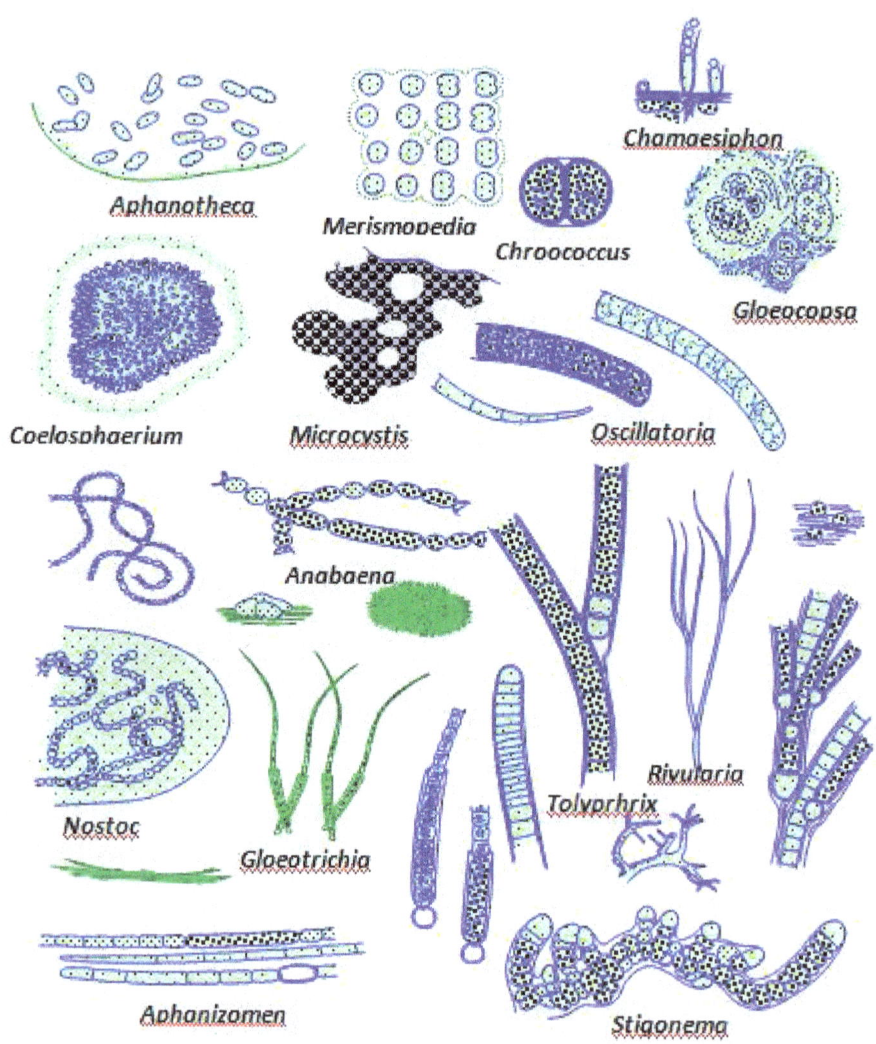

Figure 1. N2-fixing cyanobacteria

2. Effect of abiotic and biotic factor's on population and survival of Cyanobacteria

Cyanobacteria, a group of prokaryotic, oxygen-evolving, photosynthetic Gram-negative bacteria, survive in a wide variety of extreme environmental conditions; they are exposed to various types of natural stresses, such as nutrient limitation, pesticides, pollution, drought, salinity, temperature, pH, light intensity and quality, etc. (Herrero *et al.*, 2001). A Protein in the cyanobacterial thylakoid membranes was identified as a sensitive protein to environmental stress conditions: under various unfavorable conditions like drought, nutrition deficiency, heat, chemical stress, ozone fumigation as well as UV-B and visible light stresses can influence the turnover of protein (Giardi *et al.*, 1997). Many species are capable of not only surviving, but thriving in conditions previously thought to be inhabitable, tolerating desiccation, high temperatures, extreme pH, high salinity and pesticides illustrating their capacity to acclimate to extreme environments (Stal 2007). The major abiotic factor affecting the distribution of algae in soils is solar radiation, moisture, temperature, nutrients, and pH, organic matter content and soil texture are less important. Generally, the higher the soil moisture, soil temperature, and sunlight penetration to the soil surface, the greater the population and activities of algae. Increased levels of both organic and inorganic nutrients also enhance the growth and activity of terrestrial algae. Soil pH also affects the activities of certain types of algae. For example, cyanobacteria thrive best in alkaline soils (pH 7.0 and above) whereas green algae do best in more acidic soils (pH 5.5 and below). In fact, in many cases, moisture levels can be excessive, creating anaerobic conditions that favor the growth of some cyanobacterial species. On these types of sites, temperature appears to be the overriding factor influencing algal growth and activity. It was found that species of *Oscillatoria* were the predominant algae in Bermuda grass greens whereas species of *Anacystis* were the dominant algae in bent grass greens. This distribution can vary too according to the time of year and geographical location. Pesticides are another factor affecting the distribution and activity of cyanobacteria. In general, most herbicides, fungicides, and soil fumigants are toxic, whereas insecticides generally are not. Nearly all of our knowledge about pesticide toxicity to cyanobacteria comes from either laboratory culture studies or from a limited number of field studies on agricultural crops (Mahmoud et al, 1992; Issa, 1999; Issa *et al.*, 2013).

The physico-chemical changes in the environment may affect particular species and induce the growth and abundance of other species, which leads to the succession of several species in a course of time (Muthukumar *et al.*, 2007). High temperatures favour both the phytoplankton productivity and blue green algae (Roger and Reynaud, 1979). Cyanobacteria grew rapidly in the rice fields that contained ample organic matters in the soil and water as well as conditions such as pH, temperature, and organic sources in various rice fields (Choudhu ry and Kennedy, 2004). Among soil properties, pH is a very important factor in growth, establishment and diversity of cyanobacteria, which have generally been reported to prefer neutral to slightly alkaline pH for optimum growth (Koushik, 1994). The cyanobacteria *Alosira ferrilissitiia* and *Calothrix brevessima* have been reported to be ubiquitous in Kerala rice fields with pH from 3.5 to 6.5. Subhashini and Kaushik (1981) reported that the pH of the alkaline soil decreased when treated with cyanobacteria. Also, Nayak and Prasanna (2007) investigated the cyanobacteria were more in number at high pH in rice fields. Cyanobacteria have been found not only to grow in highly saline-alkali soils, but also improve the physico-chemical properties of the soil by enriching them with carbon, nitrogen and available phosphorus (Kaushik, 1994).

Many species of cyanobacteria not only fix carbon in CO_2 through photosynthesis, but they can also fix atmospheric nitrogen. Both of these processes also play an important role in humus formation. In natural soils, cyanobacteria produce considerable amounts of polysaccharide that helps to aggregate soil colloids and improve soil structure while at the same time improving water infiltration and percolation. Subsurface soil cyanobacteria are also known to associate with plant roots, producing hormones that stimulate root growth and enhance the activities of other beneficial root-associated microorganisms (Issa *et al.*, 1994). In fact in many of the rice growing regions of the world some cyanobacterial species are inoculated into soils to enhance rice yields by as much as 36%. Soil cyanobacteria also commonly interact with other microorganisms in soil. Many soil cyanobacteria excrete avariety of antimicrobial compounds that affect the activities of other microorganisms, including plant pathogens. In this case, a species of *Nostoc* was used for the biological control of a seedling disease of millet. Upon the death of nearly all algae, they serve as an important food source for many important bacteria and fungi in soils (Issa, 1999). While living, soil algae serve as food sources for protozoa, earthworms, nematodes, and micro arthropods. A number of associations of algae with othermicroorganisms in soil can result in enhanced algal growth resulting in detrimental effects on turfgrass growth and quality. On the other hand, a number of cyanobacteria species have been shown to inhibit root growth of a number of crop plants by producing antibiotic substances that also inhibit bacterial growth. It is well known in the floriculture industry that algal proliferation (usually cyanobacteria) on subirrigation mats and on roots of potted plants in greenhouses can lead to reductions in plant growth and quality.

In the Arctic, cyanobacteria are the primary source of newly fixed nitrogen (Hobara *et al.* 2006; Solheim *et al.* 2006) and form many associations with vegetation including epiphytic and endophytic facultative associations with bryophytes (Turetsky, 2003) and the lichen symbioses and soil surface colonies that are components of Biological Soil Crusts (Belnap *et al.* 2001). Bryophyte-associated cyanobacteria are an important source of N_2 within many terrestrial ecosystems, for example, a high abundance of feather moss-cyanobacterial associations occur innorthern boreal forests, where they contribute 1.5 to 2.0 kg N $ha^{-1}yr^{-1}$ (DeLuca *et al*.2002; Houle *et al*.2006; Lagerström *et al*.2007; Zackrisson *et al*.2009).While variation is often high within and between bryophyte species, the highest rates of N_2-fixation in arctic landscapes are often associated with cyanobacteria bryophyte associations (Alexander and Schell1973; Henry and Svoboda1986; Solheim et al.1996). Cyanobacterial symbioses with lichens are also a major source of fixed N_2 as they often have N_2-fixation rates exceeding that of other cyanobacterial symbioses (Schell and Alexander1973; Kallio and Kallio 1975; Crittenden and Kershaw 1978; Gunther 1989; Hobara *et al.* 2006). Finally, the prevalence of Biological Soil Crusts in many arctic ecosystems ensures that the cyanobacteria associated with those crusts are major contributors to arctic N_2 inputs (Alexander *et al*.1978). Although many environmental factors could potentially determine the microbial community present in these multidimensional ecosystems, changes in the diversity of cyanobacteria in rice fields was correlated to salinity Low salinity favored the presence of heterocystous cyanobacteria, while very high salinity mainly supported the growth of non-heterocystous genera. High nitrogen content in the low salt soils is proposed to be a result of reduced ammonia volatilization in comparison to the high salt soils. Cyanobacterial mats are dense, stratified microbial agglomerations that develop well in hypersaline habitats because of the limited grazing activities (Javor and Catenholz 1984; Cohen 1989; Farmer 1992). These mats are composed of different physiological groups of microbes such as photoautotrophic, photoheterotrophic, chemoautotrophic, and heterotrophic organisms (van Gemerden 1993; Stal 1995). Oxygenic photosynthesis is mainly performed by cyanobacteria in the top few millimeters of the mats, resulting in the development of strong oxygen gradients and the production of organics that are utilized by heterotrophic bacteria (Jonkers et al. 2003).

3. Effect of adverse soil condition on heterocyst formation and nitrogenase activity in hetreocystous cyanobacteria

Many free-living blue-green algae (cyanobacteria) fix atmospheric nitrogen and since they are photosynthetic, they do not compete neither with crop plants nor with heterotrophic soil microflora for carbon and energy. Nitrogen-fixing ability has not only been shown by heterocystous Cyanobacteria (*Nostoc, Anabaena, Aulosira,* etc.) but also by several non-heterocystous unicellular (*Gloeocapsa, Aphanothece, Gloeothece,* etc.) and filamentous (*Oscillatoria, Plectonema,* etc.) cyanobacteria (Table 2). In non heterocystous forms, the oxygenic photosynthesis was found to be separated from nitrogen fixation either temporally or spatially. In temporal separation, nitrogen fixation predominantly occurs during the dark period and photosynthesis during the light; in these forms in terms of energy the anaerobic dark conditions are not very favourable for the process of nitrogen fixation. In spatial separation, the central non-photosynthetic cells get engaged in nitrogen fixation, whereas, the outer green cells are photosynthetically active. The species with bio fertilizer potential are the heterocystous, filamentous forms belonging to the order Nostocales and Stigonematales in which the nitrogenase activity and oxygenic photosynthesis are separated spatially and nitrogenase activity is usually light-dependent. Species of *Nostoc, Anabaena, Tolypothrix, Aulosira, Cylindrospermum, Scytonema,* and several other genera are widespread in rice fields and contribute significantly to their fertility. Cyanobacteria can contribute about 20-30 kg N ha^{-1} $season^{-1}$ as well as organic matter to the soil which is quite significant for the economically weak farmers who are unable to invest on costly chemical nitrogen fertilizer. Often blooms of free-living cyanobacteria are favoured in tropical regions and inoculation of paddy fields with cyanobacteria is traditionally applied in most of the Asian countries. Besides rice, other crop plants like vegetables, wheat, sorghum, maize, cotton, sugarcane, etc. also respond to cyanobacterial biofertilizer (Abd-Alla and Issa, 1994; Abd-Alla *et al.*, 1994). In sub-tropical regions *Azolla,* a fern within the leaf cavity of which is found the heterocystous cyanobacterium-*Anabaena azollae,* is the traditional biofertilizer (Kimura, 2000; Kirk, 2004)

Biological nitrogen fixation, and specifically the nitrogenase enzyme, is notorious for its sensitivity to molecular oxygen. Moreover, high oxygen stress causes proteolysis of nitrogenase subunits (Durner *et al.*, 1996), suppresses nitrogenase synthesis, and leads to a shortage of respiratory substrates and reductants necessary for nitrogen fixation and assimilation (Gallon, 1992). Inhibitory effects of moderate levels of oxygen, or short exposure times, *in vivo* may be reversed, leading to an increase in nitrogen fixation rates (Yakunin *et al.*, 2001) and, in some diazotrophs, post-translational modification of the Fe protein from an inactive to active form (Zehr *et al.*, 1993). Furthermore, diazotrophic cyanobacteria, which provide the bulk of fixed nitrogen to the surface oceans, are the only diazotrophs that actively produce oxygen *via* photosynthesis and must contend with further restrictions on the nitrogen (Berman-Frank *etal.*, 2003). Thus, nitrogenase in the real-world operates at only a fraction of its potential activity, yet is a major elemental taxation on diazotrophic cyanobacteria both for scarce trace elements, such as iron, and in the costs of protein synthesis. These taxes have, in turn, led to a global limitation of fixed nitrogen in the oceans (Falkowski, 1997). In addition to contributing nitrogen, cyanobacteria benefit crop plants also by producing various growth promoting substances, like gibberellins, auxins like indole-3-acetic acid, indole-3-propionic acid, etc., vitamin B12, free amino acids like serine, arginine, glycine, aspartic acid, threonine, glutamic acid, etc., extra- and intra-cellular polysaccharides like xylose, galactose, fructose, etc. Such substances have several beneficial effects like improved soil structure, stimulation of growth of crop plants as well as useful bacteria, chelation of heavy metals (El-Enany and Issa, 2000). Cyanobacteria are primary colonizers and many have been shown to possess the property of tricalcium phosphate solubilization. Abundantly available rock phosphate, being insoluble, is unavailable to crop plants. Some of the cyanobacteria like, *Tolypothrix, Scytonema, Hapalosiphon*, etc. have been reported to solubilize rock phosphate.

Unicellular group	Unicellular strains growing on BG II medium without nitrogen *(Aphanothece,Gloeothece,)*
Anabaena group:	Heterocystous strains with a thin sheath, without branching, do not form mucilaginous colonies of definite shape *(Anabaena,Nodularia, Cylindrosperinurn, Anabaenopsis* etc.)
Nostoc group:	Heterocystous strains with a thick sheath, without branching, forming mucilaginous colonies of definite shape *(Nostoc)*
Aulosira group:	Heterocystous strains with a thick sheath, usually without branching, do not form diffuse colonies on agar medium *(Aulosira)*
Scytoneina group:	Heterocystous strains with false branching, without polarity, forming velvet-like patches on agar medium *(Scytonema)*
Calothrix group:	Heterocystous strains with false branching, with polarity, forming velvet-like patches on agar medium *(Calothrix, Tolypotkrix, Hassalia)*
Gloeofrichia group:	Heterocystous strains, with polarity, forming mucilaginous colonies of definite shape *(Gloeotrichia, Rivularia)*
Fischerella group:	Heterocystous strains with true branching *(Fischerella, Westiellopsis, Stigonerna)*

a All features refer to strains grown from soil or water sample dilutionsplated on agarized BGII medium without nitrogen.

Table 2. Table2. Definition of the taxa of N_2-fixing cyanobacteria

4. Nitrogenase

The enzyme complex nitrogenase (E.C.1.18.6.1) consists of a dimeric Fe-protein (the dinitrogenase reductase) functioning as an electron carrier to the tetrameric MoFe-protein (the dinitrogenase) which reduces molecular nitrogen to ammonia. Both enzymes are highly oxygen-sensitive. The intrinsically anaerobic character of the nitrogenase complex requires special adaptation in cyanobacteria which produce oxygen in a plant-type photosynthesis. Filamentous heterocystous cyanobacteria provide such an anaerobic environment by creating a diffusion barrier for gases, enhanced respiratory activity and the lack of the oxygenic photosystem I1 (Scherer *et al.*, 1988).Reductant supply of nitrogenase *via* ferredoxin is provided by photosynthates transported from vegetative cells to the heterocysts and ATP is generated by photosystem-l activity or oxidative phosphorylation in the heterocysts (Stewart and Rowell, 1986). Under a light-dark regime most heterocystous strain described so far preferentially fix nitrogen in the light (Khamees *et al.*, 1987). Natural blooms also, dominated by heterocystous cyanobacteria exhibit higher nitrogenase activity in the light than in the dark (Horne, 1979). The low activity of nitrogenase in darkness was assigned to the inability of metabolism to sufficiently generate reductants under these conditions (Ernst and Bohme, 1984). In many respects this modification resembles the ADP-ribosylation of Fe-protein of nitrogenase observed in *Rhodospirillaceae* after transfer from light to darkness or after addition of ammonia (Kanemoto and Ludden, 1984).Heterocysts have thick multilayered wall preventing the entry of oxygen, high rate of respiration which utilizes the defused oxygen, and they lack photosystem II so that there is no photosynthetic evolution of oxygen. The scheme of a heterocyst with adjacent vegetative cells is shown below. The outer and inner layers of the heterocyst envelope consist of polysaccharides and glycolipids, respectively. In this scheme the pore region is not drawn to scale and shown enlarged to accommodate metabolite exchange between the cells. Cell wall and cell membranes are not drawn separately. Heterocysts import carbohydrates from vegetative cells, with glutamine moving in the opposite direction. In a cell-free system derived from heterocysts, the following substrates supported nitrogenase activity: glycogen, maltose, sucrose (less active), glucose and fructose; glucose 6-phosphate (G6P) and other intermediates of the oxidative pentose-phosphate cycle (PPC), including dihydroxyacetone phosphate (DAP), glyceraldehyde 3-phosphate (GAP) and fructose-1,6-bisphosphate (FBP), were particularly active. Glycolytic substrates, such as phosphoenolpyruvate (PEP) and pyruvate (Pyr) were inactive or inhibitory in acetylene reduction by the heterocyst extract. In the dark, reductant for nitrogen and oxygen is generated by the activity of the oxidative PPC and possibly by isocitrate dehydrogenase. NADPH thus formed donates electrons via ferredoxin: NADP reductase (FNR) to a heterocyst-specific ferredoxin (FdxH) and then to the two components of nitrogenase (Fe-protein and FeMo-protein) as indicated. NAD(P)H and hydrogen are also electron donors to the respiratory electron transport (RET) generating the necessary ATP for the nitrogenase reaction. In the light, ATP is formed by cyclic photophosphorylation mediated by photosystem I (a PSI-dimer, as indicated). Ferredoxin could be also photoreduced by PSI at the expense of hydrogen and NAD(P)H as electron donors (Figure 2). Wyatt and Silvey (1969) for the first time reported that non-heterocystous cyanobacteria have also the ability to fix nitrogen in which all the vegetative cells contain nitrogenase but due to thepresence of oxygen the enzyme gets inactivated. Nitrogen fixation in these organisms is light stimulated process. Cyanobacteria fix nitrogen only under combined nitrogen deficient conditions and in the presence of combined nitrogen source the enzyme nitrogenase remains repressed which, similar to oxygen effect, is a reversible inhibition. Inoculation of rice fields with cyanobacteria reduces the nitrogen losses through metabolization of the applied combined nitrogen forms. The metabolized combined nitrogen as well as the biologically fixed nitrogen becomes available gradually through exudation and decomposition of these algae. The nitrogen fixation by cyanobacteria depends upon the various biotic and abiotic factors. Nitrogenases are highly sensitive to oxygen and hydrogen production catalyzed by the nitrogenase / hydrogenases can only function under anaerobic conditions because of its extreme sensitivity to oxygen. Some cyanobacteria have solved this problem by developing specialized thick walled cells known as heterocysts which maintain low oxygen tension inside, thereby facilitating nitrogenase activity, which produces hydrogen during N_2 fixation (Issa 1995). A wide range of nitrogenase activity has been reported in cyanobacteria (Table 3).

Name of the species	**Nitrogenase activity (nmol C_2H_4 mg dry wt.$^{-1}$ h^{-1})**	**Reference**
Anabaena sp. Strain CA	0.17	Antal et al. 2005
Anabaena sp. Strain N9AR	0.13	Antal et al. 2005
Phormidium valderianum	2.2	Kiran, 2007
Nostoc calcicola	9.021	Bolton, 1996
Anabaena cylindrica	5.579	Bolton,1996
Anabaena oryzae	5.076	Bolton,1996
Nostoc commune	6.346	Anjana et al. 2012
Nostoc muscorum	8.5	Issa, 1995

Table 3. Nitrogenase activity of various cyanobacterial species

5. Effect of severe conditions on nitrogenase activity in non heterocystous cyanobacteria

Cyanobacteria are oxygenic phototrophic microorganisms, usually living in aerobic and oxygen-supersaturated environments (Stanier and Cohen-Bazire 1977). Many cyanobacteria, filamentous as well as unicellular species, synthesize the enzyme nitrogenase and are able to

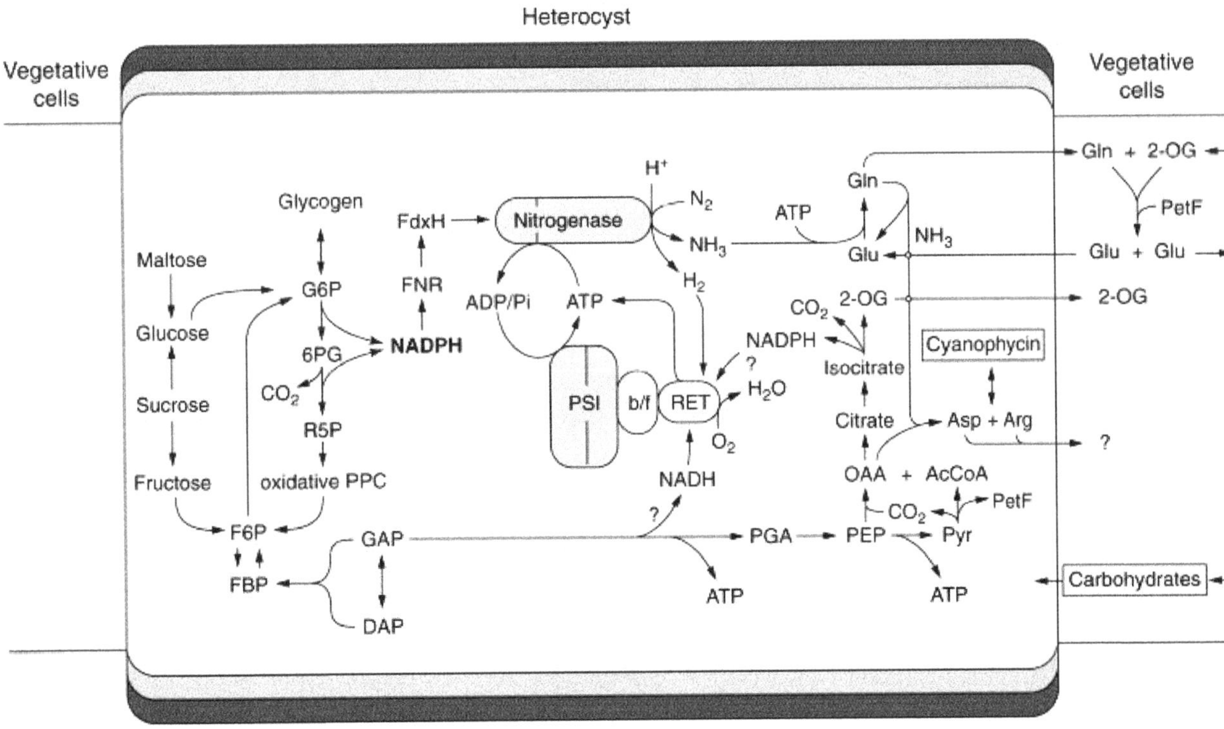

Figure 2. Heterocyst metabolism and nitrogen fixation. Abbreviations: AcCoA, acetyl coenzyme A; Arg, arginine; Asp, aspartate; b/f, cytochrome b6f complex; F6P, fructose 6-phosphate; PetF, vegetative cell type ferredoxin; Glu, glutamate; Gln, glutamine; OAA, oxaloacetate; 2-OG, 2-oxoglutarate; 6PG, 6-phosphogluconate; PGA, 3-phosphoglycerate; Pi, inorganic phosphate; R5P, ribose 5-phosphate. (Böhme,1998)

fix molecular nitrogen (Stewart 1980). This phenomenon seems to be in contradiction with the extreme sensitivity of nitrogenase towards molecular oxygen (Robson and Postgate 1980).Therefore, nitrogen-fixing cyanobacteria have developed mechanisms to protect nitrogenase from inactivation by oxygen. Thus far, these mechanisms are largely unknown. Photosynthetic oxygenevolution and nitrogen fixation cannot occur simultaneouslyin one single cell (Mitsui *et al.* 1986).Several filamentous cyanobacteria develop heterocysts.Heterocysts are non-dividing cells which have lost thecapacity of oxygenic photosynthesis and which contain theenzyme nitrogenase (Fay et al. 1968); hasorganisms thus have solved the problem by spatial separation of the incompatible processes of oxygen evolution and nitrogen fixation. In non-heterocystous filamentous and unicellular nitrogen-fixing cyanobacteria nitrogenase and photosynthesis apparently occur in the same cell. It has been suggested that in such organisms nitrogen fixation is separated from oxygenic photosynthesis temporary (Stal and Krumbein 1985a). When grown under light-dark cycles, non-heterocystous cyanobacteria show nitrogenase activity only during the dark period (Huang and Chow 1986). However, when cultures were synchronized or previously adapted to light-dark cycles, also in continuous light a cyclic pattern of nitrogenase activity can be observed (Grobbelaar et al. 1986). The strategy by which non-heterocystous cyanobacteria protect nitrogenase from deterioration by atmospheric and photosynthetic evolved oxygen, it has been shown fora variety of non-heteroeystous cyanobacteria, that when grown underlight-dark cycles, nitrogenase activity predominantly occurs during the dark period. However, all these organisms are able to grow in continuous light at the expense of molecular nitrogen, showing nitrogenase activity under such conditions. Using synchronized cultures of *Synechococeus* sp., Mitsui *et al.* (1986) showed that N_2-ase and oxygen evolution followed a reciprocal pattern, even in continuous light. They also showed that the capacity of oxygen evolution decreased to virtually zero at the maximum of nitrogenase activity and in one case even became negative (respiration exceeded possible oxygen production). These authors, however, measured photosynthesis at far higher (more than 6 times) light intensity than applied for growth and acetylene reduction. This phenomenon is most pronounced in continuous light. However, this observation cannot be taken as evidence for the coexistence of N_2-ase and photosynthesis in one single cell. Weare and Benemann (1974) provided evidence that a cyclic degradation and resynthesis of phycobiliprotein regulated photosynthetic activity. When phycobiliprotein of the cell was low, oxygen evolution ceased and nitrogenase was induced. On the other hand, Giani and Krumbein (1986) found that this phenomenon strongly depended on light intensity. At low light intensity (500 lux or less) phycobiliprotein was constant. From epifluorescence microscopy and microfluorimetry it was concluded that under nitrogen-fixing conditions, the oxygenic photosystem was still intact. However, oxygen evolution was not measured and the possibility that photosynthesis was switched off could not be excluded. In *Oscillatoria,* the phycobiliprotein content is lower in nitrogen-fixing cells than in nitrate grown cultures (Stal and Krumbein 1985b). This may result in a lower activity of the photosynthetic apparatus. However, phycobiliprotein and total protein remained constant during growth on N_2. The question arises whether this is also the case in the light (Maryan et al. 1986). Both in the dark and in the light a very high rate of respiration was observed here. The nitrogenase activity in the light, however, was about six times the dark maximum rate. Thus, light clearly stimulated nitrogenase activity. Therefore it was concluded that, in the light, respiration more likely fulfilled a protection function than that it provided energy for nitrogenase. In the dark, respiration provides both energy and a protective role. Respiration of cyanobacteria in the light has been shown for several species and the highest rates have been found in nitrogen-fixing species (Scherer and Böger 1982). However, in none of these organisms, respiration exceeded photosynthesis.

Trichodesmium spp. is non-heterocystous cyanobacteria found in tropical and subtropical seas which are important in mediating a flux of reduced nitrogen from the atmosphere to the ocean. The organism fixes nitrogen when grown with N_2 as the sole inorganic nitrogen source (Ohki and Fujita 1988). The nitrogen-fixing system of this algae is regulated at two levels: (1) the synthesis of enzyme is regulated at a transcriptional or post-transcriptional level by the presence of urea, and (2) the activity of the Fe protein is correlated with a shift in electrophoretic mobility, which is believed to be a post-translational modification (Ernst et al. 1990a, b). *Trichodesmium* lacks both akinetes and heterocystsand shows no ability to produce hormogonia. *Trichodesmium* has therefore been classified as anundifferentiated filamentous cyanobacterium exclusivelycomposed of photosynthetic vegetative cells, defined as cyanobacteria group III (Rippka etal. 1979). However, it is now becoming increasingly evidentthat *Trichodesmium* is differentiated, secludingthe nitrogen-fixing enzyme nitrogenase into a low number of cells (Fredrikssonand Bergman 1995). *Trichodesmium* istherefore the first detected non-heterocystous cyanobacterium with cells specialised for nitrogen fixation. Cyanobacteria develop specialised nitrogen-fixing cells in order to solve their problem of managing theco-existence of oxygen labile nitrogen fixation andoxygen producing photosynthesis (Fay 1992, Gallon1992). Heterocysts provide the ability to perform aerobic nitrogen fixation through the exclusion of atmosphericoxygen and oxygenic photosynthesis. *Trichodesmium* practises a different, hitherto unknown typeof specialisation to accommodate the oxygen sensitive nitrogenase (Fredriksson and Bergman 1995). The present study further characterises the nitrogenase containing cells in *Trichodesmium* by comparing theultrastructure of these cells with those lacking nitrogenase. Indeed, nitrogenase-containing cells exhibited structural modifications indicative of additional changes in gene expression. The functional implication of these changes is interesting considering *Trichodesmium's* unusual ability to perform oxygenlabile nitrogen fixation under fully aerobic and oxygen producing photosynthetic conditions withoutheterocyst formation (Saino and Hattori 1978). The differentiation of cells specialised for nitrogen fixation also questions the taxonomic affiliation of *Trichodesmium* within cyanobacteria.

The crystal structure of the enzyme one can see that nitrogenase is multisubunit enzyme. The FeMo protein is the site for N_2 reduction. The other subunit is the Fe protein, encoded by the highly conserved *nif*H gene used for sequencing and identification. The Fe protein has a Fe_4S_4 complex where ATP is hydrolyzed, providing the necessary electrons to the Fe-Mo active site. Oxygen inhibition of the enzyme occurs in this subunit where the molecule can interact with the Fe_4S_4 complex (see Figure 3). Conversion of N_2 to ammonia is no easy process; Schindelin *et al.* (1997) concluded that once ATP binds, the Fe protein goes through a substantial conformational change in order to efficiently feed electrons to the FeMo redox site. This site also goes through some reorientation, maximizing the energy put into the system. For every electron that is fed to the FeMo protein active site, two ATP are hydrolyzed. The overall reaction formula is,

$$N_2 + 8\ H^+ + 8\ e^- + 16\ ATP \rightarrow 2NH_3 + H_2 + 16\ ADP + 16\ Pi.$$

Nitrogen assimilation in natural populations of *Trichodesmium* spp. proceeds via the glutamine synthetase/glutamate synthase (GS/GOGAT) pathway. GS is necessary for NH_4 assimilation regardless of the primary form of N being used. High rates of GS transferase activity relative to rates of total N uptake have been observed in natural and cultured populations of *Trichodesmium* spp. (Mulholland et al., 1999; Mulholland and Capone, 1999). Rates of both GS transferaseand GS biosynthetic activity (which approximates *in vivo* forward reaction activity) in *Trichodesmium* spp. increase in the afternoon during the period when rates of N_2 fixation are highest. The ratio of GS transferase:biosynthetic activity decreases during the period of maximum N_2 fixation, indicating that the proportion of the GS poolthat is biosynthetically active increases during the day. The biosynthetic capacity of GS is sufficient to allow *Trichodesmium*spp. Colonies to turnover their cell N at least three times per day, suggesting that N assimilation does not limit the rate of N utilization by cells, even during midday when N_2 fixation rates are highest. Cells appear to have sufficient capacity to assimilate all of theintracellular N substrates derived from N_2 fixation and N uptake incultures growing on media with or without added N (Mulholland et al., 1999). Excess GS activity is characteristic of cells limited by N orusing N_2 as their N source. A positive correlation be-

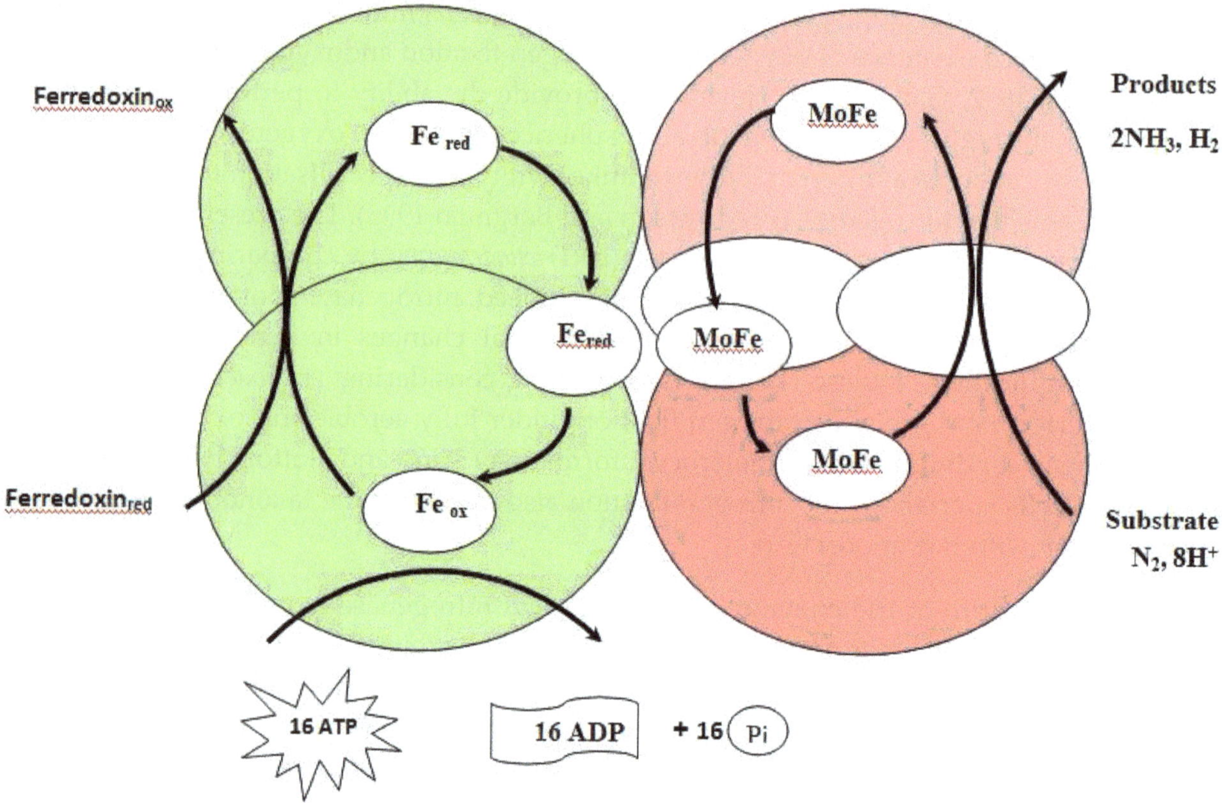

Figure 3. The structure of nitrogenase

tween GS andnitrogenase enzyme abundance and distribution has been observedin a variety of heterocystous and non-heterocystous cyanobacteria including *Trichodesmium* spp. Thus, there might be both a constitutive pool of GS, regulated for the general assimilation of N derived from various N sources, and a nitrogenase-linked pool co-regulated specifically with nitrogenase under low N conditions. Both regulatory mechanisms appear to be important. A global N-regulating gene, *ntcA,* has been identified in a *Trichodesmium* spp. isolated from the Red Sea.

6. Symbiotic cyanobacteria

Symbiotically competent cyanobacteria have some excellent features that make them particularly significant in any attempt to extend the list of N_2-fixing symbioses to include plants of commercial interest, such as cereals. Unlikerhizobia, most symbiotic cyanobacteria carry their own mechanism for protecting nitrogenase from inactivation by oxygen (heterocysts). Cyanobacteria have anunmatched host range (fungi sponges, protists and angiosperms), are not restricted to roots but may form symbiosis withvarious plant parts, and do not need to be located intracellularly within the host plant (Adams et al., 2006; Bergman et al., 2007). Cyanobionts generally supply their hosts with fixed nitrogen, although they can also provide fixed carbon to non-photosynthetic hosts. The major plant hosts are bryophytes, cycads, the angiosperm *Gunnera*, the water-fern *Azolla*, and fungi (to form lichens). Although all cyanobacteria are photoautotrophs, many are also facultative heterotrophs and so are not restricted to the areas of the plant that receive light, and can be found in roots, stems, leaves, and thalli. This review will concentrate on the cyanobacteria–bryophyte symbioses, focusing in particular on the importance of pili and gliding motility in plant infection (Meeks 2003). Plant cyanobionts all have two major characteristics in common: (i) the ability to differentiate both specialized nitrogen-fixing cells known as heterocysts (Zhang et al., 2006) and (ii) short, motile filaments known as hormogonia, which lack heterocysts and provide a means of dispersal for otherwise immotile cyanobacteria (Meeks, 1990,1998). Heterocysts usually occur singly in a semi-regular spacing within filaments of vegetative cells (Golden and Yoon, 2003; Zhang et al., 2006). The infective agents in most plant symbioses are hormogonia and some, perhaps all, plants produce chemical signals that trigger their formation and chemoattractants that guide them into the plant tissue (Figure 4). The plant cyanobionts are members of the genus *Nostoc*, which is commonly found free-living in nature (Dodds *et al.*, 1995; Rai *et al.*, 2002). However, in the laboratory, other hormogonium-developing cyanobacterial genera, such as *Calothrix* and *Chlorogloeopsis*, may infect liverworts (West and Adams, 1997). Members of the genus *Nostoc* are primarily non-motile, but a characteristic of the genus is the ability to produce specialized motile filaments known as hormogonia which serve as a means of dispersal as well as plant infection (Meeks and Elhai, 2002). Hormogonia development is triggered by a variety of environmental factors, including plant-derived chemical signals. The development of hormogonia in heterocystous cyanobacteria results from a round of rapid, synchronous cell divisions which result in a decrease in cell size (Meeks and Elhai, 2002). This is followed by fragmentation of the filament at the heterocyst–vegetative cell junctions, releasing short, motile hormogonia. Hormogonia lack heterocysts and are a temporary stage in the *Nostoc* life-cycle, soon returning to vegetative growth and developing heterocysts once more. For hormogonia to locate the symbiotic tissue of a plant host they must attach to the surface and both extracellular polysaccharides and pili (fimbriae) are thought to play a role in this process (Adams, 2000). Type IV pili are required for gliding in some unicellular cyanobacteria (Bhaya, 2004), and the cell surface of hormogonia of the symbiotically competent *Nostoc punctiforme* is covered with pili (Duggan *et al.*, 2007). Plant hosts increase the likelihood of infection by cyanobacteria by both stimulating the formation of hormogonia in potential cyanobionts and by guiding the hormogonia to the symbiotic tissues by chemotaxis. Hormogonia formation is stimulated by hormogonia-inducing factors (HIFs). HIF production has been found in the hornwort *Anthoceros punctatus* (Meeks, 2003), as well as cycads and the angiosperm *Gunnera* (Rasmussen *et al.*, 1994; Ow *et al.*, 1999). *Anthoceros punctatus* HIF is a small, heat-labile product released by the hornwort when starved of combined nitrogen (Meeks and Elhai, 2002; Meeks, 2003). The liverwort *Blasia* also releases HIF when nitrogen-starved (Adams, 2002). *Nostoc punctiforme* mutants with increased sensitivity to *Anthoceros* HIF, also show a greater initial frequency of infection of the hornwort than the wild type (Cohen and Meeks, 1997).

The infection of hornworts *via* the stomata-like opening to the slime cavity has interesting parallels with the likely method of entry of cyanobacteria into the primitive, extinct land plant *Aglaophyton major*. This symbiosis is only known from fossil evidence, but an

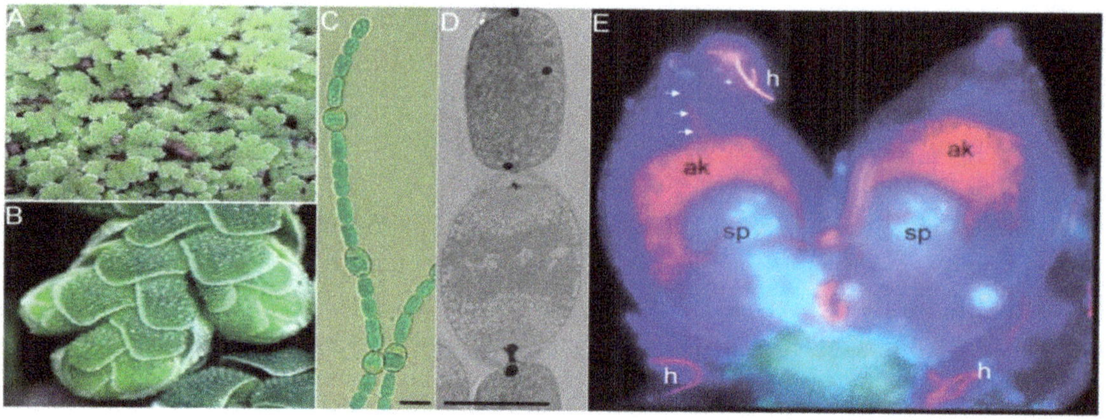

Figure 4. The partners in the Azolla symbiosis. A) Fronds of the Azolla filiculoides Lam. plant. B) Close up of an Azolla branch showing the apex and the alternating 'stacked' dorsal leaves, each containing a cavity in which the cyanobiont (NoAz) filaments reside. C), D), E) Light micrograph of the cyanobiont. Pairs of megasporocarps (blue) develop at the underside of the cyanobacterial colonized Azolla leaves. Filaments of the motile cyanobacterial cell stage (red), the hormogonia (h), are attracted to the sporocarps, gather at the base and subsequently move towards the tip, before entering the sporocarps via channels (white arrows). Once inside the sporocarp the hormogonia differentiate into individual thick walled resting spores (or akinetes; ak), seen as the intensively red fluorescing small inoculum on top of the megaspores (sp). (Ran et al., 2010)

Archaeothrix-type filamentous cyanobacterium is thought to have entered the plant via stomatal pores (Taylor and Krings, 2005). The cyanobacteria are thought to have initially colonized the substomatal chambers and then spread throughout the outer cortical tissue, where they can be seen in fossil specimens of the plant. This is somewhat similar to the infection process in the extant hornwort *Leiosporoceros dussii* in which the cyanobacteria are found in mucilage-filled 'canals' (Villarreal *et al.*, 2005; Villarreal and Renzaglia, 2006). Once the cyanobacterium has entered the host plant a number of morphological, developmental, and physiological changes occur. The development of hormogonia is repressed, whereas the development of heterocysts is greatly stimulated. The rate of cell division is reduced, ensuring that the cyanobiont does not outgrow the host. The rate of CO_2 fixation is greatly reduced, whereas nitrogen fixation is stimulated and ammonium assimilation down-regulated (Figure 5).

The nitrogen fixation rates for cyanobacteria symbiotically associated with bryophytes are several-fold higher than for the same free-living cyanobacteria. This increase is due to a greatly elevated heterocyst frequency, which may be 6–10-fold higher than in the free-living state (As little as 20% of the nitrogen fixed is retained by the cyanobiont, the remainder being transferred as ammonia to the host (Meeks and Elhai, 2002). The primary route of ammonia assimilation in cyanobacteria is the GS-GOGAT (glutamine synthetase–glutamate synthase) pathway. The level of GS protein in *Anthoceros*-associated *Noctoc* is similar to that in free-living cyanobacteria, but GS activity is reduced implying that activity is regulated by an unknown, and presumably plant-regulated, post-translational modification of the enzyme.

Close examination of an Azolla leaf reveals that it consists of a thick, greenish (or reddish) dorsal (upper) lobe and a thinner, translucent ventral (lower) lobe emersed in the water. It is the upper lobe that has an ovoid central cavity, the "living quarters" for filaments of Anabaena. Probably the easiest way to observe *Anabaena* is to remove a dorsal leaf lobe and place it on a clean slide glass with a drop of water. Then apply a cover slip with sufficient pressure to mash the leaf fragment. Under 400X magnification the filaments of *Anabaena* with larger, oval heterocysts should be visible around the crushed fern leaf. The thick-walled heterocysts often appear more transparent and have distinctive "polar nodules" at each end of the cell. The "polar nodules" may be the same composition as cyanophycin granules (co-polymer of arginine and aspartic acid). Cyanophycin granules occur in many cyanobacteria and may serve as a nitrogen storage product.

Although Azolla can absorb nitrates from the water, it can also absorb ammonia secreted by Anabaena within the leaf cavities. Rice is the single most important source of food for people and Azolla plays a very important role in rice production. For centuries Azolla and its nitrogen-fixing partner, *Anabaena,* have been used as "green manure" in China and other Asian countries to fertilize rice paddies and increase production. Republic of China has 3.2 million acres of rice paddies planted with Azolla. This provides at least 100,000 tons of nitrogen fertilizer per year worth more than $50 million annually. Extensive propagation research is being conducted in China to produce new varieties of Azolla that will flourish under different climatic and seasonal conditions. According to some reports, Azolla can increase rice yields as much as 158 percent per year. Rice can be grown year after year, several crops a year, with little or no decline in productivity; hence no rotation of crops is necessary. In addition to nitrogen fixation, Azolla has a number of other uses. Several California aquafarms grow Azolla in large vats of circulating fresh water. Apparently fish and shrimp relish the Azolla. In fact, Azolla was grown for fish food and water purification at the Biospere II project in Arizona (a 2.5 acre glass enclosure simulating an outer space greenhouse). Fresh Azolla and duckweed (Wolffia) can also be used in salads and sandwiches, just as alfalfa and bean sprouts are used. Dried, powdered Wolffia and Azolla make a nutritious, high protein powder similar to the popular alga (cyanobacterium) *Spirulina* that is sold in natural food stores. Azolla has also proved useful in the biological control of mosquitos.The mosquito larvae are unable to come up for air because of the dense layer of Azolla on the water surface. Azolla grows very quickly in ponds and buckets, and in makes an excellent fertilizer (green manure) and garden mulch.

7. Future challenges—Prospects

The nitrogen cycle of Earth is one of the most critical yet poorly understood biogeochemical cycles. Current estimates of global N_2 fixation are approximately 240 Tg N y^{-1} with a marine contribution of 100–190 Tg N y^{-1}. Of this, a single non-heterocystous genus, *Trichodesmium* sp. contributes approximately100 Tg N y^{-1} (Capone pers. comm.). Geochemical evidence suggests that, on a global scale, nitrogen fixation does not always keep pace with denitrification on time scales of centuries to millenia (Falkowski and Raven, 1997), yet it remains unclear what process (es) limits nitrogen fixation in the oceans. More importantly, given the potential for heterocystous cyanobacteria to outcompete organisms such as *Trichodesmium*, it

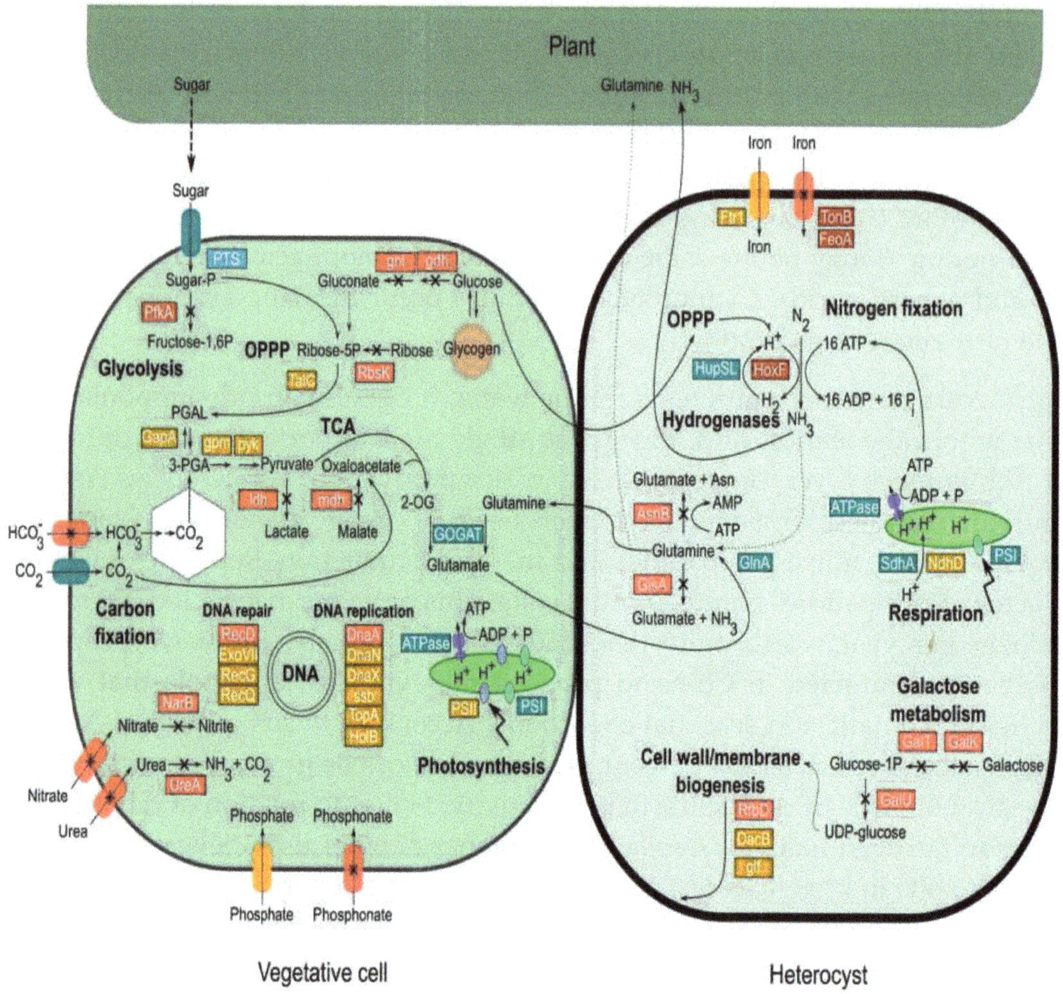

The left cell represents a vegetative cell while the right a nitrogen-fixing heterocyst. Red color indicates pseudogenes lacking functional counterpart in the No Azgenome. Orange indicates pseudogenes where a functional counterpart is present elsewhere in the genome. Fully functional gene(s) are illustrated (blue) only if their function is linked to other processes in the figure.The localization of pathways in vegetative cells or heterocysts is representative only for nitrogen fixation (heterocysts) and PSII activity(vegetative cells).Note that only amin or part of the nitrogen fixed in heterocysts is incorporate dusing the GS-GOGAT pathway and used for synthesis of amino acids,while most is exported to the plant as NH3.Sugar is provided by the plant *via* the sugar phosphotransferase system (PTS). Function has been lost int he glycolyticpathway as the pfkA gene, encoding 6-phosphofructokinase, is a pseudogeneand sugar metabolism in the Azolla cyanobiont probably proceeds via the Oxidative Pentose Phosphate Pathway (OPPP). Extensive loss of function is evident among genes involved in uptake and transport of nutrients and NoAzhas lost the capacity to both import and metabolise alternative nitrogen sources, (Ran et al., 2010).

Figure 5. Schematic illustration of important metabolic and geneticin formation pathways in NoAz.

is unclear why the apparent tempo of evolution of marine diazotrophic cyanobacteria is so slow. Diazotrophic cyanobacteria have effectively become the "gate keepers" of oceanic productivity, yet despite the rapid radiation of eukaryotic oxygenic photoautotrophs throughout the Phanaerozoic eon marine cyanobacteria seem like living fossils (Berman-Frank et al., 2003). Finally, some Questions needs answering, Are there N_2-fixing picoplankton? What limits the growth of N_2-fixing microorganisms in the open ocean? Is N_2 fixation associated with zooplankton?

Author details

Ahmed A. Issa[1], Mohamed Hemida Abd-Alla[1] and Takuji Ohyama[2]

1 Botany and Microbiology Department, Faculty of Science, Assiut University, Assiut, Egypt

2 Biological chemistry Department, Faculty of Agriculture, Niigata University, Niigata, Japan

References

[1] Abd-Alla, M.H., Mahmoud A.L. E. & A.A. Issa, 1994 . Cyanobacterial biofertilizer improved growth of wheat. Phyton 34, 11-18.

[2] Abd-Alla, M.H. & Issa, A.A. 1994. Suitability of some local Agro. Industrial wastes as carrier materials for Cyanobacterial inoculant's. Folia Microbiol, 39, 576-578.

[3] Adams, D. G. 2000. Heterocyst formation in cyanobacteria. Curr. Opin. Microbiol. 3, 618–624.

[4] Adams, D.G. 2002. Cyanobacteria in symbiosis with hornworts and liverworts. In: Rai AN, Bergman B, Rasmussen U, eds. Cyanobacteria in symbiosis. Dordrecht: Kluwer Academic Publishers, 117–135.

[5] Adams, D.G, Bergman B., Nierzwicki-Bauer S.A., Rai A.N, & Schüβler A. 2006. Cyanobacterial–plant symbioses. In: Dworkin M, Falkow S, Rosenberg E, Schleifer K-H, Stackebrandt E, eds. The prokaryotes: a handbook on the biology of bacteria. 3rd edn. Vol. 1. Symbiotic associations, biotechnology, applied microbiology, New York, NY: Springer, 331–363.

[6] Alexander, V., & Schell, D. M., 1973. Seasonal and spatial variation of nitrogen fixation in the Barrow, Alaska, tundra. Arctic and Alpine Research, 5, 77–88.

[7] Alexander, V.M., Billington, M., & Schell, D.M., 1978. Nitrogen fixation in arctic and alpine tundra. In Tieszen L.L. (ed) Vegetation and production ecology of an Alaskan Arctic tundra. New York: Springer-Verlag, 539-558.

[8] Anjana, K., Kiran B., Mona, S. & Kaushik, A. 2012. Biological photohydrogen production by Cyanobacteria : future prospects as a fuel. J. Environ. Res.Devel. 6 ,779-783.

[9] Antal, T.K. & Lindblad P.2005: Production of H_2 by sulphur-deprived cells of the unicellular cyanobacteria *Gloeocapsa alpicola* and *Synechocystis* sp. PCC 6803 during dark incubation with methane or at various extracellular pH. J. App. Microbiol. 98, 114-120.

[10] Bolton, J.R., 1996. Solar photoproduction of hydrogen: A review, *Solar Ene.* 57, 37- 50

[11] Belnap, J., Büdel, B., & Lange, O.L. 2001. Biological soil crusts: Characteristics and distribution. In: Belnap J, Lange OL (eds) Biological Soil Crusts: Structure, Function, and Management. Springer, New York, pp 3–30

[12] Bergman, B., Gallon, J. R., Rai, A. N., & Stal, L. J. 1997. N2 fixation by non-heterocystous cyanobacteria. FEMS Microbiol. Rev.19, 139-185.

[13] Bergman, B., Rasmussen U., & Rai A.N. 2007. Cyanobacterial associations. In: Elmerich C, Newton WE, eds. Associative and endophytic nitrogen-fixing bacteria and cyanobacterial associations. Dordrecht: Kluwer Academic Publishers.

[14] Berman-Frank, I., Lundgren, P., & Falkowski, P. G. 2003. Nitrogen fixation and photosynthetic oxygen evolution in cyanobacteria, Res. Microb. 154, 157–164.

[15] Berry, J.P., Gantar, M., Perez, M.H., Berry, G. & Noriega, F.G., 2008. Cyanobacterial toxins as allelochemicals with potential applications as algaecides, herbicides and insecticides. Marine Drugs 6, 117-146.

[16] Bhaya, D. 2004. Light matters: phototaxis and signal transduction in unicellular cyanobacteria. Mol Microb. 53, 745–754.

[17] Böhme ,H.1998 . Regulation of nitrogen fixation in heterocyst-forming cyanobacteria.Trends in Plant Science 3, 346–351.

[18] Castenholz, R.W. 1973 Ecology of blue-green algae in hot springs. In: N.G. Carr and B.A. Whitton [Eds] *The Biology of Blue-Green Algae.* Blackwell Scientific Publications, Oxford, 379-414.

[19] Castenholz, R. W., & Waterbury, J. B. 1989. Group 1. Cyanobacteria. In R. G. E. Murray, D. J. Brenner, M. P. Bryant, J. G. Holt, N. R. Krieg, J. W. Moulder, *et al.* (Eds.), *Bergey's manual of systematicbacteriology* (Vol. 3, pp. 1710-1799). Baltimore: Williams and Wilkins.

[20] Castenholz, R. W. 2001. Phylum B. X. Cyanobacteria. Oxygenic photosynthetic Bacteria. In D. R. Boone and R. W. Castenholz (Eds.), *Bergey's manual of systematic bacteriology* (Vol. 1, pp. 473-599). NewYork: Springer.

[21] Choudhury, A. T. M. A. & L. R. Kennedy. 2004. Prospects and potentials for systems of biological nitrogen fixation in sustainable rice production. Soil Biol.Biochem. 39, 219- 227.

[22] Cohen,Y. 1989. Photosynthesis in microbial mats and its relation to the sulfur cycle: a model for microbial sulfur interactions. – In. Cohen, Y. and Rosenber G. E. (eds): Microbial mats: physiological ecology of benthic microbial communities. pp. 22–36, American Society for Microbiology, Washington, D.C.

[23] Cohen, M.F., & Meeks J.C. 1997. A hormogonium regulating locus, hrmUA, of the cyanobacterium *Nostoc punctiforme* strain ATCC 29133 and its response to an extract of

a symbiotic plant partner *Anthoceros punctatus.* Mol Plant–Microbe Interactions 10, 280–289.

[24] Crittenden, P. D., & Kershaw, K. A., 1978 . Discovering the role of lichens in the nitrogen cycle in the boreal-arctic ecosystem. The Bryologist, 81, 258–267.

[25] DeLuca, T.H., Zackrisson, O., Nilsson, M-C., & Sellstedt, A., 2002. Quantifying nitrogenfixation in feather moss carpets of boreal forests. nature,419, 917-920.

[26] Dodds, W.K., Gudder D.A., & Mollenhauer D. 1995. The ecology of *Nostoc.* J Phycol 31, 2–18.

[27] Dor, I. & Danin, A. 1996 Cyanobacterial desert crusts in the Dead Sea Valley, Israel. Arch. Hydrobiol. Suppl. 117, Algol Studies, 83, 197-206.

[28] Duggan, P.S, Gotardello P, & Adams D.G. 2007. Molecular analysis of genes in Nostoc punctiforme involved in pilus biogenesis and plant infection. J Bacteriol 189, 4547 4551.

[29] Durner, J., Bohm, I., Knorzer, O. C., & Boger, P. 1996. Proteolytic degradation of dinitrogenase reductase from *Anabaena variabilis* (ATCC 29413) as a consequence of ATP depletion and impact of oxygen. J Bacteriol 178, 606–610.

[30] El-Enany, A.E. & Issa, A.A. 2000. Cyanobacteria as a biosorbent of heavy metals in sewage water. Envir Toxicol Pharmacol 8, 95-101.

[31] Ernst, A. & Rohme, H. 1984. Control of hydrogen-dependent nitrogenase activity by adenylates and electron flow in heterocysts of *Anabaena variabilis.* Biochim Biophys Acta 767, 362-268.

[32] Ernst, A., Reich S., Böger P. 1990a . Modification of dinitrogenase reductase in the cyanobacterium *Anabaena variabilis* due to C-starvation and ammonia. *J Bacteriol* 172,748–755

[33] Ernst, A., Liu Y-D., Reich S., & Böger P. 1990b. Diuranal nitrogenase modification in the cyanobacterium *Anabaena variabilis.* Bot Acta 103,183–189

[34] Falkowski, P. G. 1997. Evolution of the nitrogen cycle and its influence on the biological sequestration of CO_2 in the ocean. Nature, 387, 272–275.

[35] Falkowski, P. G., & Raven, J. A. 1997. *Aquatic photosynthesis* (Vol. 256). Malden, MA: Blackwell Science.

[36] Farmer , J. D. 1992, Grazing and bioturbation in modern microbial mats. – In SchopF , J. W. and Klein , C. (eds): The Proterozoic biosphere: a multidisciplinary study. pp. 247–251, Cambridge university press, Cambridge.

[37] Fay, P., Steward W.D., Walsby A.E. & Fogg G.E. 1968. Is the heterocyst the site of nitrogen fixation in the blue-green algae? Nature 220,810–812

[38] Fay, P. 1992. Oxygen relations of nitrogen fixation in cyanobacteria. Microbiol Rev, 56, 340-373.

[39] Fay, P. 1965. Heterotrophy and nitrogen fixation in *Chlorogloea fritschii.* J. Gen. Microbiol 39, 11-20.

[40] Fredriksson, C., & Bergman, B. 1995. Nitrogenase quantity varies diurnally in a subset of cells within colonies of the non-heterocystous cyanobacteria *Trichodesmium* spp. Microbiol UK, 141, 2471-2478.

[41] Gallon, J. R. 1992. Tansley Review No. 44. Reconciling the incompatible: N_2 fixation and O_2. New Phytol 122, 571-609.

[42] Gallon, J. R., Jones, D. A., & Page, T. S. 1996. *Trichodesmium,* the paradoxical diazotroph. Arch. Hydrobiol, 117 (supplement: Algological Studies 83), 215-243.

[43] Golden, J.W., &Yoon, H.S. 2003. Heterocyst development in *Anabaena.* Cur Opin Microbiol 6, 557–563.

[44] Gunther, A. J., 1989. Nitrogen fixation by lichens in a subarctic Alaskan watershed. The Bryologist 92, 202-208.

[45] Giani, D., & Krumbein, W. E. 1986. Growth characteristics of non-heterocystous cyanobacterium *Plectonema boryanum* with N2 as nitrogen source. Archives Microbiol., 145, 259-265.

[46] Giardi, M. T. Masojidek, J. & Godde, D. 1997. Effects of Abiotic Stresses on the Turnover of the D1reaction Center II Protein, Plant Physiol., 101, 635-642.

[47] Grobbelaar, N., Huang, T. C., Lin, H. Y., & Chow, T. J. 1986. Dinitrogen-fixing endogenous rhythm in *Synechococcus* RF-1. FEMS Microbiol. Lett., 37, 173-177.

[48] Huang, T.-C., & Chow, T.-J. 1986. New type of N_2-fixing unicellular cyanobacterium. FEMS Microbiol.Lett., 50, 127-130.

[49] Henry, G. H. R., & Svoboda, J. 1986. Dinitrogen fixation (acetylene reduction) in high arctic sedge meadow communities. Arctic and Alpine Research, 18, 181–187.

[50] Herrero, O. A., Muro-Pastor A. M. & Flores, E. 2001. Nitrogen Controlling Cyanobacteria, J. Bacteriol., 183, 411-425.

[51] Hobara, S., McCalley, C., Koba, K., Giblin, A. E., Weiss, M. S., Gettel, G. M., & Shaver, G. R. 2006. Nitrogen fixation in surface soils and vegetation in an Arctic tundra watershed: a key source of atmospheric nitrogen. Arctic, Antarctic, and Alpine Research, 38, 363–372.

[52] Horne, A. J. 1979. Nitrogen fixation in Clear Lake, California. 4. Diel studies on *Aphanizomenon* and *Anabaena* blooms. Limnol. Oceanogr. 24, 329-341.

[53] Houle, D., Bilodeau G. S., Paquet S, Planas D., & Warren A. 2006. Identification of two genera of N_2-fixing cyanobacteria growing on three feather moss species in boreal forest of Quebec, Canada. Can J Bot 84,1025–1029

[54] Humm, H.J. & Wicks, S.R. 1980. Introduction and Guide to the Marine Bluegreen Algae. John Wiley & Sons, New York, 194 pp.

[55] Issa, A.A. 1995. Aspects of growth and nitrogenase activity of the cyanobacterium *Nostoc muscorum* in continuous culture. Cryptogamie Alogolgie. 16, 247-253.

[56] Issa, A.A. 1999. Interference of glyphosate with the shikimate pathway by cyanobacteria in chemostat culture. Microbios.100, 47-52.

[57] Issa, A.A., Abd-Alla, M.H. & Mahmoud, A.L.E. 1994. Effect of biological treatments on growth and some metabolic activities of barley plants grown in saline soil. *Microbiol. Res. 149: 1-4.*

[58] Issa, A.A., El-Enany, A. E. & Abdel-Basset, R. 2002. Modulation of the photosynthetic source: sink relationship in cultures of the cyanobacterium *Nostoc rivulare*. Biologia plant.45, 212-225.

[59] Issa, A.A., Adam, M.S. & Fawzy, M.A. 2013. Alternations in some metabolic activities of *scenedesmus quadriquada* and *Merismopedia glauca* in response to glyophsate herbicide. J. Biol Earth Sci. 3,17-23.

[60] Javor , B.J. & Castenholz , R.W. 1984. Invertebrate Grazers of microbial mat, Laguna Gurrero Negro, Mexico. – In: Coheny., Castenholz , R.W. and Halvorson , H. O. (eds): Microbial Mats: Stromatolites. – pp. 85–94, Alan R. Liss, Inc, New york.

[61] Jonkers, H.M. LudWiG, R., De Wit, R., Pringault, O., Muyzer, G., Nienmann, H., Finke, N. & De Beer, D. 2003. Structural and functional analysis of a microbial mat ecosystem from a unique permanent hypersaline inland lake: 'La Salada de Chiprana' (NE Spain). – FEMS Microbiol. Ecol. 44: 175–189.

[62] Kallio, S. &Kallio, P. 1975. Nitrogen fixation in lichens at Kevo, North Finland. In: Wielgolaske FE (ed) Fennoscandian Tundra Ecosystems, part 1. Springer, New York, pp 292–304

[63] Kann, E. 1988. Zur AutØkologie benthischer Cyanophyten in reinen europäischen Seen und Fliessgewässern. Arch Hydrobiol. Suppl. 80, Algological Studies, 50-53, 473-495.

[64] Kanemoto, R. H. & Ludden. P. W. 1984. Effects of ammonia, darkness, and Phenazine methosulfate on whole-cell nitrogenase activity and Fe-protein modification in *Rhodospirillum rubrum*. J. Bacteriol. 158, 713-720.

[65] Kaushik, B.D. 1994. Algalization of rice in salt-affected soils. Annales of Agricultural Research 14: 105-106.

[66] Khamees. H. S. K. Gallon, J. R., & Chaplin. A. E. 1987. The pattern of acetylene reduction by cyanobacteria grown under alternatinglight and darkness. Br. Phycol. J. 22: 55-60.

[67] Kiran, B., 2007. Heavy metal sequestration from industrial effluent using native and immobilized algal biomass and optimization of metal recovery, Ph.D Thesis.

[68] Kol, E. 1968. Kryobiologie. I. Kryovegetation. In: H.J. Elster and W. Ohle (Eds) Die Binnengewsser, B and XXIV. E. Schweizerbart'sche Verlagsbuchhandlung, Stuttgart, 216 pp.

[69] Laamanen, M. 1996. Cyanoprokaryotes in the Baltic Sea ice and winter plankton. Arch Hydrobiol Suppl 117, Algological Studies, 83, 423-433.

[70] Lagerstrøm, A., Nilsson M.C., Zackrisson, O. & Wardle, D.A. 2007.Ecosystem input of nitrogen through biological fixation in feather mosses during ecosystem retrogression. Funct Ecol 21,1027–1033

[71] Mahmoud, A. L. E ,. Issa, A. A. & Abd-Alla M. H. 1992. Survival and Efficincy of N_2-Fixing Cyanobacteria in Soils under Water Stress. Journal of Islamic Academy of Sciences 5:4, 275-278, 1992

[72] Maryan, P. S., Eady, R. R., Chaplin, A. E., & Gallon, J. R. 1986. Nitrogen fixation by the unicellular cyanobacterium *Gloeothece*. Nitrogenase synthesis is only transiently repressed by O_2. FEMS Microbiol Lett 34, 251-255.

[73] Meeks, J.C. 1990. Cyanobacterial–bryophyte associations. In: Raian, ed. CRC handbook of symbiotic cyanobacteria. Boca Raton,FL: CRC Press, 43–63.

[74] Meeks, J.C. 1998. Symbiosis between nitrogen-fixing cyanobacteria and plants. BioScience 48, 266–276.

[75] Meeks, J.C. 2003. Symbiotic interactions between *Nostoc punctiforme*, a multicellular cyanobacterium, and the hornwort Anthoceros punctatus. Symbiosis 35, 55–71.

[76] Meeks, J.C. & Elhai J. 2002. Regulation of cellular differentiation in filamentous cyanobacteria in free-living and plant-associated symbiotic growth states. Microbiol Mol Biol Rev 66, 94–121.

[77] Mitsui, A., Kumazawa, S., Takahashi, A., Ikemoto, H. & Arai, T. 1986. Strategy by which nitrogen-fixing unicellular cyanobacteria grow photoautotrophically. Nature 323, 720–732.

[78] Mulholland, M.R., Ohki, K. & Gapone D. G. 1999. N_2-utilization and metabolism relative to patterns of N_2 fixation in cultures of *Trichodesmium* NIBB1067. J Phycol 35, 977–988

[79] Mulholland, M.R. & Capone, D.G. 1999. N_2 fixation, N uptake and N metabolism in natural and cultured populations of *Trichodesmium* spp. Mar Ecol Prog Ser 188, 33–49

[80] Muthukumar, C., Muralitharan, G. & Vijayakumar, R. 2007. Cyanobacterail biodiversity from different freshwater ponds of Thanjavur, Tamilnadu (India). Acta Botanica Malcitana 32,17-25.

[81] Nayak,S. and Prasanna, R. 2007. Soil pH and its role in cyanobacterial abundance and diversity in rice field soils. Appl Ecol Environ Res 5, 103-113.

[82] Ohki, K., & Fujita, Y. 1988. Aerobic nitrogenase activity measured as acetylene reduction in the marine non-heterocystous cyanobacterium *Trichodesmium* spp. grown under artificial conditions. Marine Biol 98, 111-114.

[83] Ow, M.C., Gantar, M., & Elhai, J. 1999. Reconstitution of a cycad cyanobacterial association. Symbiosis 27, 125–134.

[84] Rai, A.N., Bergman, B., & Rasmussen, U. 2002. Cyanobacteria in symbiosis. Dordrecht: Kluwer Academic Publishers.

[85] Ran L., Larsson J., Vigil-Stenman T., Nylander J.A.A., Ininbergs K., Zheng W-W., Lapidus A., Lowry S., Haselkorn R., &ergman B. 2010. Genome Erosion in a Nitrogen-Fixing Vertically Transmitted Endosymbiotic Multicellular Cyanobacterium. Plos One 5, 1-11.

[86] Rasmussen, U., Johansson, C. & Bergman B. 1994. Early communication in the *Gunnera–Nostoc* symbiosis – plant induced cell differentiation and protein synthesis in the cyanobacterium. Mol Plant Microb Inter 7, 696 702.

[87] Reed, R.H., Chudek, J.A., Foster, R. & Stewart, W.D.P. 1984. Osmotic adjustment in cyanobacteria. Arch Microbiol 138, 333-337.

[88] Rippka, R., Deruelles, J., Waterbury, J. B., Herdman, M., & Stanier, R. Y. 1979. Generic assignments, strain histories and properties of pure cultures of cyanobacteria. J Gen Microbiol 111, 1-61.

[89] Robson, R.L. & Postgate, J.R. 1980. Oxygen and hydrogen in biological nitrogen fixation. Annu Rev Microbiol 34,183–207

[90] Roger, P.A. & Reynaud, P.A. 1979. Ecology of blue green algae in paddy fields – In: International Rice Res. Institute, Los Banos, Philippines,

[91] Saino, T., & Hattori, A. 1978. Diel variation in nitrogen fixation by a marine blue-green alga, *Trichodesmium thiebautii*. Deep Sea Res 25, 1259-1263.

[92] Schell, D. M., & Alexander, V., 1973. Nitrogen fixation in arctic coastal tundra in relation to vegetation and micro-relief. Arctic 26, 130–137.

[93] Scherer, S., & Böger, P. 1982. Respiration of blue green algae in the light. Arch Microbiol 132, 329-332.

[94] Scherer, S., Almon. H., and Boger, P. 1988. Interaction of photosynthesis, respiration and nitrogen fixation in cyanobacteria. Photosynth Res 15, 95-114.

[95] Skulberg, O.M. 1996. Terrestrial and limnic algae and cyanobacteria In: A. Elvebakk and P. Prestrud [Eds] *A Catalogue of Svalbard Plants, Fungi, Algae and Cyanobacteria.* Part 9, Norsk Polarinstitutt Skrifter 198, 383-395.

[96] Solheim, B., Zielke, M., Bjerke, J. W., & Rozema, J., 2006. Effects of enhanced UV-B radiation on nitrogen fixation in arctic ecosystems. Plant Ecol 182, 109–118.

[97] Solheim, B., Endal, A., and Vigstad, H., 1996. Nitrogen fixation in arctic vegetation and soils from Svalbard, Norway. Polar Biol, 16, 35–40.

[98] Stanier, R. Y. & Cohen-Bazire, G. 1977. Phototrophic prokaryotes: the cyanobacteria. Annu Rev Microbiol 31, 225–74

[99] Stal, L. J., & Krumbein, W. E. 1985a .Isolation and characterization of cyanobacteria from a marine microbial mat. Bot. Mar., 28, 351-65.

[100] Stal, L. J., & Krumbein, W. E. 1985b. Nitrogenase activity in the non-heterocystous cyanobacterium *Oscillatoria* sp. grown under alternating light-dark cycles. Arch Microbiol 143, 67-71.

[101] Stal, L. 2007. Cyanobacteria: Diversity and versatility. In: J Sedbach (ed) Algae and Cyanobacteria in Extreme Environments Cellular Origin, Life in Extreme Habitats and Astrobiology Volume 11, 2007, pp 659-680. Springer, A.A. Dordrecht, Netherland.

[102] Stewart, W.D.P. 1980. Some aspects of structure and function in N_2 fixation cyanobacteria. Annual Review of Microbiology 34: 497-536.

[103] Stewart,W. D. P. & Rowell, P. 1986. Biochemistry and physiology of nitrogen fixation with particular emphasis on nitrogen fixing phototrophs. Plant Soil 90, 167-191.

[104] Subhashini, D., & Kaushik, B.D. 1984. Amelioration of saline sodic soils with blue-green algae. Aust. J. Soil Res. 19, 361–366.

[105] Taylor, T.N., & Krings M. 2005. Fossil microorganisms and land plants: associations and interactions. Symbiosis 40, 119–135.

[106] Turetsky, M. R. 2003. The role of bryophytes in carbon and nitrogen cycling. The Bryologist, 106, 395–409.

[107] van Gemerden , H. 1993. Microbial mats: A joint venture. Mar. Geol. 113: 3–25.

[108] Van Landingham, S.L. 1982. Guide to the Identification, Environmental Requirements and Pollution Tolerance of Freshwater Blue-Green Algae (Cyanophyta). United States Environmental Protection Agency, Cincinnati, Ohio, 341 pp.

[109] Villarreal, J.C., Duff R.J., & Renzaglia K.S. 2005. Anatomical and ultrastructural innovations in Leiosporoceros dussii (Steph.) Ha°ssel. XVII International Botanical Congress, Vienna, Austria, 450.

[110] Villarreal, J.C. & Renzaglia, K.S. 2006. Structure and development of *Nostoc* strands in *Leiosporoceros dussii* (Anthocerotophyta): a novel symbiosis in land plants. American J. Botany 93, 693–705.

[111] Weare, N. M., & Benemann, J. R. 1974. Nitrogenase activity and photosynthesis in *Plectonema boryanum*. J. Bacterial. 119, 258-265.

[112] Weber, B., Wessels, D.C.J. & Büdel, B. 1996. Biology and ecology of cryptoendolithic cyanobacteria of a sandstone outcrop in the Northern Province, South Africa. Algol Studies, 83,565-579.

[113] West, N. & Adams, D.G. 1997. Phenotypic and genotypic comparison of symbiotic and free-living cyanobacteria from a single field site. Appl Environ Microbiol 63, 4479–4484.

[114] Whitton, B.A. 1992. Diversity, ecology and taxonomy of the cyanobacteria. In: N.H. Mann and N.G. Carr (Eds) Photosynthetic Prokaryotes. Plenum Press, New York, 1-51.

[115] Whitton, B.A. 1973. Freshwater plankton, In: G.E. Fogg, W.D.P. Stewart, P. Fay and A.E. Walsby [Eds] *The Blue-Green Algae.* Academic Press, London, 353-367.

[116] Wyatt, J.T. & Silvey J.K.G. 1969. Nitrogen fixation by *Gloeocapsa*. Sci 165, 908-909.

[117] Yakunin, A. F., Fedorov, A. S., Laurinavichene, T. V., Glaser, V. M., Egorov, N. S., Tsygankov, A. A., Zinchenko, V. V., & Hallenbeck, P. C. 2001. Regulation of nitrogenase in the photosynthetic bacterium *Rhodobacter sphaeroides* containing draTG and nifHDK genes from *Rhodobacter capsulatus*, Canadian J Microbiol 47, 206–212.

[118] Zackrisson, O., DeLuca T.H., Gentili F., Selstedt A., & Jderland A. 2009. Nitrogen fixation in mixed *Hylocomium splendens* moss communities. Oecologia 160,309–319

[119] Zehr, J., Wyman, M., Miller, V., Duguay, L., & Capone, D. G. 1993. Modification of the Fe protein of nitrogenase in Natural populations of *Trichodesmium thiebautii*, Appl Environ Microbiol 59, 669–676.

[120] Zhang, C.C., Laurent, S., Sakr, S., Peng, L. & Bedu, S. 2006. Heterocyst differentiation and pattern formation in cyanobacteria: a chorus of signals. Mol Microbiol 59, 367–375.

7

Effects of Rhizobium Inoculation on Nitrogen Fixation and Growth of Leguminous Green Manure Crop Hairy Vetch (*Vicia villosa* Roth)

Takashi Sato

1. Introduction

A green manure crop is used primarily as a soil amendment and a nutrient source for subsequent crops [1]. Legumes have been utilized as biofertilizer in agriculture, because legumes can fix atmospheric dinitrogen (N_2) and accumulate a large amount of nitrogen in their organs. Accordingly, incorporating of a leguminous green manure crop provides a large amount of nitrogen and carbon source to the soils for subsequent crops. There are many leguminous green manure crops and they have been utilized all over the world [2, 3]. Table 1 shows dry weight and nitrogen accumulation of leguminous green manure crops. Legumes including clover, vetch and field pea are cultivated as winter cover crops [4, 5, 6]. They are sometimes planted together with winter cereals such as oats, rye or wheat. Legumes such as cowpea, sesbania, crotalaria or velvet beans may be grown as summer green manure crops to add nitrogen as well as organic matter to the soils [7, 8, 9, 10]. Chinese milk vetch (Renge in Japanese) and so on have been cultivated mainly southwest area in Japan, which locate in a warm climate zone [12, 13].

Hairy vetch [*Vicia villosa* Roth] is a leguminous cover crop and used for nutrient management as green manure and weeding in southwest area of Japan. Vetches (plants of the genus *Vicia*) are distributed throughout the temperate zones. There are many species of vetch, and they are important for agriculture. The species in commercial use, including hairy vetch, are all native to Europe or western Asia. Hairy vetch is a winter annual plant, and some hairy vetch cultivars can grow over winter even in north area of Japan. Recently, hairy vetch is used for increasing soil fertility and improvement of soil structures in an upland field converted from paddy field (Fig. 1).

Legumes	Dry weight (t ha^{-1})	N accumulation (kg ha^{-1})
Soybean [*Glycine max* (L.) Merr.]	2.8~5.8	31~174
Sunn hemp [*Crotalaria juncea*]	0.9~11.1	23~279
Siratro [*Macroptilium atropurpureum*]	2.4~5.5	62~178
Velvet bean [*Mucuna pruriens*]	1.7~9.3	53~183
Sesbania [*Sesbania rostrata*]	3.2~4.6	71~88
Mung bean [*Vigna radiata*]	1.1~5.5	26~88
Cowpea [*Vigna unguiculata*]	0.6~8.5	15~154
Alfalfa [*Medicago sativa*]	0.5~5.7	21~174
Barrel medic [*Medicago truncatula*]	1.0~4.5	37~131
Field pea [*Pisum sativum* L.]	3.2~7.6	107~230
Crimson clover [*Trifolium incarnatum* L.]	1.4~7.3	35~200
Red clover [*Trifolium pratense* L.]	0.3~3.7	13~115
White clover [*Trifolium repens* L.]	0.6~25	17~592
Chinese milk vetch [*Astragalus sinicus* L.]	3.4~4.9	101~158
Hairy vetch [*Vicia villosa* Roth]	1.5~10.0	58~257

Results showed by Cherr et al. [1] were modified.

Table 1. Dry weight and nitrogen accumulation of leguminous green manure crops.

Figure 1. Photos of shoot (A) and root (B) of hairy vetch planted in a upland field converted from paddy field in Akita, Japan.

2. Nitrogen utilization of hairy vetch

Hairy vetch absorbs and utilizes soil nitrogen mainly in the forms of NH_4^+ and NO_3^- by the roots, and can fix atmospheric nitrogen by the root nodule that is a symbiotic organ with *Rhizobium leguminosarum* bv. *viciae* (Fig. 2). Hairy vetch planted in a field can accumulate 100~200 kg nitrogen ha^{-1} year^{-1} in the plant. The accumulated nitrogen consists of about 90 % of nitrogen fixation by the nodules and about 10 % of absorbed soil nitrogen by the roots estimated from the results of ^{15}N natural abundance technique. The nitrogen concentration of the organs is 3~4 % with CN ratio about 10-12, and total dry matter ranges 200-500 g m^{-2} under field conditions in Japan.

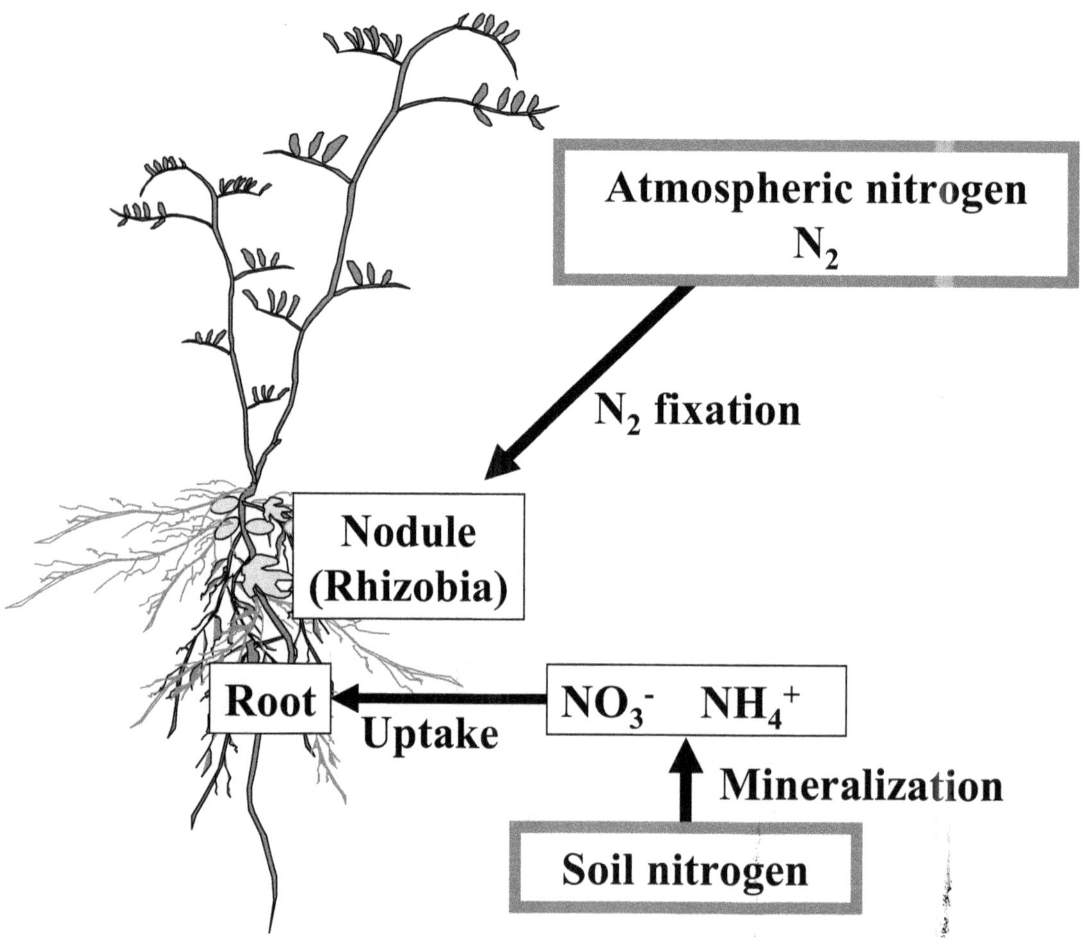

Figure 2. Nitrogen flow and utilization of a legume plants (Hairy vetch).

In early growth stage, hairy vetch plants sometimes show very poor plant growth and exhibit nutritional disorder that the leaves and stems turn to red color. From the results we investigated, the poor plant growth and the nutritional disorder is caused by nitrogen deficiency. Most of the soil nitrogen exists as organic matter, and the soil nitrogen is mineralized to inorganic nitrogen (NH_4^+ and NO_3^-) by microorganisms. When temperature is low in autumn, the rate of the soil nitrogen mineralization becomes slow because of low activity of the microorganisms in the soil. For that reason, hairy vetch cannot easily utilize soil nitrogen under low temperature conditions. On the other hand, it is considered that nitrogen deficiency is related to the poor nodule formation and low nitrogen fixation activity of the rhizobia in the nodules. It is assumed that there is no rhizobia compatible with hairy vetch in the soil or some indigenous rhizobia show low nitrogen fixing activity under low temperature conditions in northeast area of Japan.

3. Isolation of a superior hairy vetch rhizobium

Rhizobium is a genus of Gram-negative soil bacteria and universally survives in a soil. Rhizobium colonizes in the leguminous plant cells within root nodules, and fixes atmospheric dinitrogen (N_2) to ammonia. Rhizobium provides ammonia as nitrogen source to the host plant, and the host plant provides the rhizobium (bacteroid) organic compounds made by photosynthesis. There is strong host specificity in the symbiosis between rhizobia and legumes.

Hairy vetch establishes symbiosis with *Rhizobium leguminosarum* bv. *viciae* and conducts N_2 fixation by the nodules. The physiological and genetic diversity of rhizobium is very complicated in the soil [14]. If hairy vetch is infected by a rhizobium that has low affinity for hairy vetch or has low N_2 fixation activity, the host plant hairy vetch shows very poor plant growth. So, for good hairy vetch growth it is important to isolate a superior hairy vetch rhizobium and inoculate to the hairy vetch with a superior rhizobium.

Several hairy vetch rhizobium had been isolated from a heavy soil of the upland field converted from a paddy field in Hachirougata polder, Akita Japan. A superior hairy vetch rhizobium was obtained from the isolates and named Y629. The superior hairy vetch rhizobium Y629 was identified *Rhizobium leguminosarum* bv. *viciae* by 16S-rRNA gene analysis.

4. Effect of inoculation of Y629 on hairy vetch growth

Seeds of hairy vetch were sown in vermiculite medium supplied with nitrogen free culture solution with inoculation of Y629 or indigenous rhizobia. The hairy vetch plants were grown in a growth chamber with high temperature treatment (day:22°C-12 h, night:18°C -12h) or low temperature treatment (day;15°C -12 h, night;7°C -12h).

Hairy vetch plants with inoculation of Y629 normally grew, whereas the plants with inoculation of indigenous rhizobia appeared nutritional disorder irrespective of temperature with poor plant growth (Fig. 3). In high temperature treatment, the nodule number with inoculation of Y629 was fewer than that with inoculation of indigenous rhizobia. However, the nitrogen fixation activity (ARA) per plant with inoculation of Y629 was about two times higher than that with inoculation of indigenous rhizobia. The nitrogen fixation activity with inoculation of Y629 was high even in low temperature condition. These results indicate that the superior hairy vetch rhizobium Y629 shows high nitrogen fixation activity, and Y629 is considered to have low temperature tolerance in a symbiotic state.

Figure 3. Hairy vetch inoculated with Y629 or indigenous rhizobia at high temperature treatment (Day:22ºC 12h, night: 18ºC 12h).

The nodulation was observed in all the plants, but some nodules with inoculation of indigenous rhizobia showed dark greenish color inside. It has been known that leghemoglobin, a red hemeprotein specifically accumulated in nodules, is an important role in nitrogen fixation in which protects nitrogenase from inactivation by molecular oxygen (O_2) [15]. The concentration of leghemoglobin in the nodule is an indicator of nitrogen fixation activity. The section of all the nodules inoculated with Y629 showed red color, which is considered as the accumulation of normal leghemoglobin in the infected region of the nodules. However, some nodules inoculated with indigenous rhizobia exhibited dark color, and those sections showing dark greenish color in the infected region of the nodules which is derived from the decomposition product of leghemoglobin (Fig. 4). The nodules showed dark color might have low or no nitrogen fixation activity. This phenomenon is considered to be an early senescence of nodules caused by the end of symbiosis. It is assumed that rhizobium harbored in the dark color nodules have low affinity for the host plant. It is interesting to note that some rhizobium can infect and form a nodule with host plant, but cannot maintain symbiosis with their host plants.

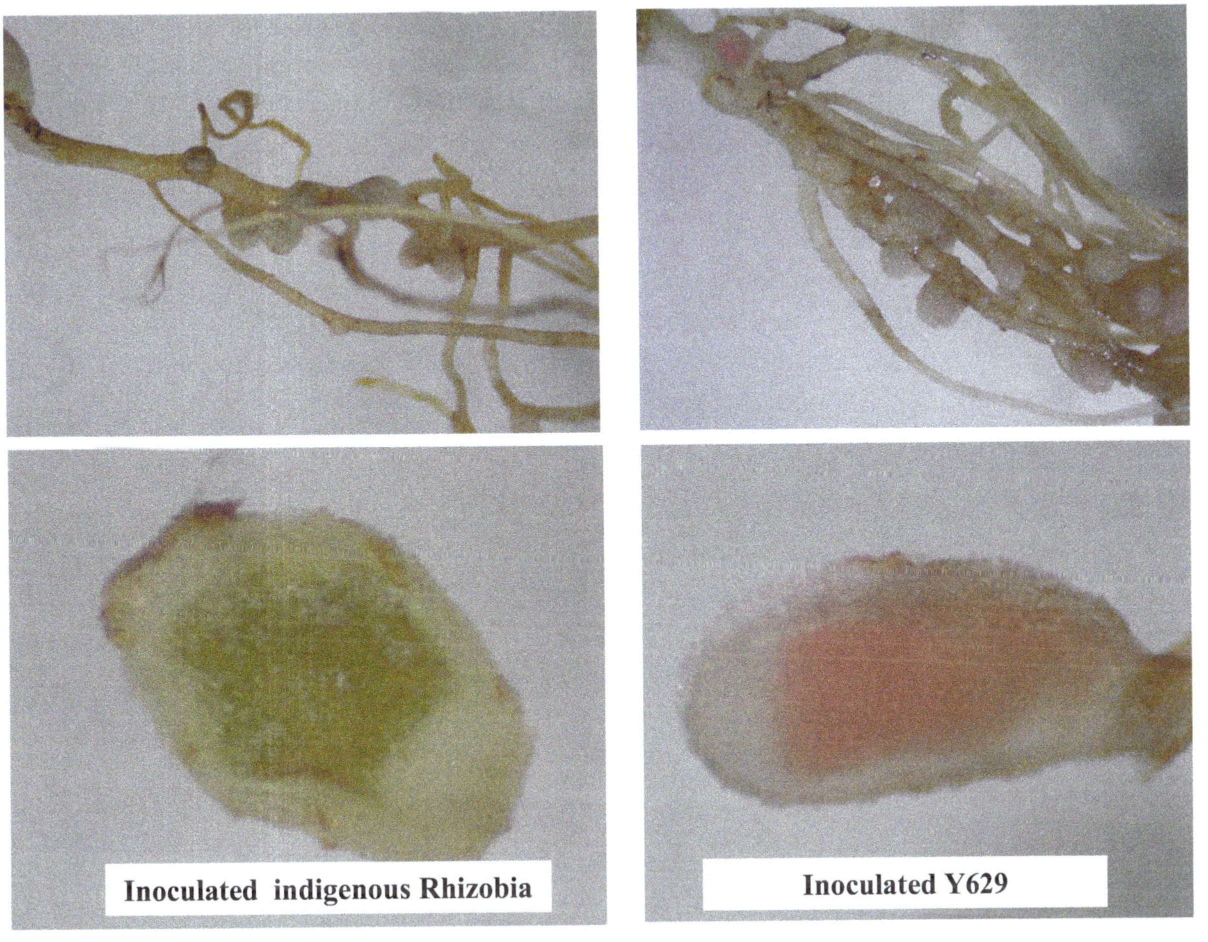

Figure 4. Nodulation and section of nodule of hairy vetch inoculated with Y629 or indigenous rhizobia at high temperature treatment (Day: 22°C 12h, night: 18°C 12h)

5. Effect of inoculation of Y629 in a field

An upland field converted from a paddy field in Hachirougata polder, Japan was used for this field experiment. The soil type is heavy clay soil and pH is 6.5. There is no hairy vetch planting history in the field. Seeds of hairy vetch were sown with or without inoculation of Y629 in early autumn. Rhizobium Y629 was cultured in YM broth at 30 ºC for 3 days. The culture solution was mixed with vermiculite and peat-moss, and then the seeds of hairy vetch were coated by Y629 mixture. The sowing density was 30 kg ha^{-1} that is about 200 seeds m^{-2}.

In the field experiment, there was no outstanding effect of Y629 inoculation on the hairy vetch growth before winter. In early spring, the stem length with inoculation of Y629 was as long as that without inoculation. However, about 10 % of the hairy vetch showed nutritional disorder in the non-inoculation treatment. It is supposed that the hairy vetch inoculated with Y629 was avoided nitrogen deficiency according to high nitrogen fixation activity by the nodules infected with Y629. In non-inoculation treatment, some hairy vetch plants were infected with indigenous rhizobia, which might have low nitrogen fixation activity or low affinity for the host plant.

The stem length of the plants inoculated with Y629 was slightly higher than those without inoculation in early summer. The dry weight of the shoots with inoculation of Y629 was significantly high compared with non-inoculation treatment. It is confirmed that the hairy vetch growth is promoted by high nitrogen fixation activity by the nodules infected with Y629 under field conditions.

There are many kinds of rhizobia which can establish symbiosis with hairy vetch in the soil of the experimental field. The infection ratio of Y629 with the inoculation method in this experiment might be below 10 % under the field condition where indigenous hairy vetch rhizobia exist in the soil. It is important that the effect of Y629 inoculation on hairy vetch growth is observed in the field condition even if indigenous rhizobia compatible with hairy vetch exist in the soil (Fig. 5).

Figure 5. Hairy vetch inoculated with or without Y629 under field condition at 180 days after sowing (15 May 2007).

6. Rhizobium inoculation using micro zeolite powder

It has been considered that infection ratio of inoculant rhizobium to host leguminous plant is very low in a field condition. In general, commercial rhizobium inoculant product is peat moss-based things, and leguminous crop seeds are usually inoculated mixing with rhizobium inoculant prior to sowing. In this study, the infection ratio of Y629 with the seed inoculation might be below 10 % under the field condition where indigenous hairy vetch rhizobia exist in the soil as described above.

On the other hand, it is necessary to inoculate rhizobium just before sowing, because the population of the inoculated rhizobium on the seed decreased immediately due to drought stress [16]. Thus, it is important for increase in infection ratio of inoculated rhizobium to have high rhizobium population on the seed and to maintain high population of rhizobium until sowing.

The micro zeolite powder of about 0.003 mm in diameter was used for the carrier material of rhizobium inoculant instead of peat moss in the next study. The Y629 YM culture solution at a density of about 10^8 cells mL^{-1} was mixed with the micro zeolite powder at a rate of 1:1 (volume ratio). Then, 100 mL of the rhizobium inoculant was mixed with 5 kg of a hairy vetch seeds, and the inoculated seeds were air dried for 1 hour (Fig. 6). The inoculated seeds were

Figure 6. Photo of hairy vetch seeds and seeds inoculated with rhizobium (Y629) using the micro zeolite powder.

sown to a field with density of 30 kg ha^{-1}. On the other hand, the inoculated seeds were stored at room temperature with dry condition for 3 month, and then sown to a vermiculite medium filled with nitrogen free culture solution.

The hairy vetch plant was collected from the field at 30 days after sowing, and the rhizobium was isolated from the nodules. The inoculation rhizobium (Y629) was identified by a genotype. The infection ratio of Y629 was about 30 % under the field condition, whereas that with peat moss-based inoculation might be below 10 % under same condition. The rhizobium population on the seeds with the micro zeolite powder inoculant was as high as with peat moss-based inoculant just after the inoculation. The hairy vetch seed takes about ten days for germination under field condition. In the field experiment in this study, the hairy vetch seeds were not conducted soil cover. Therefore, the rhizobium on the seed surface was got drought stress until germination. The rhizobium with the micro zeolite powder might alive on the seed surface in spite of under drought stress, although the reason has not clear yet. The pH of Y629 culture solution and the micro zeolite powder are about 3 and 10, respectively. It was supposed that the inoculant was neutralized by mixing the rhizobium culture solution and the micro zeolite powder, and the rhizobium was able to maintain population due to improvement of condition to alive on the seed surface. Furthermore, the rhizobium with the micro zeolite powder maintained population enough to form the nodules for 3 month after the inoculation under room temperature with dry condition. The rhizobium inoculant with the micro zeolite powder may be considered to make the rhizobium drought stress tolerant.

7. Flow inoculation of rhizobium in paddy and upland rotation system

Hairy vetch has been used for nutrient management as green manure to increase in soil fertility and to amend soil physical properties in paddy field and upland field. The rhizobial seed inoculation is troublesome when the cultivation area is very large and the seed amount is a lot. Furthermore, the host plants may not formed nodules even if the seeds are inoculated with rhizobium, when the method of the inoculation was inappropriate or the climatic conditions are not appropriate for rhizobium survival in the soil. Thus, it is necessary to develop new and easy method for rhizobium inoculation in paddy and upland rotation system.

In a paddy field (1.25 ha), 20 L of the Y629 YM broth culture solution at a density of about 10^8 cells mL^{-1} was applied with the irrigation water in summer (Fig. 7). The irrigation water was supplied to the field with flow quantity of about 100 m^3 per hour. The application speed of the rhizobium solution was about 5 L per hour. After the flow inoculation, conventional farm management was carried out until hairy vetch sowing. The hairy vetch seeds were sown without seed inoculation to the field with density of 30 kg ha^{-1} at 50 days after the flow inoculation.

The hairy vetch with the flow inoculation formed nodules as well as seed inoculation. The inoculant rhizobium (Y629) population in the soil increased drastically compared with that without flow inoculation treatment. It is considered that the inoculant rhizobium (Y629) could propagate in the soil of the paddy field. The infection ratio of Y629 was about 50 % even if indigenous rhizobia compatible with hairy vetch exist in the soil. In addition, the hairy vetch

Figure 7. Photos of inoculant setting on the water intake of the paddy field (Left) and applying of rhizobium (Y629) culture solution (Right) with the flow inoculation.

plant growth with the flow inoculation was promoted compared with the seed inoculation treatment. It is supposed that the rhizobium infection to the hairy vetch root with the flow inoculation was faster than that with seed inoculation because the inoculant rhizobium population was very high in the soil surface. Consequently, the flow inoculation of rhizobium was considered to be effective to improve the infection ratio of inoculant rhizobium by simple treatment.

8. Conclusion

Hairy vetch is useful for improving soil structure due to their deep root system, increasing soil fertility and weeding for subsequent crops as described above. Hairy vetch is planted in late summer or autumn, and grows until late autumn. The seedling can survive under the snow in winter and then grow vigorously from spring to early summer. If there is no compatible rhizobium strains suitable for hairy vetch in a soil, inoculation with superior rhizobium such as Y629 bring a significant result on promoting hairy vetch growth. The seed inoculation of rhizobium Y629 using the micro zeolite powder is able to keep rhizobium population on the seed surface after the inoculation possibly by protecting from drought stress. The flow inoculation of rhizobium Y629 is simple and effective method to improve the infection ratio of inoculant rhizobium and to promote the hairy vetch growth under field condition.

Author details

Takashi Sato

Faculty of Bioresource Sciences Akita Prefectural University, Kaidobata-Nishi, Shimoshinjo-Nakano, Akita, Japan

References

[1] Cherr C. M., Scholberg J. M. S., McSorley R. Green Manure Approaches to Crop Production: A Synthesis. Agron. J., 2006; 98, 302-319.

[2] Mappaona, Yashida S., Kitou M. Yield Respone of Cabbage to Several Tropical Green Manure Legumes Incorporated into Soil. Soil Sci. Plant Nutr., 1994; 40, 415-424.

[3] Badaruddin M., Meyer D. W. Green-Manure Legume Effects on Soil Nitrogen, Grain Yield, and Nitrogen Nutrition of Wheat. Crop Sci., 1990; 30, 819-825

[4] Biederbeck V.O., Bouman O.T., Looman J., Slinkard A.E., Bailey L.D., Rice W.A., Janzen, H.H. Productivity of four annual legumes as green manure in dryland cropping systems. Agron. J., 1993; 85, 1035-1043.

[5] Ross S.M., King J.R., Izaurralde R.C. O'Donovan J.T. Weed suppression by seven clover species. Agron. J., 2001; 93, 820–827.

[6] Karpenstein-Machan M., Stuelpnagel, R. Biomass yield and nitrogen fixation of legumes monocropped and intercropped with rye and rotation effects on a subsequent maize crop. Plant Soil, 2000; 218, 215–232.

[7] Choi B., Daimon, H. Effect of hairy vetch incorporated as green manure on growth and N uptake of sorghum crop. Plant Prod. Sci., 2008; 11, 211-216.

[8] Kouyate Z., Franzluebbers K., Juo A.S.R., Hossner, L.R. Tillage, crop residue, legume rotation, and green manure effects on sorghum and millet yields in the semiarid tropics of Mali. Plant Soil, 2000; 225:141–151.

[9] Ladha J.K., Dawe D., Ventura T.S., Singh U., Ventura W., Watanabe, I. Long-term effects of urea and green manure on rice yields and nitrogen balance. Soil Sci. Soc. Am. J., 2000; 64, 1993–2000.

[10] Carsky R.J., Oyewole B., Tian,G. Integrated soil management for the savanna zone ofWest Africa: Legume rotation and fertilizer N. Nutr. Cycling Agroecosyst, 1999; 55, 95–105.

[11] Steinmaier N., Ngoliya A. Potential of pasture legumes in low-external-input and sustainable agriculture (LEISA): I. Results from green manure research in Luapula Province, Zambia. Exp. Agric., 2001; 37, 297–307.

[12] Yoshioka K., Maruyama, Y. Characterization and symbiotic nitrogen fixation of rhizobium that nodulates Chinese milk vetch (*Astragalus sinicua*, L.). Soil. Soil Sci. Plant Nutr., 1990; 36, 83-90.

[13] Kim S.Y., Oh S.H., Hwang W.H., Choi K.J., Oh, B.G. Optimum Soil Incorporation Time of Chinese MilkVetch (*Astragalus sinicus* L) for its Natural Re-seed-ing and Green Manuring of Rice in GyeongnamProvince, Korea. J. Crop Sci. Biotech., 2008; 11, 193-198.

[14] Palmer K. M., J. Young, P. W. Higher Diversity of *Rhizobium leguminosarum* biovar *viciae* populations in arable soils than in grass soils, Appl. Environ. Microbiol., 2000; 66, 2445–2450

[15] Appleby C.A. Leghemoglobin and Rhizobium Respiration. Annu. Rev. Plant Physiol., 1984; 35, 443-478. Varia. Soil Sci. Plant Nutr. 55, 235-242.

[16] Rosalind D., Rodney J. R., Ivan R. K. Legume seed inoculation technology—a review. Soil Biology and Biochemistry, 2004; 36, 1275–1288.

8

Impact of *Harsh* Environmental *Conditions* on Nodule Formation and Dinitrogen Fixation of Legumes

Mohamed Hemida Abd-Alla, Ahmed A. Issa and
Takuji Ohyama

1. Introduction

Dinitrogen (N_2) comprises almost 80% of the atmosphere. The triple-bonded molecule cannot be used directly by higher plant. The element nitrogen, or "azote," meaning "without life," as Antonie Lavoisier called it about 200 years ago, has proved to be anything but lifeless, since it is a component of food, poisons, fertilizers, and explosives (Schoot Uiterkamp 1990). The atmosphere contains about 10^{15} tonnes of N_2 gas, and the nitrogen cycle involves the transformation of some 3 × 10^9 tonnes of N_2 per year on a global basis (Postgate 1982). However, transformations (e.g., N_2 fixation) are not exclusively biological. Lightening probably accounts for about 10% of the world's supply of fixed nitrogen (Sprent and Sprent 1990). The fertilizer industry also provides very important quantities of chemically fixed nitrogen. World production of fixed nitrogen from dinitrogen for chemical fertilizer accounts for about 25% of the Earth's newly fixed N_2, and biological processes account for about 60%. Nitrogen availability is probably the second most limiting factor in agricultural production, second only to water availability (Date 2000). The legume-*Rhizobium* symbiosis is the single most important source of biologically fixed nitrogen in agricultural systems (Graham and Vance 2000). Since it is a biological process, it does not depend on external sources of energy, except for free and renewable sunlight, and has few detrimental ecological effects (Phillips 1999).

2. Importance of biological nitrogen fixation by legumes

The legume-*Rhizobium* symbiosis allows many species to obtain their nitrogen nutrition from N_2. This has been known since the Greeks discovered the value of legumes in enhancing soil fertility, some 2300 years ago. Since 1888 it has been known that this is due to the nodules on legume roots and the bacteria in them (Date 2000). This nitrogen fixation is very hard to quantify in economic (Lanyon 1995) or ecological terms (Hoglund and Brock 1987), but is estimated to be responsible for between 45% (Danso 1995) and 85% (Vance et al. 1988) of the biological nitrogen fixation in agricultural areas. Biological nitrogen fixation (BNF) has the advantage of being environmental friendly and therefore would be ideal for sustainable agriculture. Biological nitrogen fixation, a key source of N for farmers using little or no fertilizer, constitutes one of the potential solutions and plays a key role in sustainable grain legumes production. Given the high cost of fertilizer in developing countries and the limited market infrastructure for farm inputs, current research and extension efforts have been directed to integrated nutrient management, in which legumes play a crucial role (Chianu et al. 2008). Certain microorganisms have the ability to use the renewable source of energy to fix atmospheric nitrogen under mild conditions, such as normal temperature and normal pressure. Nitrogen fixation is a key process in which molecular nitrogen is reduced to form ammonia, which is the form of nitrogen that is used by living systems for the synthesis of many bioorganic compounds. Biologically-fixed nitrogen could be directly "absorbed" by plants and keep the environment almost "untouched". Crop rotation with legumes has been recognized to increase soil fertility and agricultural productivity since ancient China and Rome (Cheng 2008). Rhizobia live in the rhizosphere of leguminous plants such as soybeans and faba bean, forming colonies that produce root nodules (Foth 1990). Biological nitrogen fixation is an oxygen dependent reaction and therefore is prevalent in legumes growing in aerated, upland soils. Currently, approximately 2 tons of industrially-fixed nitrogen are needed as fertilizer for crop production to equal the effects of 1 ton of nitrogen biologically-fixed by legume crops. Therefore, biologically-fixed nitrogen influences the global nitrogen cycle substantially less than industrially-fixed nitrogen. One day, this situation needs to be changed. Research in this field has pivotal importance and would be significantly beneficial.

Inoculation with compatible and appropriate rhizobia may be necessary where a low population of native rhizobial strains predominates and is one of the solutions which grain legume farmers can use to optimize yields. It is critical for sustained yield in farmlands deficient in native rhizobia and where N supply limits production. Research on use of *Rhizobium* inoculants for production of grain legumes showed it is a cheaper and usually more effective agronomic practice for ensuring adequate N nutrition of legumes, compared with the application of N fertilizer (Chianu et al 2008). The major findings are: (1) complete absence of or very weak institutions, policy and budgetary support for biotechnology research and lack of its integration into wider agricultural and overall development objectives, (2) limited knowledge of inoculation responses of both promiscuous and specifically nodulating legume varieties as well as the other factors that inhibit BNF, hence a weak basis for decision-making on biotechnology issues, (3) limited capacity and lack of sustainable investment, (4) poorly developed marketing channels and infrastructure, and limited involvement of the private sector in the distribution of inoculants, and (5) limited farmer awareness about and access to (much more than price) inoculants. The lessons learned include the need: (1) to increase investment in *Rhizobium* inoculation technology development, and strengthen policy and institutional support, (2) for public private partnership in the development, deployment and dissemination of BNF technologies, (3) to develop effective BNF dissemination strategies (including participatory approach) to reach farmers, and (4) for greater emphasis on capacity building along the BNF value chain. According to an FAO report, production of N fertilizer for 2007 was 130 million tons of N, and this should further increase in the coming years (FAO 2008). This extensive use has certain drawbacks. A proportion of added fertilizer is lost as a result of denitrification and leaching of soil by rainfall and irrigation. In addition, leaching leads to water pollution is caused by eutrophication. As a consequence, extending application of biological nitrogen fixation by any means is an important issue.

The legume-*Rhizobium* symbiosis is the single most important source of biologically fixed nitrogen in agricultural systems (Graham and Vance 2000); although under some conditions free-living and associative systems are important (Phillips 1999). Since it is a biological process, it does not depend on external sources of energy, except for free and renewable sunlight, and has few detrimental ecological effects (Phillips 1999). Conversely, chemical nitrogen fixation depends on fossil fuels (generally natural gas), both as an energy source and as a reductant source (Braswell et al. 1997). In addition, extensive use of nitrogen based fertilizers tends to be inefficient, with absorption by the crop ranging from 20 to 50 % of applied nitrogen (Graham and Vance 2000). Much of the nitrogen not absorbed by crops can find its way into ground water, lakes and rivers, where it poses both health and environmental problems (Graham and Vance 2000). Since the legume-bacteria symbiosis is so important in agriculture, even a slight increase in its efficiency, in terms of total nitrogen fixation, would have a major impact a global nitrogen inputs, and increase the efficiency of production for the corresponding cropping system (Phillips 1999). The interaction between legumes (Leguminosae) and bacteria in the family Rhizobiaceae leads to the development of a nitrogen-fixing symbiosis (Ohyama et al. 2009). Several environmental factors can adversely affect the performance of symbiotic nitrogen fixation by legumes. These factors can intervene at the following levels: survival of rhizobia in the soil, the infection process, nodule growth and nodule function. Also, these factors can affect N_2 fixing performance indirectly through their effects on host plant growth.

3. Effect of harsh environmental conditions on survival of rhizobia in agroecosystems and process of nodulation and nitrogen fixation of legumes

Rhizobia is the common name given to a group of small, rod-shaped, Gram-negative bacteria that collectively have the ability to produce nodules on the roots of leguminous plants and belong to the family Rhizobiaceae, which are part of the α-proteobacteria. In early studies, the taxonomy of rhizobia was based on the rate of growth of isolates on laboratory media and their selective interaction with their plant hosts. It was soon established that no strain could nodulate all plants, but that each could nodulate some legumes though not others (Long 1989). This led to the concept of cross-inoculation groups, with organisms grouped according to the hosts they nodulated. Within the genus *Rhizobium* several strains nodulate a common host, but are distinct according to genetic and/or phenotypic properties and are therefore classified as distinct species (e.g. *R. tropici* and *R. etli*). For a time this was the basis on which rhizobia were identified. However, developments in molecular biology and advances in bacterial taxonomy (Graham et al. 1991) have resulted in a rhizobial taxonomy based on a wide range of characteristics and to the distinction of new genera and species. Rhizobia were confined to the Alphaproteobacteria, namely in the order Rhizobiales. The Alphaproteobacteria class comprises 11 genera that include bacteria able to induce nodule formation in legume plants: *Agrobacterium, Allorhizobium, Azorhizobium, Bradyrhizobium, Devosia, Mesorhizobium, Methylobacterium, Ochrobactrum, Phyllobacterium, Rhizobium* and *Ensifer* (Euzéby 1997). *Agrobacterium* genus has been proposed by Young et al. (2001) to be emended and reclassified as *Rhizobium*. The members of the genus *Sinorhizobium* were transferred to the genus *Ensifer* by decision of the Judicial Commission of the International Committee on Systematics of Prokaryotes (Tindall 2008). Three additional genera belonging to the *Betaproteobacteria* class were recently added to the list of genera containing rhizobia species, namely, *Burkholderia* (Moulin et al. 2001), *Cupriavidus* (Chen et al. 2001,2003) and *Herbaspirillum* (Valverde et al. 2003). It is noteworthy to mention that these 14 genera also include a variety of non-symbiotic bacteria. More recently a member of the γ-proteobacteria has been found that also nodulates legumes (Benhizia et al. 2004).

When studying any living organism, it is important to know how each species grows and responds to certain conditions that can be found in their natural environment. Ascertaining how bacteria respond to environmental signals, or stressful conditions, is a vital part to understanding how those microbes live, thrive and survive. Every bacterium has optimum conditions that make this process easier, however in order to survive in a changing environment (or some other form of stress) the bacteria must be able to adapt. This adaptation is a stress response. Two types of stress responses operate in microorganisms: the general stress response and specific stress responses. The general stress response is normally controlled by a single or a few master regulators (Bremer and Krämer 2000) and provides cross-protection against a wide variety of environmental cues, regardless of the initial stimulant (Hecker et al. 1996; Hecker and Völker 1998). This response is effective in allowing the cell to survive, but it may not be enough to let the cell grow under the stressful conditions (Bremer and Krämer 2000). Under prolonged stress conditions cells employ specific stress responses, which utilize highly integrated networks of genetic and physiological adaptation mechanisms (Bremer and Krämer 2000). Usually, there is also a complex relationship between cellular response systems and global regulators, adding another level of control to the cell's emergency stress response and long-term survival reactions (Hengge-Aronis 1999). Although the above description is usually what happens, not all general responses occur immediately on stressful stimuli as some activate on entry into stationary phase; likewise some specific stress response are induced as soon as stress is detected.

3.1. Temperature

3.1.1. Survival of Rhizobia

Soil environmental conditions are critical factors to the persistence and survival of rhizobia in the soil. The changes in the rhizospheric environment can affect both growth and saprophytic competence, which will influence competitiveness and persistence (Dowling and Broughton 1986). In arid and semiarid regions of the tropics, the soil temperatures near the surface can be very high. In Egyptian sandy soils, the temperature near the soil surface was 59 °C at the air temperature 39°C. However, the soil temperature decreased rapidly with depth, being moderate 35 °C, at 15 cm. It appears, however, that rhizobia are more resistant to high temperatures in soil than in laboratory medium (AbdelGadir and Alexander 1997). The study performed with Indian desert soils suggested that not the high soil temperature but the low organic matter and poor soil moisture were the major factors that reduced the numbers of different micro-organisms (Rao and Venkateswarlu 1983). Indeed, in drought-affected *Acacia senegal* soils the numbers of culturable rhizobia were significantly reduced, approximately from 10^7 to 10^6 cfu g^{-1}. In conclusion, one can assume that during the dry season, the water deficit together with the high soil temperature will considerably decrease rhizobial numbers or cause a lack of rhizobia in the surface soils (0-10 cm). In order to improve the yield of legumes in more adverse environments, stress-tolerant cultivars should be combined with stress tolerant rhizobia. Studies on rhizobia biodiversity are an important approach to find more tolerant strains, even when non-adverse environments are sampled, since populations often contain tolerant strains to non-acting stresses, as resilience to respond to future problems (Giller et al. 1997). The ability of rhizobia to persist in the absence of their host plant is perhaps more dependent of their ability to endure adverse environmental factors than during symbiosis, where the nodule represents a protective environment.

Environmental stresses especially during the period between inoculation of seeds and germination impose severe problems to establishment of a successful symbiosis (Weaver and Holt 1990; Weaver et al. 1985). Every bacterium has its own optimum conditions, under which it grows at its best. For most rhizobia, the optimum temperature range for growth is 28 – 31 °C, and many are unable to grow at 37°C (Zahran 1999). Not only do the bacteria themselves have an optimum temperature range, but the processes within them do as well. Survival of *Rhizobium leguminosarum*, the microsymbiont of faba bean, may be affected mainly by extremes in temperature. Differences in adaptation to high temperatures have been demonstrated from different climatic zones. Eaglesham and Ayanaba (1984) reported that more than 90 % of cowpea rhizobia isolated from hot dry Sahelian Savanna in Niger were able to grow at 40 °C while the rhizobia isolated form cooler humid regions of West Africa did not grow at this temperature (Bowen and Kennedy 1956; Munevar and Wollum 1981). Some rhizobial strains isolated from nodules in arid environments are able to grow at 40 °C or even higher (Eaglesham et al. 1981). Kluson et al. (1986) reported differences in the tolerance of bradyrhizobia cultivated in liquid medium and in soil to temperatures within the range of 20 to 35 °C. Gewaily et al. (1991) similarly reported that survival of *Rhizobium leguminosarum* was significantly reduced at 40 °C when inoculated in sterile and non sterile soils.

Baldani and Weaver (1992) attributed heat tolerance of *Rhizobium leguminosarum* biovar *trifolii* strains to cryptic plasmids. These plasmids induce the synthesis of heat shock proteins upon exposure of bacteria above normal growth temperatures (Sen et al. 1990). Abd-Alla and Abdel-Wahab (1995a) reported that Strain RCR 1001 was more resistant to heat and nodulated faba bean better than other tested strains. Although strain adaptation to high temperature has been reported, it has been associated with decreased effectiveness in establishing symbiosis, mostly due to plasmid loss (Hungria and Franco 1993). Additionally, heat stress has been shown to induce the synthesis of heat shock proteins in *Rhizobium* (Michiels et al. 1994) and to modify lipopolysaccharide patterns and bacterial mobility (Zahran et al. 1994). Studies on rhizobia biodiversity are an important approach to find more tolerant strains, even when non-adverse environments are sampled, since populations often contain tolerant strains to non-acting stresses, as resilience to respond to future problems (Giller et al. 1997). Temperature is often pointed out as the major factors in determining the bacterial community diversity (Fierer and Jackson 2006; Staddon et al. 1998).

The optimum temperature for rhizobia growth is 25-30 °C (Zhang et al. 1995), however in both saprophytic and symbiotic life rhizobia are often subject to temperatures out of this range. Most studies in rhizobial temperature stress tolerance focus soybean and common-bean microsymbionts. Soybean isolates grow weakly at 40 °C and no isolate was able to grow at 45°C (Chen et al. 2002). Rhizobia nodulating *P. vulgaris* can survive to 47 °C, but their symbiotic effectiveness is loss at high temperatures; while other isolates tolerant to 40 °C were able to remain infective at that temperature (Karanja and Wood 1988). Nandal et al. (2005) reported that mutants tolerant to high temperature (43 °C), obtained from a thermo-sensitive *Rhizobium* sp. strain, exhibited a different protein profile from that of the wild type at high temperature, namely the mutant strains showed overexpressed proteins and new proteins. A protein of 63-75 kDa was overproduced in all mutant strains, which probably corresponds to DnaK. In chickpea rhizobia, a 60 kDa protein that could correspond to GroEL was found to be consistently overproduced when isolates were submitted to heat stress (Rodrigues et al. 2006). Strains isolated from a chickpea wild relative (*Cicer anatolicum*), collected from high altitudes, were successful in nodulating chickpea at low temperatures, what represents an alternative source of chickpea nodulating rhizobia with potential use as inoculants (Ogutcu et al. 2008). The changes in the population occupancy of some bradyrhizobia tested were in accordance with the geographical distribution of indigenous soybean-nodulating bradyrhizobia (Saeki et al. 2006). Saeki et al. (2010) reported that the occupancies of bradyrhizobia in soil microcosms after long-term incubation may change depending on the incubation temperature. They suggest that soil temperature may be one important environmental gradient determining bradyrhizobial niches from northern to southern regions of Japan. Temperature stress is generally divided into two classes: heat shock and cold shock. Bacterial heat shock is the more characterised of the two (Phadtare et al. 2000). The heat shock response is very similar to the acid stress response, in that many proteins with a similar mode of action are synthesised. Heat shock proteins contribute to heat tolerance by conferring heat protection on the bacteria but do not alter the internal temperature of the cell (Yura et al. 2000). Like acid shock protein, there are two main types of heat shock protein: chaperones and proteases. These work in the same way as the cid shock proteins. Heat shock proteins, and their regulation, structure and function, have been studied in great detail. Their function appears to be highly conserved between both prokaryotes and eukaryotes (Netzer and Hartl 1998). Some of these proteins are also vital under normal (non-heat shock) growth conditions (Münchbach et al. 1999; Lentz 2004).

Cold shock is essentially the opposite of heat shock. Instead of proteins misfolding and denaturing, cells undergoing cold shock have to contend with a loss of membrane and cytosol fluidity and with the stabilisation of secondary structures of RNA/DNA (Phadtare et al. 2000). RNA/DNA stabilisation leads to a decrease in the efficiency of translation, transcription and replication. Bacterial cold shock response is an immediate and transient response to the temperature downshift. This is followed by low temperature adaptation that allows continued growth at low temperatures (Panoff et al. 1997). Generally, bacteria overcome loss of fluidity by increasing the amount of unsaturated fatty acids in the membrane phospholipids (Phadtare et al. 2000). Cold shock response also leads to the production of many cold shock proteins (CSPs). Just like ASPs and HSPS, these too are mainly chaperones and proteases (Phadtare et al. 2000). However, instead of protecting against the misfolding of proteins, the CSP chaperones are primarily used to bind to RNA/DNA to prevent stabilisation and allow translation and transcription to proceed as usual (Phadtare et al. 2000). CspA is an RNA chaperone and a major CSP found in many bacteria (Jiang et al. 1997). A CspA homologue is present in *Sinorhizobium meliloti* and is induced following a temperature downshift from 30 to 15 °C, along with the three rRNA (*rrn*) operons. It is unknown what function the genes and products of the *rrn* operons or CspA have in response to cold shock, as mutations made in these genes showed no change in cell phenotype at 15 °C compared to the wild-type (O'Connell et al. 2000; Gustafson et al. 2002). TypA is also required for growth at low temperatures and is believed to act as a regulator by controlling the phosphorylation of proteins (Kiss et al. 2004). Expression of several heat shock operons, mainly coding for small heat shock proteins, is under the control of repression of heat shock gene expression in various rhizobial species (Nocker et al. 2001). The molecular bases of temperature stress tolerance in rhizobia were investigated, by comparing the expression of chaperone genes dnaKJ and groESL in thermotolerant and thermosensitive isolates. Tolerance to cold, heat and heat shock was evaluated for 53 mesorhizobia (Alexandre and Oliveira 2011). The analysis of the dnaK and groESL expression by Northern hybridization, using isolates from three species groups, showed an increase in the transcripts levels with heat, but not with cold stress. Laranjo and Oliveira (2011) reported that under heat stress a protein of approximately 62 kDa was newly detected in *Mesorhizobium chacoense.* The over expression of a 62 kDa protein in *Mesorhizobium huakuii* and *Mesorhizobium septentrionale* was controlled groEL expression and increased upon heat shock. Interestingly, a 62 kDa protein, the NdvCprotein, has been described as important in osmoregulation, associated with symbiotic systems in *B. japonicum* (Bhagwat et al. 1996). Further studies are required to elucidate the function of overexpressed genes in order to clarify their role in environmental stress tolerance of rhizobia, as well as their contribution to symbiotic effectiveness (Alexandre and Oliveira 2012).

3.1.2. Rhizobium- Legume molecular signalling exchange

Molecular signalling in the rhizosphere controls the nature of relationships between plants and other soil organisms. For instance, legume-derived signals, such as betaines and isoflavonoids, function by chemoattracting rhizobia and trigger the events that lead to endosymbiosis. The molecular signalling exchange between both microsymbiont and legume is fundamental for an effective legume-*rhizobium* symbiosis and can determine the specificity of this symbiotic relationship. Production of Nod factors or lipo-chito-oligosaccharide signalling molecules by the prokaryotic partner is activated by the release of plant phenolic signals, mainly flavonoids, into the rhizosphere. The phenolic flavonoid compounds partly determine the specificity of the symbiotic relationship as each *Rhizobium* species responds to specific flavonoids. Another determinant of host symbiont specificity is attributed to the different Nod factors substituents attached to the oligosaccharide backbone (Dénarié et al. 1996; Oldroyd 2001). Production of Nod factors or lipo-chito-oligosaccharide signalling molecules by the prokaryotic partner is activated by the release of plant phenolic signals, mainly flavonoids, into the rhizosphere. The phenolic flavonoid compounds partly determine the specificity of the symbiotic relationship as each *Rhizobium* species responds to specific flavonoids. Another determinant of hos symbiont specificity is attributed to the different Nod factors substituents attached to the oligosaccharide backbone (Dénarié et al. 1996; Oldroyd 2001). Most rhizobia species interact with only a few select legumes, but some have been shown to have a broad host range (Pueppke and Broughton 1999). For example, the strain *Ensifer* sp. NGR234 is able to nodulate over 120 plant genera, including the non-legume *Parasponia andersonii*. This feature may depend up on the family of Nod factors secreted, which are more diverse than in all other rhizobia known (Schmeisser et al. 2009) and in the concentration of Nod factors released by the NGR234 that is much higher than usual. More recently, *M. opportunistum* WSM2075 was isolated from *Biserrula pelecinus* root nodules, but the symbiotic genes of this organism provide a broader range of hosts for nodulation, including also *Astragalus adsurgens, A. membranaceus, Lotus peregrinus* and *Macroptilium atropurpureum* (Nandasena et al. 2009). On the other hand, there are legumes species that can be nodulated by several rhizobia species and others that are very restrict for nodulation and only accept as microsymbionts a reduced number of species. For example, *Phaseolus vulgaris* is known as a promiscuous host, since it can be nodulated by rhizobia belonging to diverse genera (such as *Bradyrhizobium, Rhizobium* and *Ensifer*) while *Cicer arietinum* is considered a restrict host, because it is nodulated only by *Mesorhizobium* species. Nevertheless, the host range depends on the legume cultivar used and conditions tested (Martinez-Romero 2003).

Temeprature play a critical role on the exchange of molecular signals between rhizobia and their host, thus reducing nodulation. Low temperature inhibits inter-organismal signaling between the two symbiotic partners. It has been shown that low temperature inhibits the biosynthesis and rhizosecretion of plant to-bacteria signal molecules (for example genistein from soybean roots, which are necessary for the induction of the nod genes of *B. japonicum* (Zhang and Smith 1997; Abd-Alla 2001, 2011). Low temperature also inhibit the induction of bacterial nodulation genes (nod gene) required for the biosynthesis of bacteriato-plant signaling molecules, lipo-chitooligosaccarides (LCOs), the so-called Nod factors (Zhang et al.

1995). The disruption of Nod factor production/excretion at low temperature incubations has been previously reported in *R. leguminosarum* bv. *trifolii* (McKay and Djordjevic 1993), and in *B. japonicum* (Duzan et al. 2005). Low temperatures delay the onset of nodulation (Pan and Smith 1998) and reduce the rate of subsequent nodule growth, resulting in effects on small final nodule size (de Lira Juniora et al. 2005). The presence of appropriates flavonoids in root exudates is a critical factor in nodule formation (Richardson et al. 1988 a,b) and dictate rhizobia legume specificity. The nodulation status of the pea (*Pisum sativum* L.) was most profoundly improved via the addition of the flavonoid naringenin (Bandyopadhyay et al. 1996), possibly via the induction of nod gene expression in *R. leguminosarum*. In the soybean-*B. japonicum* symbiosis, Kosslak et al. (1987) reported that genistein was the most effective inducer for the expression of the nodYABC operon of *B. japonicum*. Exogenous application of genistein results in the short circuiting of plant-bacterium signaling, and has been confirmed as an effective means to mitigate the adverse effects of low temperature on nodulation and nitrogen fixation (Zhang and Smith 1997). Pre-incubation of *B. japonicum* with genistein hastened the onset of nitrogen fixation, and increased the number and size of the nodules and plant growth. Additionally, this beneficial effect of genistein increased with decreasing temperature (Zhang et al. 1995). Firmin et al. (1993) and Ovtsyana et al. (1999) observed that nodule formation on cv. Afghanistan inoculated with *R. leguminosarum* bv. *viciae* is at least partly controlled by Nod-factors that carry a modification on the reducing terminus (an acetyl group encoded by the acetyl transferase NodX) as well as by a single genetic locus sym2A in the host-plant. *R. leguminosarum* bv. *viciae* strains producing Nod factors lacking this modification cannot nodulate these pea plants at low temperatures. Somehow, blockage of nodulation is overcome at higher temperatures suggesting a temperature-sensitive gene-for-gene relationship between nodX and sym2A (Kozik et al. 1995; Lie 1984; Olsthoorn et al. 2000). There is, however, little information available about the effect of high temperature on *Rhizobium*- Legume molecular signalling exchange, additional studies are required.

3.1.3. Rhizobia-legume nodulation and nitrogen fixation

The interaction between *Rhizobium* and host legume results in the formation of root nodules. The relation between rhizobia and legume is a selective one: individual species of rhizobia have a distinct host range, from narrow to broad, allowing nodulation of a particular set of leguminous species. Also, a particular leguminous species can be infected by a certain range of rhizobia. Development of functional nodules requires temporally and spatially controlled activity of genes and gene products of both partners (Oldroyd et al. 2001). The *Rhizobium*-legume interactions are controlled by signal exchange between the two partners. Legumes secrete specific flavonoid-type molecules (Gagnon and Ibrahim 1998). These compounds are main components in the chemical communication between symbiotic legumes and different species of nitrogen-fixing bacteria leading to nodule formation and N_2 fixation (Werner 2001). Flavonoid perception attracts the specific rhizoba to the root hairs and activates *nodulation* (*nod*) gene expression, via the bacteria activator NodD (Lindström et al. 2002). NodD activates transcription of *nod* boxes promoters, and represents the first level of host-specific recognition (Schultze and Kondorosi 2008). The *nod* genes expression lead to the production of strain-specific lipo-chito-oligosaccharides, also called as Nod factors (Spaink 2000). Nod factors are considered the second level of host-specific recognition (Perret et al. 2000). Nod-factors induce various plant responses, including root-hair deformation, cortical cell-division, 'pseudo-nodule' and nodule-formation (Broughton et al. 2003). On production of Nod factors, the rhizobia then surround and attach to the tip of root, causing the root to start to curl (Yao and Vincent 1969). Rhizobia trapped in a curled hair, or between a hair and another cell, proliferate and begin to infect the outer plant cells, which in turn stimulates plant cells to produce infection threads (Callaham and Torrey 1981). Rhizobia are released from infection threads into the cytoplasm of plant cells. The Rhizobia are surrounded by plasma membrane of plant origin and then briefly replicate their DNA and divide before stopping both processes (Robertson et al. 1978). The membrane-enveloped rhizobia continue to divide within the host cells before they differentiate into bacteroids and start to fix nitrogen (Roth and Stacey 1989a, b). The endosymbiotic forms of the rhizobia are referred to as bacteroids and begin to fix nitrogen by the action of the enzyme nitrogenase (Xi et al. 2000). Atmospheric nitrogen is converted into ammonia by bacteroids and is subsequently assimilated into the plant following its conversion to glutamine by glutamine synthase. Within the nodule interior and the neighboring plant cells, essential nutrients are exchange between bacteroids and plant cells through a symplastic and apoplastic pathway (Abd-Alla et al. 2000 a).

Temperature regulates the metabolism of the plant and the bacteria, as well as the plant–bacteria association (Young et al. 2006). Sometimes, the sensitivity of the host toward low temperature affects nitrogen fixation severely, leading to an abrupt cut-off at temperatures where the bacterial cells can still grow and metabolize. Root hair infection is much more temperature sensitive than nodule development (Hafeez et al. 2000). Naeem et al. (2008) suggested that suboptimum temperatures affect the growth of bacteroids even within the nodule and also affect the cell-to-cell movement of the bacteroids. Most of the work related to low soil temperature effects on signal exchange and nodulation has been conducted by Smith group on soybean. The root zone ternperatures blow 17.5 °C cause reduction and delays nodule formation and the onset of nitrogen fixation (Zhang and Smith 1996). Soil temperatures below 25 °C are prevalent in Canada until mid-June/mid-July (Zhang and Smith 1994). This delay reduces the time during which there is nitrogen fixation by more than 10 days (Zhang and Smith 1995). This leads to decline in total nitrogen fixation and lower crop yield (Lynch and Smith 1994), and can be at least partially corrected by the addition of genistein (Zhang and Smith 1997; Pan and Smith 1998). It is known, however, that low soil temperatures affect nodule initiation (Cordovilla et al 1999) and function (Rice et al. 1995).A greater concen in the tropical zone, but also important during the peak of summer in temperate zones., is excessive soil temperature and its ability to be an important limitation to nodulation and nitrogen fixation (Hungria and franco 1993; Boddey et al. 1997; Graham and Vance 2000). Hungria and Vargas (2000) note that maximum soil ternperature in the tropics regularly exceed 40 °C at 5 cm below the soil surface and 50 °C at 1 cm. Such high temperatures can limit nodulation. The range for optimal growth of legurnes dependent on nitrogen fixation have been recognized to be narrower than for nitrogen-fed legumes since at least (Hungria and Vargas 2000). Elevated temperature inhibited plant growth by diminishing net photosynthesis and nitrogen fixation (Aranjuelo et al. 2007). The optimum temperature for nodule formation is ranged between 25 and 33 °C (Pankhurst and Sprent 1976), although decline in nitrogenase activity have been recorded above 28 °C (Hungria and Franco 1993). High temperature might affect N_2 fixation directly or indirectly. Direct inhibition by temperature is a consequence of decreased nodule development (Dart and Mercer 1965; Piha and Munns 1987), activity (Meyer and Anderson 1959; Piha and Munns 1987; Hernandez-Armenta et al. 1989) and accelerated nodule aging. Indirect inhibition is related to temperature effects on root hair formation depression, decrease of nodulation sites (Frings 1976; Jones and Tisdale 1921), and modified adherence of bacteria to root hairs (Frings 1976). Moreover, Aranjuelo et al. (2007) observed that high temperature negatively affected bacteroid enzyme activity regardless of water availability, although the inhibitory effect on the bacteroid fraction was even greater. The decrease in malate dehydrogenase activity was observed in plants exposed to elevated temperature suggests that malate should also be reduced. Malate depletion implies that less carbon is redirected to mitochondria and that there is less energy available for N_2 fixation. The observed decrease in enzyme activity could be related to a reduction of the total soluble protein of the bacteroid. The large (70%) reduction in soluble protein could affect nitrogenase protein responsible for N_2 fixation activity. These results show that the nodule bacteroid fraction was more sensitive to drought and elevated temperature than the nodule plant fraction. Relatively high temperature has also been shown to influence infection, N_2-fixation activity, and legume growth (Mohammadi et al. 2012) and has a strong influence on specific strain and cultivar interactions (Arayankoon et al. 1990). Reduced nitrogenase activity in such nodules can also be explained on the basis of these effects. It has been reported that there is considerable variation in the rhizobia strains in terms of survival and nodulation ability at high temperature (Michiels et al. 1994). Karanja and Wood (1988) found that a high percentage of the strains that persisted at 45 °C lost their infectiveness. They attributed these losses to plasmid curing. It appears that every legume and *Rhizobium* combination has an optimum temperature relationship, which is around 30 °C for clover and pea, between 35 to 40 °C for soybean, peanut and cowpea, and between 25 to 30 °C for common bean (Long 2001).

3.2. Desiccation

3.2.1. Survival of Rhizobia

One of the most severe and widespread problems facing the crops production is the degradation of soil quality due to desiccation and salinity. In fact, almost 40% of the world's land surface is affected by salinity-related problems (Vriezen 2007). These two harsh environmental conditions can have a dramatic impact on the endogenous soil bacteria (Fierer et al. 2003; Han et al. 2005). Water, and its availability, is one of the most vital environmental factors to affect the growth and survival of micro-organisms (Potts 1994). Desiccation is one of the most common stresses soil microorganisms have to face so often. The responses of bacterial cells to desiccation can be: shrinkage of the bacterial cytoplasm and capsular layers, increase in intracellular salt levels, crowding of macromolecules, damage to external layers (pili, membranes), changes in ribosome structure, and decrease in growth. Reactive oxygen species can also damage proteins and DNA, leading to accumulation of mutations (Potts 1994). Death of rhizobial cells during desiccation was suggested to associate with changes in membrane integrity (Bushby and Marshall 1977). It has been hypothesised that during dehydration, the removal of water hydrogen bond to the phospholipid head groups of the membrane decreases the spacing between adjacent lipids. Then, the membrane is converted from the liquid crystalline into the gel phase already at room temperature. Subsequent rehydration results in a further phase transition of the membrane back to the liquid crystalline phase. As consequence, the membrane barrier is disrupted, leading to leakage of membranes (Potts 1994: Welsh 2000). Desiccation tolerance is the ability of cells to undergo nearly absolute dehydration through air drying, without being killed. This is the most severe water deficit stress since the removal of the cell-bound water imposes such structural, physiological and biochemical stresses that cells must adapt or die (Billi and potts 2002).

Of particular importance to the crop production is the impact of these harsh environmental conditions on the persistence and survival of rhizobia. The ability of rhizobia, to survive desiccation depends on their ability to cope with radiation stresses, reactive oxygen species, certain salts and solutes, and temperature extremes (Potts 1994; Billi and potts 2002; Ramos et al. 2001). Desiccation stress can be differentiated into three main stages: drying, storage and rehydration. These phases can be manipulated in several ways, namely, by the severity and the speed of drying and rehydration and by the duration of storage. The consequences of drying are fourfold: (i) the accumulation of salts and solutes, (ii) hyperosmotic stresses, (iii) the distribuance of metabolism when a certain water activity has been reached, and (iv) the accumulation of damage when the aqueous monolayer is removed from macromolecules. The accumulation of damage during storage is comparable to that caused by ionizing and UV radiation and damage by reactive oxygen species (Mattimore and Battista 1996; Shirkey et al. 2000) when organisms are not metabolically active and thus unable to repair any damage (Potts 1994). Finally, during rehydration, hypoosmotic stresses and the appearance of reactive oxygen species affect survival (Stead and Park 2000). A low survival of rhizobia occurs when drying was rapid (Bushby and Marshall 1976; Antheunissen and Arkestein-Dijksman 1979). These results were confirmed by Mary et al. (1985, 1986) who used *Sinorhizobium meliloti* RCR2011 to test survival after slow and fast drying with and without the addition of salts. It was recorded that fast drying process resulted in a decrease in survival of rhizobia in mineral soil (Chao and Alexander 1984). The reduction of water content causes the accumulation of salts and other compounds that produce osmotic and salt stress. High concentration of these compounds may be toxic and decrease the viability of cells (Steinborn and Roughley 1975; Vriezen et al. 2006). Conversely, the accumulation of certain compounds, including osmoprotectants and compatible solutes, may increase desiccation survival (Fougere and Le Rudulier 1990; Gouffi and Blanco 2000; Madkour et al. 1990). Osmoprotectants are exogenous solutes that stimulate bacterial growth in an environment with high osmolality, whilst compatible solutes are specific organic osmolytes that accumulate in high amounts within a cell to counter a hyper-osmotic gradient, but do not conflict with cellular functions (Miller and Wood 1996). Several compatible solutes have also been shown to stabilise enzyme stability in cells under stressful conditions (Poolman et al.

2002). Some compounds can function as osmoprotectants and compatible solutes, whilst some can only function as one of these groups.

Many osmoprotectants are transported into the cytoplasm where they act as, or are converted into compatible solutes (Hirsch 2010). Compatible solutes can be collected in high concentrations (Bremer and Krämer 2000). Since only a limited number of compounds meet the required criteria, the same compatible solutes are employed against hyper-osmosis throughout various bacteria (Braun 1997). Different compatible solutes work more effectively than others within their bacteria; e.g. glycine betaine is more effective in *S. meliloti* and *E. coli* than it is in *Bacillus subtilis* (Botsford and Lewis 1990); whilst proline is a compatible solute in *E. coli* but not in rhizobia (Gloux and LeRudulier 1989). Also, the strength of hyper-osmolarity can determine how the bacteria respond and what osmoprotectants are used (Breedveld et al. 1990; Gouffi et al. 2000). In a similar way, the compound used to bring about hyper-osmosis can stimulate a stronger stress response compared to others; e.g. generally sodium chloride (NaCl) induced hyper-osmosis causes a stronger stress response then sucrose induced hyper-osmosis, due to the ionic nature of NaCl (Gloux and Le Rudulier 1989). Compatible solutes can either by synthesized *de novo*, when required by the bacteria, or they are accumulated from the environment, depending on the situation. Under conditions where osmotic upshift is severe and immediate, cells do not have the time required to synthesise compatible solutes and so must acquire them from their environment. Reduction of water activity below 0.53 inhibited the function of RNA polymerase (Brown 1990). The storage phase is characterized by a slow decline in viable counts in rhizobia after slow drying. Mary et al. (1985, 1986) noted that the better survival of *Sinorhizobium* during storage under desiccation conditions from 22% to 67% than at 3% and 83.5% relative humidity. Similar trend have been recorded in *Bradyrhizobium* (Mary et al. 1994; Boumahdi et al. 1999). Cell death occurs at 83.5% relative humidity due to failure of intracellular enzymes. Antheunissen et al. (1981) showed that when dried occur at slow rate, rhizobia can survive desiccation for up to 4 years. These long-term storage studies are rare, but they show that in the family Rhizobiaceae, sinorhizobia can survive desiccation for years. A decline in viable cells during long-term storage under desiccation conditions can be explained by the accumulation of oxygen- and radiation-induced damage (Vincent et al. 1961; Mary et al. 1994). The rate of rehydration has important consequences for repairing of accumulated damage and survival of bacteria. Combined process of drying and fast rehydration resulted in rupture of the cell envelope. Cells ruptured on the subpolar region which is also the region from which flagella originate. A suggestion is made that the point where a flagellum emerges from the cell is a point of weakness (Bushby and Marshall 1977; Salema et al. 1981). Slow rehydration of *S. meliloti, R. leguminosarum*, and *Pseudomonas putida* from dried inoculant formulations provide higher viable counts than did rapid rehydration (Kosanke et al. 1991). Chen and Alexander (1973) reported that a higher percentage of drought-tolerant than drought-sensitive bacteria was able to grow at low water activities. When these bacteria were grown in media with high salt concentrations, bacteria generally became more tolerant of prolonged drought and they persisted longer. In general, rhizobia do not, or cannot, synthesise their own solutes so use uptake systems to accumulate them (Gloux and Le Rudulier 1989). However, Streeter and Gonez (2006) reported that many rhizobia accumulate various carbohydrates such as trehalose, a disaccharide made up of two glucose molecules joined together by an α, α-1,1 linkage, is employed by many organisms to protect membranes and proteins from desiccation stress. Rhizobia also accumulate trehalose, among other carbohydrates, and also betaine and proline, in response to desiccation. Trehalose and sucrose are the only carbohydrates that are synthesized de novo in response to stress. Bacteria can produce trehalose from glucose 6-phosphate and UDP-glucose (the OtsA–OtsB pathway), from glycogen-like α(1-4)-linked glucose polymers (the TreY–TreZ pathway) and from maltose (the TreS pathway). The TreYZ pathway is common to many rhizobia (Streeter and Bhagwat 1999), whereas the OtsAB and TreYZ pathways are found in *R. leguminosarum* bv. *trifolii* strain NZP561 (Mcintyre et al. 2007). Cultured *Bradyrhizobium japonicum* USDA110 and *B. elkanii* were found to have three enzymes for trehalose synthesis: trehalose synthase (TS), maltooligosyltrehalose synthase (MOTS), and trehalose-6-phosphate synthetase (Streeter and Gomez 2006). Trehalose at relatively high concentrations is also present in *B. japonicum* bacteroids residing within nodules, suggesting that these differentiated nitrogen-fixing cells are under stress. Addition of trehalose to culture medium of *B. japonicum* at the time of desiccation stress had a significant positive effect on survival (Streeter 2003). Although it may not be practical to use trehalose as a carbon source in inoculant production, it may be possible to engineer greater trehalose accumulation in rhizobia to enhance their survival in response to dryness. Trehalose may protect desiccated cells by their ability to form glasses under dry conditions, in this way maintaining the native conformation of proteins and other macromolecules (Ramos et al. 2001). Trehalose levels also increase in *R. leguminosarum* bv. *trifolii* TA1 cells as they encounter osmotic stress (Streeter 1985; Breedveld et al. 1993). *R. leguminosarum* bv. *trifolii* strain NZP561 accumulates trehalose upon entry into stationary phase (Mcintyre et al. 2007), but in this rhizobial strain, trehalose synthesis is constitutive and modified post transcriptionally rather than induced as in other rhizobia. Mutations in otsA or treY individually in strain NZP561 did not dramatically affect trehalose accumulation, but double otsA treY mutants did not accumulate trehalose and were more sensitive to desiccation. They were also less competitive with regard to occupying nodules than were wild-typestrains (Mcintyre et al. 2007). The transcriptional analysis of the genome of *B. japonicum* subjected to desiccation stress indicated that genes critical for pilus are upregulated (Cytryn et al. 2007). Pili, especially type IV pili, are often important for biofilm formation (Shime-Hattori et al. 2006; Jurcisek and Bakaletz 2007). It is extremely likely that desiccation-stressed *B. japonicum* cells show some of the same patterns of gene expression, as do cells in biofilms. Similarly, *S. meliloti* cells grown under salt and osmotic stress (Dominguez-Fererras et al. 2006) upregulate some of the same genes uncovered in transcriptome arrays of biofilm cells of other bacteria (An and Parsek 2007). Biofilms are one of the many ways that bacteria use to protect themselves from desiccation and it is well known that exopolysaccharide, an important component of the biofilm matrix, protects bacteria from drought stress. Loss-of-function EPS mutants of many bacteria show impaired biofilm formation (Yildiz and Schoolnik 1999; Danese et al. 2000; Whiteley et al. 2001; Matsukawa and Greenberg 2004), as do *S. meliloti* exoY loss-of-function mutant cells (Fujishige et al. 2006) and exopolysaccharide mutants of *M. tianshanense* (Wang et al. 2008). Nevertheless, it is not known whether exopolysaccharide-deficient mutants that are incapable of biofilm formation are less capable of surviving desiccation stress under field conditions. Vanderlinde et al. (2010) demonstrated the important role for exopolysaccharide in desiccation tolerance in *Rhizobium leguminosarum* bv. *viciae* 3841 and identify a novel genetic element (ABC transporter) required for biofilm formation. Identification of a novel genetic element required for desiccation tolerance, and proper biofilm formation. An increased tolerance to desiccation can enhance the survival of rhizobacteria within the soil considerably (Rokitko et al. 2003). This is of particular interest with nitrogen-fixing rhizobia because desiccation is a major cause of the poor on-seed survival rates of commercial inoculants and the subsequent poor performance of rhizobial inoculants in the field (Deaker et al. 2004). To further our understanding of how bacteria persist and survive in the soil environment, and improve on-seed survival of rhizobial inoculants, a complete understanding of the mechanisms used for desiccation tolerance is necessary.

3.2.2. Rhizobium-Legume molecular signalling exchange, noduation and nitrogen fixation

To the best of our knowledge, there is no information available addressing the impact of drought on molecular signals exchanges between the two partners. Further research is needed to clarify the impact of drought on *Rhizobium*-legume molecular signaling exchange. However, several studies both on herbaceous and woody legumes have shown that drought cause deleterious effect on nodulation and nitrogen fixation (Singleton and Bohlool 1984; Zahran and Sprent 1986; Arayangkoon et al. 1990; Marcar et al. 1991; Purwantari et al. 1995; Hatimi 1999). If the stress conditions are long or/and strong enough, formation of nodules will cease completely. It is generally believed that the infection process is the most sensitive phase during nodule development (Singleton and Bohlool 1984). Once nodules have formed, they are less affected by stress factors than the initial nodule formation process (Purwantari et al. 1995). Studies with herbaceous legumes having the root hair infection mode indicated that drought and salinity stresses cause changes in root hair morphology and decrease the numbers of markedly curling hairs. In mildly water-stressed *Vicia faba* nodules, the loss of turgor induced a reduction in the peribacteroid space, the vesicle membrane became sinous, and the cell cytoplasm appeared granular. Under more severe water stress, a total disappearance of peribacteroid spaces was associated with the rupture of vesicle membranes. The bacteroid content appeared to be very heterogeneous (Guerin et al. 1991). A shortage of water supply can slow the growth of the nodule and accelerate its senescence. Nitrogenase activity is decreased significantly, accompanied by the decrease in respiratory activity of the soybean and common bean nodules (Weisz et al. 1985; Gerosa-Ramos et al. 2003). A limitation in metabolic capacity of bacteroids and oxidative damage of cellular components are contributing factors to the inhibition of nitrogenase activity in alfalfa nodules (Naya et al. 2007). In addition, the transport of fixed nitrogen out of the nodule is decreased, possibly due to an insufficient supply of photosynthates in stems and leaves under stress (Huang et al. 1975).

The major pathway of water movement into nodules is through vascular connections with the root. The surface of the nodule is the main area for gaseous exchange and is therefore more adapted to water loss than uptake. Nodules always need an efficient water supply to export the product of fixation. Water-stressed plants transpire at a lower rate than unstressed plants. A lower rate of water movement out of the nodule during drought stress may restrict export of N_2 fixation products, thus inhibiting nitrogenase activity via a feed-back mechanism (Serraj et al. 1999). Water shortage may induce oxidative stress in nodules. This leads to general decrease of antioxidant activities that are associated with nodule senescence (Hernandez-Jimenez et al. 2002; Porcel et al. 2003). It was found that water stress imposed during vegetative growth was more detrimental to nodulation and nitrogen fixation than that imposed during the reproductive stage (Pena-Cabrales and Castellanos 1993). The mechanisms of desiccation tolerance could be attribitted to the ability to limit to a minimum the cell metabolism, the increased catalase activity and the presence of specific plasmids for drought tolerance. Some experiences have shown that working with legumes and rhizobia strains selected for desiccation tolerance, a symbiotic interaction can take place (Soria et al. 1996). Abd-Alla and Abdel-Wahab (1995b) observed that nodulation and nitrogenase activity were significantly decreased by increasing drought stress. Leghaemoglobin and protein contents of nodule cytosol were also severely inhibited by drought stress. This decline was attributed to the induction of protease activity. However, carbohydrate contents of the nodule cytosol increased significantly. This accumulation was attributed to a sharp decline in invertase activity and low use of sugar by the bacteroids. This study indicated that harmful effects of water deficits can be alleviated by increasing K^+ supplementation (Abd-Alla and Abdel-Wahab 1995b). Water stress is quickly reflected as changes in hormonal content (Hsiao 1973). The nodules are an active site of synthesis of auxins and cytokinins. Therefore, it is likely that nodules, besides the supply of organic N, are a source of cytokinins that makes the plant more tolerant to water stress (Phillips and Torrey 1973).

Many legumes are very sensitive to excess water. Nodule development and function are usually more affected than the infection itself, and some effects such as decreased nitrogenase activity may be even more intense than in the case of water deficit. Reduced to zero, the contribution of O_2 to the nodule appears to be the main problem of the effect of waterlogging. The diffusion of O_2 within the nodules is in part regulated by a physical barrier located in nodular parenchymal cells (Andres et al. 2012). Under stress conditions, the diffusion resistance increases by identifying a lack of O_2 inside the nodule, leading to inhibit its activity (Day and Copeland 1991). The ability of aerobic bacteria to utilize nitrogenous oxides, as terminal electron acceptors, enables them to survive and grow during periods of anoxia. This may be advantageous for the survival of rhizobia in soils (Zablotowicz et al. 1978).

3.3. Salinity stress

3.3.1. Survival of Rhizobia

The world's demand for food is increasing at such a rate that the ability to meet anticipated needs in the next several decades is becoming questionable. Irrigated agriculture presently accounts for about one-third of the world's production of food and fibre; it is anticipated that it will need to produce nearly 50 percent by the year 2040 (Rhoades et al. 1999). This will likely be difficult, because extensive areas of irrigated land have been and are increasingly becoming degraded by salinization and waterlogging resulting from over-irrigation and other forms of poor agricultural management (Ghassemi, et al. 1995). Soil salinity affects about 800 Mha of arable lands worldwide (Munns and Tester 2008), and this area is expanding. Salinity affects agricultural production in arid and semiarid regions, where rainfall is limited and is not sufficient to transport salts from the plant root zone (Tester and Davenport 2003). The damaging effects of salt on organisms are caused not only by osmotic forces, but also by toxic levels of sodium and chloride. Salinity stress is important factor limiting the productivity of leguminous crops. Soil salinity reduces survival and growth of rhizobia in the soil and inhibits rhizobia-legume symbiosis, resulting in lower productivity of legumes (Abd-Alla 1992; Abd-Alla et al. 1998). Rhizobia are known to be more salt tolerant than their respective plant partners. Maximal limit of tolerance to salinity is superior in rhizobia as compared to their host plant which frequently constitute the limiting factor in saline soils (Kassem et al. 1985). In some cases bacteria that are highly salt tolerant as free-living cells produce ineffective nodules, showing reduced symbiotic efficiency and/or low rates of nitrogen fixation (Chien et al. 1992, Abd-Alla and Abdel-Wahab 1995a). Some bacterial strains of various species are able to grow within the range of 300–700 mM NaCl (Mpepereki et al. 1997). Many authors reported that some rhizobial strains can persist and survive in the saline soils (ElSheikh and Wood 1989, 1990a; Ishaq et al. 1989), although the majority of rhizobia are not capable of tolerating the harmful effects of high osmolarity (Soussi et al. 2001). The ability of rhizobia to tolerate salt stress depends also on the species and even the strain of rhizobia studied (Bernard et al. 1986). Fast growing rhizobia are generally considered to be more tolerant to saline stress than bradyrhizobia (ElSheikh and Wood 1990b) and strains isolated from saline soils are typically more tolerant (Hua et al. 1982). Nevertheless, many fast growing rhizobia are very salt sensitive and some rhizobia isolated from saline soils are sensitive (Zahran 1999). The ability of rhizobia to adapt to fluctuations in the osmolarity of their surrounding is of fundamental importance for their survival. Osmotolerant rhizobia use a variety of accumulated or non-accumulated osmoprotectants, including betaines, amino acids and sugars as a strategy to counter the high oscillations of their environment osmolarity (Bernard et al. 1986; Gouffi et al. 1999). Osmoregulation is the main strategy employed by rhizobia to cope with salt stress (Ghittoni and Bueno 1996), including altered polysaccharide production upon salt treatment (Lloret et al. 1998). Survival and growth in saline environments are the result of adaptive processes, such as ion transport and compartmentation, osmotic solute synthesis and accumulation, which lead to osmotic adjustment and protein turnover for cellular repair (Munns and Termaat 1986; Paul and Cockburn 1989).

Bacteria grown in saline environments must maintain positive turgor across the membrane by allowing influx of salt or solutes or by exclusion of salts via the production of compatible solutes or other organic osmolytes. Bacteria prefer uptake over *de novo* synthesis of organic osmolytes when present in surrounding environment. Exogenous osmolytes that improve cell growth under adverse osmotic conditions are referred to osmoprotectans (Galinski 1995). Under elevated salinity *S. meliloti* can accumulate K^+ ions (Miller and Wood 1996) and an organic anion, glutamate, and synthesize the following compatible solutes: N-acetylglutaminylglutamine amide (NAGGN), trehalose, and glycine betaine, if the medium contains its precursor, choline (Hua et al. 1982; Botsford and Lewis 1990; Smith et al. 1994; Talibart et al. 1994, 1997). Recent data indicate that the set of endogenous osmolytes produced by rhizobia can vary at least according to the species level and NAGGN has been found only in *S. meliloti* (Smith and Smith 1989; Smith et al. 1994; Talibart et al. 1994, 1997). The types of the accumulating osmolytes also depend on the stress level and on the growth phase of the cell culture. In *S. meliloti*, glutamate accumulated at low salt concentrations, but at higher levels, glutamate and NAGGN were observed. All three osmolytes, glutamate, NAGGN, and trehalose accumulated only at extremely high NaCl concentrations (Smith et al. 1994). When glycine betaine was exogenously supplied in the growth medium, it accumulated as the major osmolyte during the lag and early exponential phases, whereas glutamate and NAGGN prevailed at the late exponential phase (Talibart et al. 1997). There was no improvement in the growth of chickpea rhizobia with the addition of glutamate to 340-400 $mmol^{-1}$ NaCl in the medium (Botsford 1984; Elsheikh and Wood 1989; Gonzalez-Gonzalez et al. 1990).

- Trehalose was the major osmolyte of the stationary phase (Smith et al. 1994, Talibart et al. 1997). A number of different environmental factors, such as the level of osmotic stress, growth phase of the culture, carbon source and osmolytes of the growth medium control the combination of naturally occurring endogenous osmolytes in rhizobial cells (Smith et al. 1990, 1994). Bacteria prefer uptake of compatible solutes over synthesis *de novo*. The transport systems for external osmolytes (osmoprotectants) are relatively unspecific, and bacteria accept components of plant and animal origin (Galinski 1995). The enhancement of growth of rhizobia from different species, resulting from added glycine betaine and other betaines under saline conditions, has been recorded (Sauvage et al. 1983, Bernard et al. 1986). *S. meliloti* and many other rhizobial species can use glycine betaine as a C and/or N source under low osmotic stress but as an osmoprotectant in high osmolarity. Contrary, Bostford (1984) observed that exogenous application of glycine betaine did not ameliorate inhibition of the growth of *S. meliloti* in NaCl. Only the slow-growing *B. japonicum*, which is considered as an osmosensitive species, is incapable to transfer glycine betaine and its precursor choline (Boncompagni et al. 1999). Under low osmotic stress glycine betaine is degraded via successive demethylation to glycine, while at elevated osmolarity the catabolism is blocked and glycine betaine is accumulated in cells (Sauvage et al. 1983; Bernard et al. 1986; Smith et al. 1988; Boncompagni et al. 1999). Furthermore, Exogenous application of calcium alleviated the inhibitory effect of NaCl on growth of rhizobia and induced NaCl tolerance (Chien et al. 1991; Howieson et al. 1992; Reeve et al. 1993). There is an increasing demand for identifying rhizobial species that could also work under stressed soil environment so that the productivity of the inoculated legumes does not suffer under derelict soils.

3.3.1.1. Rhizobum-legume signalling exchange

Nodulation (nod) factors are products of nod gene exerssion of rhizobia. Nod factors are signals known to be used in the communication with leguminous plants in the process of root nodule formation. Nod factors are major host-specificity determinants (factors that determine which rhizobium nodulates with a specific host plant) and trigger the nodulation program in a compatible host (Mergaert et al. 1997). The influence of salt (NaCl) stress on the expression of enzymes involved in Nod factor production, and on the huge variety of Nod factors produced by growing *R. tropici* have been investigated (Estevez et al. 2009). In surprising study by Guasch-Vidal et al. (2013) observed that, in the absence of flavonoid inducers, high concentrations of NaCl induced nod genes and the production of Nod factors. The higher transcriptional activity at the nod region in the presence of NaCl, as revealed by the increased the β-galactosidase activity of a *nodP:lacZ* fusion, that lead to a higher and detectable production of Nod factor. However, a high concentration of salinity had adverse effect on Nod factor activity and decreased root hair deformation in soybean (Duzan et al. 2005). Evidence existed that introduction of exogenous *nod* gene inducers increases nodulation and nitrogen fixation of some legume species (Abd-Alla et al. 2013a). Pre-treatment of *Bradyrhizobium japonicum* with genistein increased nodulation and nitrogen fixation of soybean and common bean (Zhang and Smith 1996; Abd-Alla 2001, 2011); and pre-induction of *Rhizobium leguminosarum* with hesperetin and naringenin was found to stimulate nodulation and plant dry matter accumulation of pea and lentil plants (Begum et al. 2001). Flavonoid inducers act in low concentrations and the pre-activation of rhizobia used as inoculants (biofertilizers) in undoubtedly economically justified. Pre-activation of strains might increase rhizobial competitiveness in the soil environment (Hungria and Philips 1993). Indeed, it has been shown that flavonoid pre-activated *B. japonicum* increased soybean nodule quantity and weight (about 30%), the seasonal level of N_2 fixation (35%) and yields (10-40%) when compared to conventional inoculants (Zhang and Smith 2002). Likewise, field pea and lentil plants displayed increased nodulation and biomass production when inoculated with *Rhizobium leguminosarum* preinduced with hesperetin (Begum et al. 2001). In common bean (*Phaseolus vulgaris* L.), a negative effect of NaCl on the expression of *nod* genes by *Rhizobium tropici* and *Rhizobium etli* and on nodulation factors' pattern was observed (Dardanelli et al. 2008). The preincubation of *B. japonicum* with the signal molecule genistein, under saline conditions, was described as a method to alleviate the stressful effects of salt on soybean–*B. japonicam* symbiosis (Miransari and Smith 2009). Efforts to overcome the inhibitory effect of salinity on nodulation of feungreeck by pre-inocubation of *Rhizobium tibeticum* with hesperetin and apaginein were successful (Abd-Alla et al. 2013a). It is supposed that flavonoids have selective value in plant-microbe interaction under salinity stress. Improving legume inoculation efficiency is extremely important to improve legume production under drastic conditions.

3.3.1.2. Rhizobium-legume nodulation and nitrogen fixation

In legumes salinity can result in a root system devoid of root hairs and a mucilaginous layer, and incapable of forming an infection thread (Singleton and Bohlool 1984; Zahran and Sprent 1986). Unsuccessful symbiosis may be attributed to failure in the infection process due to poor establishment of *Rhizobium* (Rai and Prasad 1983). The reduction of nodulation in soybean under saline conditions was attributed to shrinkage of the root hairs (Tu 1981). Yousef and Sprent (1983) showed that NaCl affected nodulation and they concluded that there may also be effects on infection. In salt-treated faba bean plants, total nodule weight per plant was reduced with increasing amounts of salt given. A reduction of 16.7, 41.2, 50.8 and 72.6% in nodule weight was caused by salinity levels of 5.8, 8.8, 11.6 and 14.6 dSm^{-1}, respectively. Total nodule weight per plant followed a similar pattern to nodule number per plant (Abd-Alla 1992). The detrimental effects of salinity on nodulation parameters were most pronounced for NOD1-3 and Williams 82, intermediate for DR-1, and less marked for PI416937. Self-grafted NOD1-3 plants showed 50 to 62% inhibition in nodulation responses (activity, number, mass) while grafting of PI416937 scions to NOD1-3 roots resulted in less than 7% inhibition by salt. Nodule number on PI416937 roots was greater when grafted to NOD1-3 scions (relative to self-grafted PI416937 plants), confirming a shoot role in autoregulation of nodule number (Abd-Alla et al. 1998). These results indicated that shoot factors are of primary importance in determining salt-tolerance of the PI416937 genotype and that hypernodulation expression in the mutant is negatively affected by salt treatment.

The salt injury on the symbiotic interaction not only inhibits the formation of the nodules, but also thereby leads to the inhibition in nitrogenase activity and reduction of the growth of the host plant. Other effects of salinity on the nodulation, includes formation of non-functional nodules with abnormal structure, and degradation of peribacteroid membrane (Bolanos et al. 2003). Tu (1981) observed that the interaction of *Bradyrhizobium japonicum* to the root hairs under increasing concentrations of NaCl in the range from 0% to 1.8% in the culture solution was affected. At a concentration of 1%, inhibition was evident in the bending of the hairs, a phenomenon markedly accentuated as rising concentrations. At 0.2% NaCl, growth was slow and at 1.2% nodulation did not occur. Results of some experiments on *Vicia faba* and *Vigna unguiculata* showed that the application of different concentrations of NaCl at the time of root hair formation significantly reduced root colonization, root hair curling, shrinking root hairs and hypodermic cells (Zahran and Sprent 1986; Georgiev and Atkins 1993). Several hypotheses have been suggested to explain the negative effects of salt on nitrogen fixation in plant legumes. It is likely that the depression in specific nitrogenase activity was due to salt reducing the protein, leghaemoglobin and carbohydrate contents of both the cytosol and the bacteroids. Salinity levels at 11.6 and 14.6 dSm^{-1} induced a 37 and 59% decline in carbohydrate and 41 and 53% in leghaemoglobin and 19 and 53% in protein content of nodules, while nodule C_2H_2 reduction (specific nitrogenase activity) showed a 55 and 61% decreases, respectively (Abd-Alla 1992). Abd-Alla and Omar (1997) reported that 0.5% NaCl significantly reduced nodulation of fenugreek and addition of wheat straw and cellulytic fungi minimize the adverse effect of salinity on nodulation.

Protein synthesis is readily inhibited by water stress because of the decreases in the level of polyribosomes (Bewley 1981). Thus, the decline of nodule soluble protein may result from a general reduction of protein synthesis and from an increased protease activity in the cytosol (Becana et al. 1986). The disappearance of nodule soluble protein induced by plant water stress can be compared to the situation naturally occurring during nodule senescence (Pladys and Rigaud 1985) or after feeding the plant with nitrate (Ohyama and Kumazawa 1981, Nishiwaki et al. 1997; Ohyama et al. 2011; Abdel-Wahab and Abd-Alla 1995a,b, 1996). In both cases, the presence of active proteases was responsible for the protein digestion, and it seems likely that these enzymes may also be induced under water restricted conditions (Guerin et al.1990; 1991). It is clear that most biochemical parameters of the bacterial and plant components of the nodules (cytosol) were affected by high levels of salinity of 100 and 125 mM NaCl (Abd-Alla 1992). Both the protein and carbohydrate contents of nodule cytosol were more sensitive to salinity than the corresponding bacteroids at the same levels of salinity. The negative effects of salinity on the oxygen diffusion barrier, which normally would reduce the O_2 flux into the nodule was also demonstrated (Soussi et al. 1998; Serraj and Drevon 1998). The decline in nitrogen fixation can be attributed to a reduced carbon supply to bacteroids, mainly in the form of malate limitation and likely a result of the salt induced inhibition of nodule carbon metabolism through the inhibition of sucrose synthase activity (Ben Salah et al. 2009). The accumulation of solutes such as proline, sucrose, and D-pinitol has been described in nodules of some legumes such as alfalfa and cowpea, and they exert an osmoregulatory function in situations of salinity (Irigoyen et al. 1992). Miransari and Smith (2007) indicate that under high levels of stress plant spent most of its energy to inhibit or alleviate the stress rather than developing a symbiosis with N-fixing bacteria. Miransari and Smith (2008) reported that genistein application into rhizobia inoculant improves plant growth through improved nodulation and nitrogen fixation in both normal and salt stress conditions. Inhbition of N_2-fixation of legumes under salt stress seems to some extent to be due to nodule senescence. More work is required to elucidate the mechanisms by which salinity interferes with normal nodule physiology and function to alleviate the inhibitory effect of salinity in crop legume production.

3.4. Soil pH

3.4.1. Survival of Rhizobia

Agricultural soils are either alkaline or acidic that effect on rhizobial growth, survival and subsequent formation of nitrogen-fixing symbiosis with a legume host. Global warming and agricultural practices cause increase in the amount of soil affected by acidity, and thus limit legume crop productivity. Worldwide, more than 1.5 Giga hectares of acid soils limit agriculture production (Graham and Vance 2000) and as much as 25% of the agriculture land is impacted by problems associated with soil acidity (Munns 1986). Each *Rhizobium* has its own optimum pH, under which it grows at its best. Although neutral conditions are generally optimum for bacteria, different species of *Rhizobium* display varying degrees of pH resistance as measured by their ability to grow (not just survive) (Glenn and Dilworth 1994). Rhizobia can be more sensitive to acidic condition than their legume host. Indeed, it is in many cases the inability of the rhizobia to persist and survive under acidic conditions that reduces the effectiveness of the symbiosis. Therefore, the selection of rhizobial strains tolerant to acidic conditions may improve the acid tolerance of the legume through an efficient symbiotic nitrogen fixation under acidity conditions. However, the relationship between low pH of soil and rhizobia competitiveness, and ability to presist under acidic stress is not always straight forward. Chen et al. (1993) reported that mutants of *R. leguminosarum* have been able to grow at a pH as low as 4.5. Moreover, Foster (2000) recorded that *S. meliloti* was viable only below to pH 5.5. Some rhizobia have wide range of pH such as *S. fredii* can grow well between pH 4 – 9.5 but *B. japonicum* cannot grow at the extremes of that range (Fujihara and Yoneyama 1993). These values are the extremes and hindered the growth and survival of bacteria between 1 and 2 pH units (Richardson and Simpson 1989). Although some acid-tolerant rhizobia strains have been selected (Wood et al. 1988; Vinuesa et al. 2005; Laranjo and Oliveira 2011), the mechanisms that employ to survive and grow under acidic conditions have not been fully elucidated and therefore the molecular basis for differences in pH toleranceamong strains of rhizobia is still not clear. Several genes, such as actA, actP, exoR, lpiA, actR, actS, and phrR, were shown to be essential for rhizobia growth at low pH (de Lucena et al. 2010). Kurchak et al. (2001) have been identified 20 genes in *R. leguminosarum* that are responsible for acid stress namely as *act* genes (acid tolerance).

Root nodule bacteria employ mechanisms for maintenance of intracellular pH (pHi) are crucial. Like other Gram-negative bacteria, root nodule bacteria show an adaptive acid tolerance response, with growth at moderately acidic conditions protecting against an extreme acid shock. Variations in acid tolerance within species of root nodule bacteria imply a genetic basis to low pH tolerance and studies of acid-sensitive mutants suggest that as many as 20 genes could be involved (Glenn and Dilworth 1994). Sensing mechanisms are composed of two components: a sensor and a regulator, and one has been found in *S. meliloti*; the genes *actR* and *actS* encode for the regulator and sensor respectively (Tiwari et al. 1996b). ActS is the membrane bound product of *actS* that, on detection of external acidity, activates ActR (product of *actR*) via phosphorylation. ActR then goes on to activate the transcription of other acid response genes within the bacterium (Tiwari et al. 1996b). The membrane bound product of *actA* is basic and responsible for maintaining internal pH at around 7, when the external pH drops below 6.5 (Tiwari et al. 1996a). They demonstrated that mutants defective in this gene are unable to maintain intracellular pH and cannot grow at a pH lower then 6. Tiwari et al. (1996a) indicated that calcium involves in acid tolerance mechanism *S. meliloti*. Riccillo et al. (2000) reported that glutathione play a key role in acid tolerance of *Rhizobium tropici*. Glutathione also provides protection against chlorine compounds in *E. coli* and against oxidants in *E. coli* and *Rhizobium leguminosarum* bv. *phaseoli* (Crockford et al. 1995). The activation of glutathione synthesis might be essential for tolerance to acid stress (Muglia et al. 2007). TypA is act as a regulator by controlling the phosphorylation of proteins and required for growth at acidic condition (Kiss et al. 2004). Reeve et al. (1998) show that, in addition to the genes like actA, acts and actR that are absolutely essential for growth of *S. meliloti* at low pH, there is phrR gene which, while not essential for growth, appears to be induced by exposure to low pH. Cunningham and Munns (1984) found that *Rhizobium* produce greater amounts of exopolysaccharides are able to survive in acidic conditions more successfully than *Rhizobium* that can only produce smaller amounts.

Many agricultural fields are alkaline with an average pH above 7.0 to 8.5. A major problem in alkaline soils is reduced nutrient availability. Alkalinity stress can also retard *Rhizobium* from growing and subsequent establishment of a viable nitrogen-fixing symbiosis with a legume host. Therefore, it makes good sense agriculturally to select rhizobial isolates that are tolerant of alkaline conditions as well as capable of nodulating legumes (Abd-Alla et al. 2013b). Although *R. leguminosarum* bv. *trifolii* has been reported to colonize soil at a higher rate and produce nodulates at a higher frequency in alkaline conditions; it is also known to grow unaffected at pH 11.5 (Zahran 1999). Homospermidine, a polyamine present in high concentrations in root nodule bacteria, is also known to accumulate in *B. japonicum* in alkaline conditions, although its function is unknown (Fujihara and Yoneyama 1993). *Rhizobium etli* strains EBRI 2 and EBRI 26 are more competitive than strain CIAT 899G in soils with high salt or alkaline conditions (Shamseldin and Werner 2004). These isolates can be use as inoculants for alkaline agriculture fields.

3.4.1.1. Rhizobium-legume molecular signalling exchange

Soil pH adversely affects several stages during the development of symbiosis, including the exchange of molecular signals between the legume and the microsymbiont (Hungria and Vargas 2000). Hungria and Stacey (1997) observed that release of isoflavonoids *nod*-gene inducers by soybean and common bean roots was less at pH 4.5 than at pH 5.8 with some nodulation genes, including *nodA,* switched off as the pH decrease (Richardson et al. 1988 a, b). At low pH, induction of nod gene expression in *R. leguminosarum* biovar *trifolii* is markedly reduced in the presence of flavone-inducer. Furthermore, inducibility of nod gene expression in *R. leguminosarum* bv. *trifolii* is also affected by a net reduction in the concentration of nod gene-inducing factors present in the root exudates of clover seedlings grown in acidic conditions (Richardson et al. 1989). Depression in soybean and common bean nodulation under acidic conditions may be partially alleviated by supplementation with flavonoid nod-gene inducers (Hungria and Stacey 1997), with studies to examine varietal differences in response or to maximize the expression of rhizobial nod-genes under acid conditions still needed. Also, acidic soil can affect the production and excretion of nodulation factors of *R. leguminosarum* bv. *trifolii* (McKay and Djordjevic 1993). Soil pH has been shown to effect on the profile of Nod factors secreted by *R. tropici* CIAT899, which is tolerant of acidic conditions (Morón et al. 2005). At least seven different classes of Nod factor structures were identified. More than 50 different Nod factors were detected at pH 4.5, and greater induction of the nod genes also occurred at this pH. This diversity of Nod factor structure may facilitate nodulation of bean at an acidic pH. Angelini et al. (2003) observed that Initial root colonization by rhizobia was advesly affected when both acid-tolerant and acid sensitive (pH 5.0) peanut rhizobia were grown at low pH. This effect could be attributed to deformation produced by acidic pH on the microsymbiont rather than on host legume. At acid pH, very low nodC gene expression was observed in acid-sensitive isolates, while a change in the flavonoids inducer effectiveness was determined in acid-tolerant isolates. There is no information available on the of effect alkalinity on the exchange of molecular signals between the legume and the microsymbiont and additional studies are needed.

3.4.2. Rhizobium-legume nodulation and nitrogen fixation

The availability of some essential nutrients such as calcium, magnesium, phosphorus, and molybdenum is low in acid soils, whereas high levels of aluminum and manganese can become toxic to plants and rhizobia (Coventry and Evans 1989). Soil pH can be distorted and this depending on the type of plant nitrogen nutrition and host legume. Thus, the N_2-fixing plants released protons, which lead to lower the soil pH. The buffering capacity of certain agriculture field can avoid major changes in pH, but in soils with low cation exchange capacity the problem may be significant. It is clear that N_2-fixing plants are more sensitive to acidity than plants of the same species that feed on mineral nitrogen (Andrews 1976).

Growth of rhizobia in acidified culture media has proved useful for selecting strains with an ability to colonize the rhizosphere and to nodulate their host plant in acid soils (Cooper 1988). A major problem is to distinguish between the effects of low pH and toxicity of some minerals, especially aluminum. In acidic soils with pH of >5.0, where heavy metal activity is relevant, the presence of available aluminum inhibits nodulation.

The information available on alkalinity is scarce compared to that on acidity. Singh et al. (1973) observed that the number of nodules on alfalfa root can be significantly decreased in culture solutions containing 0.1% Na_2CO_3 and $NaHCO_3$. Apparently, soil alkalinity significantly inhibited nodulation and nitrogen fixation of faba bean inoculated with *R. leguminosarum* bv. *viciae* STDF-Egypt 19 (HM587713). However, dual inoculation of *R. leguminosarum* bv. *viciae* STDF-Egypt 19 (HM587713) and AMF improve the nodulation status and nitrogen fixation of faba bean grown under alkalinity stress (Abd-Alla et al. 2013b). AMF colonization generally enhances rhizobial nodulation and N_2 fixation (Barea and Azcon-Aguilar 1983; Hayman 1986; Fitter and Garbaye 1994; Chalk et al. 2006). Many authors have now reported the enhancement of rhizobial nodulation in roots colonized with AM fungi (Meghvansi and Mahna 2009; Badri et al. 2010; Wang et al. 2011; Sakamoto et al. 2013).

3.5. Pesticides

3.5.1. Suvrival of Rhizobia

Over the past 60 years pesticides have been used increasingly in the environment. Pesticides are massive applied to agriculture land to produce a larger yield and high quality of crop as well as reduce the input of labor and energy into crop production (Ayansina 2009). Worldwide, about 3 billion kg of pesticides are applied each year with a purchase price of nearly $40 billion per year (Pan-UK 2003). These compounds pose great threats to the varied agroecosystems. Although the application of these chemicals has been banned or restricted in many countries especially the developed ones, some developing countries are still using these compounds because of their low cost and versatility in industry, agriculture and public health (Tanabe et al. 1994, Sarkar et al. 1997). These pesticides can enter the soil environment either by direct application or via plant root exudates. This input of pesticides can affect many soil organisms in different manners. Information from previous studies indicated that some soil bacteria are able to tolerate or degrade some pesticides as sole carbon or nitrogen source (Dick and Quinn 1995; Liu et al. 1991; Sarnaik et al. 2006) bacteriostatic and lethal effects can also occur (Fox et al. 2007). Some pesticides were toxic and inhibited rhizobial growth and survival (Graham et al. 1980; Mallik and Tesfai 1983; Singh and Wright 2002; dos Santos et al. 2005).

The effect of pesticides on rhizobia will depend upon the species and strain (Sawicka and Selwet 1998; Singh and Wright 2002; Zablotowicz and Reddy 2004), type of pesticide applied, and the dose applied (Kaur et al. 2007; Zawoznik and Tomaro 2005).

A *Rhizobium* sp. strain, named PATR, was isolated from an agricultural soil and found to actively degrade the herbicide atrazine (Bouquard et al. 1997). They showed that the hydroxylation of atrazine due to presence of constitutive enzyme consist of four 50-kDa subunits.

Drouin et al. (2010) showed that fungicides had the highest deleterious effects on the rhizobia, followed by herbicides and then insecticides. Tolerance or resistance of rhizobia against pesticides is a complex process and is controlled by physiological/genetic level of microorganism. These microorganisms developed such tolerance or resistance have the ability to degraded or change the configuration of pesticide (Kumar et al. 1996; Andres et al. 1998a, Ortiz-Hernández and ánchez-Salinas 2010). The resistance or tolerance against pesticides might be attributes to physiological activities that induce the microbial metabolism for the formation of a new metabolic pathway to bypass a biochemical reaction inhibited by a specific pesticide (Bellinaso et al. 2003). Dominant resistance of rhizobia to specific pesticide depends upon genetic modifications, inherited by the subsequent generation of microbes (Johnsen et al. 2001; Herman et al. 2005). It well established that pesticides not only have toxic impact on rhizobial growth and survival but also have deleterious effect on their ability to produce plant growth promoting substances. Further studies are required to addressing the toxicological effects of pesticides on survival of rhizobial and their plant growth promoting activates at molecular level to fortify the effective implementation of this approach to protect the soil ecosystem from pesticide hazard.

3.5.2. Rhizobium- Legume molecular signaling exchange

Published studies regarding pesticides' effects on *Rhizobium*-legume molecular signaling exchange are usually very scattered and done with few pesticides on a limited number of same genus (Fox et al. 2001, 2004). They reported using *in vitro* assays that 30 different pesticides and environmental contaminants specifically disrupted crucial symbiotic signaling between flavonoid phytochemicals and *S. meliloti* NodD receptors. Fox et al. (2007) reported *in vivo* evidence that a subset of organochlorine pesticides and pollutants inhibit symbiotic signaling between alfalfa and *S. meliloti*, resulting in delayed symbiotic recruitment, reduced symbiotic nitrogen fixation, and a decline in alfalfa plant yield. Pesticides may also disrupt symbiosis by altering the array of flavonoid phytochemicals a plant produces or by reducing the overall flavonoid secretion pattern, thereby disrupting plant–rhizobial signaling. Pesticides may affect the plants in a manner that the bacteria-induced root hair deforming factors, similar to Nod Rm-1 and related compounds (Truchet et al. 1991), are incapable to influence morphogenic activity of the plants. Andres et al. (1998b) reported that alfalfa and soybean seed and root exudates treated with fungicide (thiram) inhibit the expression of rhizobia nodulation genes. Temporal and chemical specificity of symbiotic signaling is crucial for coordinating the actions of host legume and microsymbiont necessary for symbiotic nitrogen fixation. In the dynamic soil environment, rhizobia are exposed to a mixture of agonistic and antagonistic phytochemicals, and the degree to which its NodD receptors are able to interpret these signals to locate its symbiotic partner. This will most likely determine the efficiency of symbiotic nitrogen fixation in that particular soil environment.

3.5.3. Rhizobium-Legume and nodulation and nitrogen fixation

Pesticides are essential for controlling plant pests, and accordingly, improve the productivity of major crops including legumes. The presence of residual pesticides in the rhizosphere may interfere with the infection process by inhibiting bacteria-induced root hair deformation and nodule formation. The infection process may be changed either by pesticide effects on the virulence of the attacking bacteria or by effect on root fibers of plant in which the infection occurs (Kumar 1981). Nodulation inhibition with pesticides may also be due to alteration of root hair morphology or to alteration in the quality and quantity of root exudation (Ratnayake et al. 1978) including isoflavonoid compounds and lectin that play an important role in the attraction and attachment of rhizobia to root hairs (Hansen 1994). Pesticides applied to leguminous crops constitute a potential hazard to growth, nodulation and nitrogen accumulation (Abd-Alla and Omar 1993). Abd-Alla et al. (2000 b) indicated that root nodule formation was inhibited in cowpea, common bean and lupin, the effect being most evident in the case of cowpea. Reduction in the number of nodules on soybean was recorded following trifluralin application at rates of 0.74 and 1.1 kg^{-1} (Kust and Struckmeyer 1971). Nodule dry mass of soybean was also decreased in five soil types treated with pesticides (Dunigan et al. 1972). Nodulation of broad bean was inhibited by the herbicides trifluralin and metribuzin (Bertholet and Clark 1985). Similar results were also obtained by Eberbach and Douglas (1991) using the herbicides paraquat and glyphosate. Abd-Alla and Omar (1993) recorded a significant reduction in nodule formation by faba bean plants grown in herbicide-treated pot soil. Curley and Burton (1975) attributed the inhibition in nodulation, concomitant with pesticide application, to the inability of rhizobia/bradyrhizobia to multiply in the presence of pesticides. Pesticides may inhibit nodulation through their effect on cellulolytic and pectolytic enzyme production by rhizobia, as occurs in other micro-organisms (Mahmoud and Omar 1995). Production of these enzymes by rhizobia is essential for root hair penetration (Hansen 1994). Reports on the effect of agrochemicals on symbiotic attributes of legumes are, however, contradictory. For example, Miettinen and Echegoyen (1996) observed a substantial decline in nodulation in legume crops grown in soil amended with imidacloprid. Pesticides may influence on the nodule development and effectivenss of nitrogen fixation through the effect inside the host plant or nodule structure (Rup 1988). Aggarwal et al. (1986) evaluated the effect of carbamate on nodulation in *Pisum sativum* and *Vigna sinensis*. They observed that the low doses of the insecticides had little effect on nodulation whereas the higher concentrations adversely affected it. In a similar study, Alonge (2000) evaluated the phytotoxicity of imazaquin on the growth of soybean plants and found that chlorophyll content in the leaves, root nodules, shoot growth, whole plant dry weight, and grain yield were reduced. Pesticides can reduce the symbiotic efficiency of nitrogen fixing bacteria and the host leguminous plants. Pesticides, which are either applied to crops or found as contaminants in the soil, significantly disrupt symbiotic nitrogen fixation and subsequently lower plant yields. The decrease in the nitrogen content of plant tissue is mainly due to the side effects of pesticides on rhizobia/bradyrhizobia.The decrease in nodule number on pesticide-treated plants may explain the decrease in tissue N (Abd-Alla et al. 2000 b). Similarly, Anderson et al. (2004) claimed that insecticides (chlorsulfuron) negatively affect on nodule formation and nitrogen fixation of chickpea. A comparable observation on the effect of insecticides (fipronil) on legumes has been reported. For example, the effect of insecticide fipronil on growth and nodulation of chickpea, pea, lentil and green gram were determined by Ahemad and Khan (2011a, b).They showed that Fipronil displayed varying degrees of toxicity to the tested legumes. The highest toxicity of fipronil was observed on shoot dry biomass, leghaemoglobin, and chlorophyll content and the seed protein in chickpea, nodule numbers and nodule biomass in pea.

3.6. Heavy metals

3.6.1. Survival of rhizobia

Heavy metals are most important inorganic pollutants such as Cu, Ni, Cd, Zn, Cr, Pb. Heavy metals enter soil from industrial operation; animal manures and sewage sludge application after these elements enter to the soil they remain for several thousands of years. Accumulations of these metals in various ecological systems cause a massive threat to the varied agroecosystems (Abd-Alla et al. 1999; Ceribasi and Yetis 2001; Cheung and Gu 2007). Recently there has been increasing concern with heavy metal contamination, not only due to their toxic impact on living organisms, but also due to their irreversibly immobilized in soil components (McGrath and Lane 1989). Some metals such as Zn, Cu, Ni and Cr are essential micronutrients for living organisms at very low level. Optimum level of physiological function can be achieved in suitable environment and adequate nutrient supplies. Some elements, such as heavy metals, though essential for organisms, are harmful if present in excess. The impact of certain metals such as Cd, Hg and Pb on biological and physiological functions of living organisms is still obscure. However, all these metals could be toxic at relative low concentrations (Gadd 1992). Soil microorganisms are very sensitive when subjected to moderate heavy metal concentrations (Giller et al. 1998). High levels of heavy metals in soil cause dramatic changes in microbial composition and their activities (Baath et al. 1998; Lakzian et al. 2002; Khan and Scullion 2002, Paudyal et al. 2007; Wani et al. 2008a; Khan et al. 2009; Krujatz et al. 2012) resulting in microbial populations with higher tolerance to metals, but with lower diversity, when compared to unpolluted soils (Baath 1992; Baath et al. 1998). These changes in ecosystem led to losses in soil fertility. Wood and Cooper (1988) reported inhibition of multiplication of rhizobial strain caused at 50 μM Al concentration. Al toxicity is great problem under acidic medium as solubility of free Al ion (Al^{3+}) increases rapidly under acidic condition (McLean 1976). Fast growing acid producing bacteria, it promotes Al to show its full toxicity. Since the distribution of various ionic species of Al is pH dependent (Martin 1991), slight change in pH may significantly affect the relative concentration of the various charged Al species and their ligands and hence the toxicity of aluminium. Further, slight change in pH alone can significantly affect the growth of root nodule bacteria (Thornton and Davey 1983; Richardson and Simpson 1989). In combination with acidity, Al is detrimental to root nodule bacteria in those cases where growth occurs to longer log periods, decreases growth rates and lowers final cell densities (Whelan and Alexander 1986). The toxic effect of aluminium is due to its effect on bacterial DNA (Johnson and Wood 1990; Martin 1988). The deleterious effects of metals on rhizobial composition within agroecosystem and different legume genotypes, however, have been contradictory (Wani et al. 2007a, b, 2008a, b). To validate this concept of conflicting effects of metals on rhizobia, Paudyal et al. (2007) conducted an experiment which revealed that rhizobia grew poorly in culture medium supplemented with even lower concentration of aluminium, while rhizobial growth was completely inhibited at 50 mM Al concentration (Wood and Cooper 1988; Chaudri et al. 1993; Broos et al. 2004). Hirsch et al. (1993), in a study, demonstrated that the persistence of *R. leguminosarum bv. trifolii* was drastically altered by long-term exposure to heavy metals, and this *Rhizobium* lost their infectability to form functional nodules. In a similar study, Chaudri et al. (2000) reported that the irrigation for long term with sewage sludge containing Zn or Cu or mixture of Zn and Cu significantly decreased growth and survival of *R. leguminosarum* bv. *viciae* and *R. leguminosarum* bv. *trifolii*, in soils. On the contrary, Angle and Chaney (1991) indicated that even in highly contaminated soils, metal activity was not high enough to exert an antagonistic influence on the soil rhizobial population or the symbiotic association between alfalfa and *R. meliloti*. Arora et al. (2010) investigated the impact of aluminium and copper, iron and molybdenum on growth and enzyme activity of fast and slow-growing rhizobial species. They indicated that *Sinorhizobium meliloti* RMP5 had a greatest tolerance to copper, iron and molybdenum compared to *Bradyrhizobium* BMP1. However, these strains were very sensitive to Al than other metals. In addition, Al was significantly depressed the various enzymatic activities of *Sinorhizobium meliloti* strains RMP5 and BMP1.

The lack of adverse effects of metals on soil populations of *R. meliloti* is contrary to several previous reports with other species of *Rhizobium*. Many studies indicated that *R. leguminosarum* biovar *trifoli*, or some portion of the symbiotic process, was adversely affected by metal contamination of soil (McGrath et al. 1988; Giller et al. 1989; Mcilveen and Cole 1974; Rother et al. 1983). Tong and Sadowsky (1994) found that fast-growing rhizobia were more susceptible to Zn and Co (20-40 μg ml^{-1}) than slow-growing (80-480 μg ml^{-1}), and less susceptible to Mo. The discrepancy is related to the differential metal tolerance of the rhizobial species (Abd-Alla et al. 2012). A higher protein expression was usually related to tolerance mechanisms in *Rhizobium* (Saxena et al. 1996, Pereira et al. 2006).

3.6.2. Rhizobium-Legume molecular signalling exchange, noduation and nitrogen fixation

During the last decades, a plethora of reports have been published concerning the impact of heavy metals on survival of rhizobium and nitrogen fixation. However, there is minor information about the effect of heavy metals on molecular signals between rhizobia and their host legumes. Recently, Pawlak-Sprada et al. (2011a, b) have found that treatment of soybean (*Glycine max*) and yellow lupine (*Lupinus luteus*) with cadmium (Cd^{2+}) or lead (Pb^{2+}) trigged an increase of phenylalanine ammonia-lyase mRNA level and phenylalanine ammonia-lyase activity. Phenylalanine ammonia-lyase is a key enzyme of the phenylpropanoid pathway. Several products of this pathway such as flavones and isoflavones are strongest inducer of rhizobial nodulation genes (Firmin et al. 1986; Redmond et al. 1986). The total amount of isoflavonoids in soybean and yellow lupine were significantly increased by about 15 % in cadmium- and 46 % in lead-treated plants. High concentration of heavy metals may reduce the isoflavonoid exudate and therefore inhibits the induction of *nod* genes. Some reports are available dealing with the toxicity of different heavy metals on nodulation and nitrogen fixation (Bhandal et al. 1990; Hung et al. 1974; Paivoke 1993a, b; Yakolevya 1984). Conversely, field studies by Heckman et al. (1984) failed to detect adverse changes in either plant growth or N_2-fixation in the sludge-amended soils. Borges and Wollum (1981) added

Cd salts to the soil and reported that N_2-fixation of soybean was not affected. The deleterious effects of high levels heavy metals on nodulation and N_2 fixation of *Rhizobium*–fab bean symbiosis are probably due to their inhibitory effects on the growth and activity of both symbionts (El-enany and Abd-Alla 1995). Wetzel and Werner (1995) reported that 75 mg kg-soil as $CdCl_2$ significantly decreased nodulation of alfalfa roots and shoots dramatically increased. Younis (2007) reported that when 50–200 mg kg^{-1} soil of Co, Cu, Cd and Zn was added deliberately to soils used for *Lablab purpureus* cultivation, these metals significantly decreased growth, nodulation and nitrogenase activity of plants in both pot and field experiments. Sepehri et al. (2006) observed that 2 mg Cd/kg soil had a variable effect on symbiotic properties of *S. meliloti* strains and consequently on *S. meliloti*–alfalfa symbiosis. A decreasing effect of cadmium concentration on root nodules and N concentration in plants inoculated with sensitive rhizobial strains in comparison with plants bacterized with tolerant strains was 68% and 41%, respectively. The effects of metals on rhizobial composition within soil or nodule environment and different legume genotypes, however, have been contradictory (Wani et al. 2007a, b, 2008a, b). Mandal et al. (2011) reported that nitrogenase activity was reduced to almost 2 fold in plants with Arsenate-treated soil but not abolished in root nodule of blackgram. Their microscopic observations of the cross section of root nodules revealed that the effective areas of N_2-fixing zone were reduced in plants grown in arsenic contaminated field. Studies examining the effect of *Mesorhizobium metallidurans* and leguminous *Anthyllis vulneraria* symbiosis in soil that was contaminated with Zn, Cd and Pb showed a significant increase in the total N levels in both plants and soil (Frerot et al. 2006; Mahieu et al. 2011). *Cupriavidus metallidurans* is another bacterial species isolated from mining areas that tolerates high concentrations of heavy metals, and this organism has been deemed a model of tolerance to metals (Mergeay et al. 2003; Vaneechoutte et al. 2004) and efficiently nodulates *Mimosa pudica* (Chen et al. 2008). Recently, a symbiotic capacity for N_2 fixation and a high tolerance to heavy metals was demonstrated for some *Cupriavidus necator* strains (Ferreira et al. 2012; Silva et al. 2012; Avelar Ferreira et al. 2013).

Metalloprotein (Cd-binding protein) was isolated from nodules of faba bean, grown for 58 days at 100 ppm Cd and 0 Cd by fractionation on a sephadex G-100. The major cadmium containing protein eluted with a molecular weight higher than 67 kD, close to the void volume of the column (200 kD). This protein constituted 38% of the total Cd-binding protein (metallothioneins). A second peak was noticed in protein fractions of a molecular weight of about 67 KD. This fraction contained about 53% of Cd-binding protein. A very small amount of cadmium (ca. 3.4%) was eluted with proteins of 8-17 kD. Apparently, the cadmium-binding protein of 67 kD it appears that Cd-binding proteins play an important role in the detoxification of excess Cd and determining resistance of faba bean-Rhizobium symbiosis to cadmium toxicity was synthesised only in the presence of appreciable cadmium concentrations (El-enany and Abd-Alla 1995). The assessment of variation in protein profile is considered a good indicator to estimate the level of stress imposed on *Rhizobium*-legume symbiosis exposed to heavy metal contamination.

4. Conclusion

Impact of *harsh* environmental *conditions play an essential role* in the control of legume-rhizobia interactions. They can arrest the growth, multiplication and survival of rhizobia in soil rhizosphere. The harsh environmental conditions may also have depressive effect on the steps involved in legume-*Rhizobium* symbiosis such as molecular signaling, infection process, nodule development and function, resulting in low nitrogen fixation and crop yield. Selection of hosts and their nitrogen-fixing endosymbionts that are tolerant to a broad range of environmental stresses is important for agriculture system. Understanding the key molecular factors and steps in rhizobia-legume interaction is of crucial importance for the development of *Rhizobium* strains and legume cultivars with high N_2-fixation potential. Prevelence and aboundance of rhizobia species vary in their tolerance to major environment factors; consequently, the selection of resistant strains is an important option. Better N_2 fixation can be achieved by selecting tolerance or resistance rhizobia from soil subjected to environmental stress. The selection and characterization of harsh conditions-tolerant strains with efficient symbiotic performance may be a strategy to improve *Rhizobium*-legume symbiosis and crop yield in adverse environments. Environmental stress severely affected on various metabolic activities of legumes including, nod gene expression, photosynthesis, synthesis of proteins, enzymes and carbohydrates. Therefore, understanding the environmental stress–rhizobia–legume interactions is urgently required for growing legumes under harsh environmental stress. Research into these areas is currently underway in several research groups throughout the world and it is anticipated that this research will provide beneficial outcomes resulting in improved sustainability and productivity in agricultural systems.

Author details

Mohamed Hemida Abd-Alla[1*], Ahmed A. Issa[1] and Takuji Ohyama[2]

*Address all correspondence to: mhabdalla2002@yahoo.com

1 Botany and Microbiology Department, Faculty of Science, Assiut University, Assiut, Egypt

2 Biological chemistry Department, Faculty of Agriculture, Niigata University, Niigata, Japan

References

[1] Abd-Alla, M. H. 1992. Nodulation and nitrogen fixation in faba bean (*Vicia faba* L.) plants under salt stress. Symbiosis 12: 311-319.

[2] Abd-Alla, M. H. 2001. Regulation of nodule formation in soybean-*Bradyrhizobium* symbiosis is controlled by shoot or/and root signals. Plant Growth Regulation 34 (2): 241-250.

[3] Abd-Alla, M. H. 2011. Nodulation and nitrogen fixation in interspecies grafts of soybean and common bean is controlled by isoflavonoid signal molecules translocated from shoot. Plant, Soil and Environment 57 (10): 453-458.

[4] Abd-Alla, M. H. & Abdel Wahab, A. M. 1995 a. Survival of *Rhizobium leguminosarum* biovar *viceae* subjected to heat, drought and salinity in soil. Biologia Plantarum 37 (1): 131-137.

[5] Abd-Alla M.H. & Abdel Wahab A.M. 1995 b. Response of nitrogen fixation, nodule activities, and growth to potassium supply in water-stressed broad bean. Journal of Plant Nutrition 18: 1391-1402.

[6] Abd-Alla, M. H. & Omar S. A. 1993. Herbicides effects on nodulation, growth and nitrogen yield of faba bean induced by indigenous *Rhizobium Leguminosarum*. Zentralblet fur Mikrobiologie 148: 593-597.

[7] Abd-Alla M.H. and Omar S.A.1997. Wheat straw and cellulolytic fungi application increases nodulation, nodule efficiency and growth of fenugreek (*Trigonella foenum-graceum* L.) grown in saline soil. Biology and fertility of Soils 26: 58-65.

[8] Abd-Alla, M. H., Voung, T. D. & Harper J. E. 1998. Genotypic differences in dinitrogen fixation response to NaCl stress in intact and grafted soybean. Crop Science 38: 72-77.

[9] Abd-Alla, M. H., Yan, F. & Schubert, S. 1999. Effects of sewage sludge application on nodulations, nitrogen fixation and plant growth of faba bean, soybean and lupin. Journal of Applied Botany 73: 69-75.

[10] Abd-Alla, M. H., Omar, S. A. & Karanxha, S. 2000 b . The impact of pesticide on arbuscular mycorrhizal and nitrogen fixing symbioses in legumes. Applied Soil Ecology 14: 191-200.

[11] Abd-Alla, M. H., Kyro, H. V., Yan, F., Shubert, S. & Peiter, E. 2000 a. Functional structure of faba bean nodules: Implications for metabolite transport. Journal of Plant Physiology 157: 335-343.

[12] Abd-Alla, M. H., Morsy, F. M., El-Enany, A. E. & Ohyama, T. 2012. Isolation and characterization of a heavy metal resistant isolate of *Rhizobium leguminosarum* bv. *viciae* potentially applicable for biosorption of Cd (II) and Co. International biodegradation and Biodeterioration 67: 48-55.

[13] Abd-Alla, M. H., El-enany, A. E., Bagy, M. K. & Bashandy, S. R. 2013a. Alleviating the inhibitory effect of salinity stress on nod gene expression in *Rhizobium tibeticum*-fenugreek (*Trigonella foenum graecum*) symbiosis by isoflavonoids treatment. Journal of Plant Interaction (http://dx.doi.org/10.1080/17429145.2013.824622).

[14] Abd-Alla, M. H., El-Enany, A. E., Nafady, N. A., Khalaf, D. M. & Morsy F. M., 2013b. Synergistic interaction of *Rhizobium leguminosarum* bv. *viceae* and arbuscular mycorrhizal fungi as a plant growth promoting biofertilizers for faba bean (*Vicia faba* L.) in alkaline soil. Microbiological Research (http://www.sciencedirect.com/science/article/pii/S0944501313001171#).

[15] Abdel Wahab, A. M. & Abd-Alla, M. H. 1995a. Effect of form and level of combined nitrogen on nitrogenase and nitrate reductase activities in *Vicia faba*. Biologia Plantarum 37: 57-64.

[16] Abdel Wahab, A. M. &. Abd-Alla, M. H. 1995b. Nodulation and nitrogenase activity of *Vicia faba* and *Glycine max* in relation to rhizobia strain, form and level of combined nitrogen. Phyton 35: 177-187.

[17] Abdel Wahab, A. M. & Abd-Alla, M. H. 1996. Effect of different rates of N-fertilizers on nodulation, nodule activities and growth of two field grown cvs. of soybean. Fertilizer Research 34: 37-41.

[18] AbdelGadir, A. H. & Alexander, M. 1997. Procedures to enhance heat resistance of *Rhizobium*. Plant and Soil 188: 93-100.

[19] Aggarwal, T. C., Narula, N. & Gupta, K. G. 1986. Effect of some carbamate pesticides on nodulation, plant yield and nitrogen fixation by *Pisum sativum* and *Vigna sinensis* in the presence of their respective rhizobia. Plant and Soil 94: 125-132.

[20] Ahemad, M. & Khan, M. S. 2011a. Pesticide interactions with soil microflora: Importance in bioremediation. In: Ahmad, I., Ahmad, F. and Pichtel, J. (ed.), Microbes and Microbial Technology: Agricultural and Environmental Applications. Springer, New York, p. 393-413.

[21] Ahemad, M. & Khan, M. S. 2011b. Comparative study of the growth parameters of legumes grown in fipronil-stressed soils. EurAsian Journal of BioSciences 5: 29-36.

[22] Alexandre, A. & Oliveira, S. 2011. Most heat-tolerant rhizobia show high induction of major chaperone genes upon stress. FEMS Microbiology Ecology 75: 28-36.

[23] Alexandre, A. & Oliveira, S. 2012. Response to temperature stress in rhizobia. Critical Reviews in Microbiology DOI: 10.3109/1040841X.2012.702097.

[24] Alonge, S. O. 2000. Effect of imazaquin applications on the growth, leaf chlorophyll and yield of soybean in the Guinea Savanna of Nigeria. Journal of Environmental Science and Health Part B 35: 321-336.

[25] An, D. & Parsek, M. R. 2007. The promise and peril of transcriptional profiling in biofilm communities. Current Opinions in Microbiology 10: 292-296.

[26] Anderson, A., Baldock, J. A., Rogers, S. L., Bellotti, W. & Gill, G. 2004. Influence of chlorsulfuron on rhizobial growth, nodule formation, and nitrogen fixation with chickpea. Australian Journal of Agricultural Research 55: 1059-1070.

[27] Andres, J. A., Correa, N. S. & Rosas, S. B. 1998a. Survival and symbiotic properties of *Bradyrhizobium japonicum* in the presence of thiram: Isolation of fungicide resistant strains. Biology and Fertility of Soils 26: 141-145.

[28] Andres, J. A., Correa, N. S. & Rosas, S. B. 1998b. Alfalfa and soybean seed and root exudates treated with thiram inhibit the expression of rhizobia nodulation genes. Phyton (Vincente Lopez) 62: 47-53.

[29] Andres, J. A., Rovera, M., Guiñazú, L. B., Pastor, N. A. & Rosas, S. B. 2012. Interactions between legumes and rhizobia under stress conditions. In: Dinesh K. M. (ed.), Bacteria in Agrobiology: Stress Management. Springer-Verlag Berlin Heidelberg, Germany, p. 77-94.

[30] Andrews, C. S. 1976. Effect of calcium, pH and nitrogen on the growth and chemical composition of some tropical and temperate pasture legumes. I. Nodulation and growth. Australian Journal of Agricultural Research 27: 611-623.

[31] Angelini, J., Castro, S. & Fabra, A. 2003. Alterations in root colonization and nodC gene induction in the peanut-rhizobia interaction under acidic conditions. Plant Physiology and Biochemistry 41: 289-294.

[32] Angle, J. S. & Chaney, R. L. 1991. Heavy metal effects on soil populations and heavy metal tolerance of *Rhizobium meliloti*, nodulation, and growth of alfalfa. Water, Air and Soil Pollution 57-58: 597-604.

[33] Antheunisse, J. & Arkestein-Dijksman, L. 1979. Rate of drying and the survival of microorganisms. Antonie Van Leeuwenhoek 45: 177-184.

[34] Antheunisse, J., de Bruijn-Tol, J. W. & VanderPol-VanSoest, M. E. 1981. Survival of microorganisms after drying and storage. Antonie Van Leeuwenhoek 47: 539-545.

[35] Aranjuelo, I., Yrigoyen, J. J. & Sanchez-Diaz, M. 2007. Effect of elevated temperature and water availability on CO_2 exchange and nitrogen fixation of nodulated alfalfa plants. Environmental and Experimental Botany 59: 99-108.

[36] Arayankoon, T., Schomberg, H. H. & Weaver, R. W. 1990. Nodulation and N_2 fixation of guar at high root temperature. Plant and Soil 126: 209-213.

[37] Arora, N. K., Khare, E., Singh, S. & Maheshwari, D. K. 2010. Effect of Al and heavy metals on enzymes of nitrogen metabolism of fast and slow growing rhizobia under explanta conditions. World Journal of Microbiology and Biotechnology 26: 811-816.

[38] Avelar Ferreira, P. A., Lopes, G., Bomfeti, C. A., de Oliveira Longatti, S. M., de Sousa Soares, C. R., Guimarães Guilherme, L. R. & de Souza Moreira, F. M. 2013. Leguminous plants nodulated by selected strains of *Cupriavidus necator* grow in heavy metal contaminated soils amended with calcium silicate. World Journal of Microbiology and Biotechnology. DOI 10.1007/s11274-013-1369-2.

[39] Ayansina, A. D. V. 2009. Pesticide use in agriculture and microorganisms, In: Khan, M. S., Zaidi, A. & Musarrat, J. (ed.), Microbes in Sustainable Agriculture. Nova Science Publishers, New York, p. 261-284.

[40] Baath, E. 1992. Measurement of heavy metal tolerance of soil bacteria using thymidine incorporation into bacteria extracted after homogenization-centrifugation. Soil Biology and Biochemistry 24: 1167-1172.

[41] Baath, E., Diaz-Ravina, M., Frostegard, A. & Campbell, C. D. 1998. Effect of metal-rich sludge amendments on the soil microbial community. Applied and Environmental Microbiology 64: 238-245.

[42] Badri, M., Soula, S., Huguet, T. & Aouani, M. E. 2010. Symbiotic performance of *Medicago ciliaris* lines in association with different *Sinorhizobium* strains. Journal of Ecobiotechnology 2: 6-11.

[43] Baldani, J. I. & Weaver, R. W. 1992. Survival of clover rhizobia and their plasmid-cured derivatives in soil under heat and drought stress. Soil Biology and Biochemistry 24: 737-742.

[44] Bandyopadhyay, A. K., Jain, V. & Nainawatee, H. S. 1996. Nitrate alters the flavonoid profile and nodulation in pea (*Pisum sativum* L.). Biology and Fertility of Soils 21: 189-191.

[45] Barea, J. M. & Azcon-Aguilar, C. 1983. Mycorrhizas and their significance in nodulating nitrogen-fixing plants. Advances in Agronomy 36: 1-54.

[46] Becana, M., Aparicio-Rejo, P., Pena, J., Aquirreotea, J. & Sanchez-Diaz, M. 1986. N_2 fixation (C_2H_2-reducing activity) and leghemoglobin content during nitrate and water stress induced senescence of *Medicago Sativa* root nodules. Journal of Experimental Botany 37: 597-605.

[47] Begum, A. A., Leibovitch, S., Migner P. & Zhang, F. 2001. Inoculation of pea (*Pisum sativum* L.) by *Rhizobium leguminosarum* bv. *viceae* preincubated with naringenin and hesperetin or application of naringenin and hesperetin directly into soil increased pea nodulation under short season conditions. Plant and Soil 237: 71-80.

[48] Bellinaso, M. L., Greer, C. W., Peralba, M. C., Henriques, J. A. & Gaylarde, C. C. 2003. Biodegradation of the herbicide trifluralin by bacteria isolated from soil. FEMS Microbiology Ecology 43: 191-194.

[49] Ben Salah, I., Albacete, A., Martinez Andujar, C., Haouala, R., Labidi, N., Zribi, F., Martinez, V., Perez-Alfocea, F. & Abdelly, C. 2009. Response of nitrogen fixation in relation to nodule carbohydrate metabolism in *Medicago ciliaris* lines subjected to salt stress. Journal of Plant Physiology 166 (5): 477-488.

[50] Benhizia, Y., Benhizia, H., Benguedouar, A., Muresu, R., Giacomini, A. & Squartini, A. 2004. Gamma proteobacteria can nodulate legumes of the genus *Hedysarum*. Systematic and Applied Microbiology 27: 462-468.

[51] Bernard, T., Pocard, J. A., Perroud, B. & Le Rudulier, D. 1986. Variations in the response of salt-stressed *Rhizobium* strains to betaines. Archives of Microbiology 143: 359-364.

[52] Bertholet, J. & Clark, W. 1985. Effect of Trifluralin and Metribuzin on faba bean growth, development and symbiotic nitrogen fixation. Canadian Journal of Plant Science 65: 9-20.

[53] Bewley, J .B. 1981. Protein synthesis. In: Paleg, L. G. and Aspinall, D. (ed.), The Physiology and Biochemistry of Drought Resistance in Plants. Academic Press, New York, p. 261-282.

[54] Bhagwat, A. A., Gross, K. C., Tully, R. E. & Keister, D. L. 1996. Betaglucan synthesis in *Bradyrhizobium japonicum*: characterization of a new locus (ndvC) influencing beta-(1leads to 6) linkages. Journal of Bacteriology 178: 4635-4642.

[55] Bhandal, I. S., Kaur, H. & Bhandal, B. K. 1990. Heavy metal inhibition of in vitro *Rhizobium*, growth, nodulation and nitrogen-fixation in *Vigna radiata*. Phyton 30: 57-63.

[56] Billi, D. & Potts, M. 2002. Life and death of dried prokaryotes. Research in Microbiology 153: 7-12.

[57] Boddey, R. M., Sa, J. C., Alves, B. J. & Urquiaga, S. 1997. The contribution of biological nitrogen fixation for sustainable agricultural systems in the tropics. Soil Biology and Biochemistry 29: 787-799.

[58] Bolaños, L., El-Hamdaoui, A. & Bonilla, I. 2003. Recovery of development and functionality of nodules and plant growth in saltstressed *Pisum sativum – Rhizobium leguminosarum* symbiosis by boron and calcium. Journal of Plant Physiology 160: 1493-1497.

[59] Boncompagni, E., Østerås, M., Poggi, M. C. & Le Rudulier, D. 1999. Occurrence of choline and glycine betaine uptake metabolism in the family Rhizobiaceace and their in osmoprotection. Applied and Environmental Microbiology 65: 2072-2077.

[60] Borges, A. & Wollum, A. 1981. Effect of cadmium on symbiotic soybean plants. Journal of Environmental Quality 10: 216-221.

[61] Botsford, J. L. 1984. Osmoregulation in *Rhizobium meliloti*: inhibition of growth by salts. Archives of Microbiology 137:124-127.

[62] Botsford, J. L. & Lewis, T. A. 1990. Osmoregulation in *Rhizobium meliloti*: production of glutamic acid in response to osmotic stress. Applied and Environmental Microbiology 56: 488-494.

[63] Boumahdi, M., Mary, P. & Hornez, J. 1999. Influence of growth phases and desiccation on the degrees of unsaturation of fatty acids and the survival rates of rhizobia. Journal of Applied Microbiology 87: 611-619.

[64] Bouquard, C., Ouazzani, J., Prome, J., Michel-Briand, Y. & Plesiat, P. 1997. Dechlorination of atrazine by a *Rhizobium* sp. isolate. Applied and Environmental Microbiology 63 (3): 862-866.

[65] Bowen, G. D. & Kennedy, M. M. 1956. Effect of high soil temperatures on *Rhizobium* spp. Queensland Journal of. Agricultural Science 18: 177-196.

[66] Braswell, B. H., Schimel, D. S., Linder, E. & Moore, B. 1997. The response of global terrestrial ecosystems to interannual temperature variability. Science 238: 870–872.

[67] Braun, A. D. 1997. Microbial Water Stress. Bacteriology Review 40: 803-846

[68] Breedveld, M. W., Zevenhuizen, L. P. T. M. & Zehnder, A. J. B. 1990. Osmotically induced oligo- and polysaccharide synthesis by *Rhizobium meliloti* SU-47. Journal of General Microbiology 136: 2511-2519.

[69] Breedveld, M. W., Dijkema, C., Zevenhuizen, L. & Zehnder, A. J. B. 1993. Response of intracellular carbohydrates to a NaCl shock in *Rhizobium leguminosarum* biovar *trifolii* TA1 and *Rhizobium meliloti* SU47. Journal of General Microbiology 139: 3157-3163.

[70] Bremer, E. & Krämer, R. 2000. Coping with osmotic challenges: osmoregulation through accumulation and release of compatible solutes in bacteria. In: Storz, G. and Hengge-Aronis, R. (ed.), Bacterial Stress Response, p. 79-97.

[71] Broos, K., Uyttebroek, M., Mertens, J. & Smolders, E. 2004. A survey of symbiotic nitrogen fixation by white clover grown on metal contaminated soils. Soil Biology and Biochemistry 36: 633-640.

[72] Broughton, W. J., Zhang, F., Perret, X. & Staehelin C. 2003. Signals exchanged between legumes and *Rhizobium*, agricultural uses and perspectives. Plant and Soil 252: 129-137.

[73] Brown, A. D. 1990. Microbial water stress physiology, principles and perspectives. John Wiley & Sons, Inc., New York.

[74] Bushby, H. V. A. & Marshall, K. C. 1976. Some factors affecting the survival of root-nodule bacteria on desiccation. Soil Biology and Biochemistry 9: 143-147.

[75] Bushby, H. V. A. & Marshall, K. C. 1977. Desiccation induced damage to the cell envelope of root-nodule bacteria. Soil Biology and Biochemistry 9: 149-152.

[76] Callaham, D. A. & Torrey, J. G. 1981. The structural basis for infection of root hairs of *Trifolium repens* by *Rhizobium*. Canadian Journal of Botany 59: 1647-1664.

[77] Ceribasi, I. H. & Yetis, U. 2001. Biosorption of Ni (II) and Pb (II) by *Phanaerochaete chrysosporium* from a binary metal system kinetics. Water SA 27: 15-20.

[78] Chalk, P. M., Souza, R. D. F., Urquiaga, S., ALves, B. J. R. & Boddy, R. M. 2006. The role of arbuscular mycorrhiza in legume symbiotic performance. Soil Biology and Biochemistry 38: 2944-2951.

[79] Chao, W. L. & Alexander, M. 1984. Mineral soils as carriers for *Rhizobium* inoculants. Applied and Environmental Microbiology 47: 94-97.

[80] Chaudri, A. M., McGrath, S. P., Giller, K. E., Reitz, E. & Suerbeck, D. R. 1993. Enumeration of indigenous *Rhizobium leguminosarum* biovar *trifolii* in soils previously treated with metal contaminated sewage sludge. Soil Biology and Biochemistry 25: 301-309.

[81] Chaudri, A. M., Allain, C. M., Barbosa-Jefferson, V. L., Nicholson, F. A., Chambers, B. J. & McGrath, S. P. 2000. A study of the impacts of Zn and Cu on two rhizobial species in soils of a long term field experiment. Plant and Soil 22: 167-179.

[82] Chen, H., Richardson, A. E. & Rolfe, B. G. 1993. Studies of the physiological and genetic basis of acid tolerance in *Rhizobium leguminosurum* biovar *trifolii*. Applied and Environmental Microbiology 59: 1798-1804.

[83] Chen, L. S., Figueredo, A., Villani, H., Michajluk, J. & Hungria, M. 2002. Diversity and symbiotic effectiveness of rhizobia isolated from field-grown soybean nodules in Paraguay. Biology and Fertility of Soils 35: 448-457.

[84] Chen, M. & Alexander, M. 1973. Survival of soil bacteria during prolonged desiccation. Soil Biology and Biochemistry 5: 213-221.

[85] Chen, W. M., Laevens, S., Lee, T. M., Coenye, T., De Vos, P., Mergeay, M. & Vandamme, P. 2001. *Ralstonia taiwanensis* sp nov., isolated from root nodules of *Mimosa* species and sputum of a cystic fibrosis patient. International Journal of Systematic and Evolutionary Microbiology 51: 1729-1735.

[86] Chen, W. M., Wu, C. H., James, E. K. & Chang, J. S. 2008. Metal biosorption capability of *Cupriavidus taiwanensis* and its effects on heavy metal removal by nodulated *Mimosa pudica*. Journal of Hazardous Materials 151: 364-371.

[87] Chen, W. M., Moulin, L., Bontemps, C., Vandamme, P., Béna, G. & Boivin-Masson, C. 2003. Legume symbiotic nitrogen fixation by beta-Proteobacteria is widespread in nature. Journal of Bacteriology 185: 7266-7272.

[88] Cheng, Q. 2008. Perspectives in biological nitrogen fixation research. Journal of Integrative Plant Biology, 50 (7): 784–796

[89] Cheung, K. H. & Gu, J. D. 2007. Mechanism of hexavalent chromium detoxification by microorganisms and bioremediation application potential: a review. International Biodeterioration and Biodegradation 59: 8-15.

[90] Chianu, J. N., Vanlauwe, B., Mahasi, J. M., Katungi, E., Aketch, C., Mairura, F.S., Chianu, Justina N & Sanginga, N. 2008. Soybean situation and Outlook analysis; the case of Kenya.TSBF-CIAT, AFNET, p. 68.

[91] Chien, C. T., Shetty, K., Mortimer, M. & Orser, C. S. 1991. Calcium-induced salt toler-

ance in *Rhizobium leguminosarum* biovar *viciae* strain C1204b. FEMS Microbiology Letters 83: 219-224.

[92] Chien, C. T., Maundu, J., Cavaness, J., Dandurand, L. M. & Orser, C. S. 1992. Characterization of salt-tolerant and salt-sensitive mutants of *Rhizobium leguminosarum* biovar *viciae* strain C12046. FEMS Microbiology Letters 90: 135-140.

[93] Cooper, J. E. 1988. Nodulation of legumes by rhizobia in acid soils. In: Vancura, V. and Kunc, F. (ed.), Developments in soil science. Elsevier Science, Amsterdam, p 57-61.

[94] Cordovilla, M. D. P., Berrido, S. L., Ligero, F. & Lluch, C. 1999. *Rhizobium* strain effects on the growth and nitrogen assimilation in *Pisum sativum* and *Vicia faba* plant growth under salt stress. Journal of Plant Physiology 154: 127-131.

[95] Coventry, D. R. & Evans, J. 1989. Symbiotic nitrogen fixation and soil acidity. In: Robson, A. D. (ed.), Soil Acidity and Plant Growth. Academic Press, Sydney, p. 103-137.

[96] Crockford, A. J., Davis, G. A. & Williams, H. D. 1995. Evidence for cell-dependent regulation of catalase activity in *Rhizobium leguminosarum* bv. *phaseoli*. Microbiology 141: 843-851.

[97] Cunningham, S. D. & Munns, D. N. 1984. The correlation of the exopolysaccharide production and acid-tolerance in *Rhizobium*. Soil Science Society of America Journal 48: 1273-1276.

[98] Curley, R. L. & Burton, J. C. 1975. Compatibility of *Rhizobium japonicum* with chemical seed protectants. Agronomy Journal 67: 807-808.

[99] Cytryn, E. D., Sangurdekar, D. P., Streeter, J. G., Franck, W. L., Chang, W., Stacey, G., Emerich, D. W., Joshi, T., Xu, D. & Sadowsky, M. J. 2007. Transcriptional and physiological responses of *Bradyrhizobium japonicum* to desiccation-induced stress. Journal of Bacteriology 189: 6751-6762.

[100] Danese, P. N., Pratt, L. A. & Kolter, R. 2000. Exopolysaccharide production is required for development of *Escherichia coli* K-12 biofilm architecture. Journal of Bacteriology 182: 3593-3596.

[101] Danso, S. K. A. 1995. Assessment of biological nitrogen fixation. Fertililizer Research, 42: 33-41.

[102] Dardanelli, M. S., Fernandez de Cordoba, F. J., Espuny, M. R., Rodrıguez Carvajal M. A., Soria Diaz, M. E., Gil Serrano, A. M., Okon, Y. & Megıas, M. 2008. Effects of *Azospirillum brasilense* coinoculated with *Rhizobium* on *Phaseolus vulgaris* flavonoids and Nod factor production under salt stress. Soil Biology and Biochemistry 40: 2713-2721.

[103] Dart, P. J. & Mercer, F. V. 1965. The effect of growth temperature, level of ammonium nitrate and light intensity on the growth and nodulation of cowpea (*Vigna sinensis* Endl. ex. Hassk.). Australian Journal of Agricultural Research 16: 321-345.

[104] Date, R. A. 2000. Inoculated legumes in cropping systems of the tropics. Field Crops Research 65: 123-136.

[105] Day, D. A. & Copeland, L. 1991. Carbon metabolisms and compartmentation in nitrogen fixing legume nodules. Plant Physiology and Biochemistry 29: 185-201.

[106] de Lira Juniora, M., Limaa, A. S. T., Arruda, J. R. F. & Smith, D. L. 2005. Effect of root temperature on nodule development of bean, lentil and pea. Soil Biology and Biochemistry 37: 235-239.

[107] de Lucena, D. K., Pühler, A. & Weidner, S. 2010. The role of sigma factor RpoH1 in the pH stress response of *Sinorhizobium meliloti*. BMC Microbiology 10: 265. doi: 10.1186/1471-2180-10-265.

[108] Deaker, R., Roughley, R. J. & Kennedy, I. R. 2004. Legume seed inoculation technology – a review. Soil Biology and Biochemistry 36: 1275-1288.

[109] Dénarié, J., Debelle, F. & Promé, J. C. 1996. *Rhizobium* lipo-chito-ligosaccharide nodulation factors: Signalling molecules mediating recognition an morphogenesis. Annual Review of Biochemistry 65: 503-535.

[110] Dick, R. E. & Quinn, J. P. 1995. Glyphosate-degrading isolates from environmental samples: occurence and pathways of degradation. Applied Microbiology and Biotechnology 43 (3): 545-550.

[111] Dominguez-Fererras, A., Pérez-Arnedo, R., Becker, A., Olivares, J., Soto, M. J. & Sanjuán, J. 2006. Transcriptome profiling reveals the importance of plasmid pSymB for osmoadaptation of *Sinorhizobium meliloti*. Journal of Bacteriology 188: 7617-7625.

[112] dos Santos, J. B., Ferreira, E. A., Kasuya, M. C. M., da Silva, A. A. & Procopio, S. O. 2005. Tolerance of *Bradyrhizobium* strains to glyphosate formulations. Crop Protection 24 (6): 543-547.

[113] Dowling, D. N. & Broughton, W. J. 1986. Competition for nodulation of legumes. Annual Review of Microbiology 40: 131-157.

[114] Drouin, P., Sellami, M., Prévost, D., Fortin, J. & Antoun, H. 2010. Tolerance to agricultural pesticides of strains belonging to four genera of Rhizobiaceae. Journal of Environmental Science and Health Part B 45: 780-788.

[115] Dunigan, E. P., Frey, J. P., Allen, L. D., McMahon, A., 1972. Herbicidal effects on nodulation of *Glycine max* (L.) Merrill. Agronomy Journal 64: 806-808.

[116] Duzan, H. M., Mabood, F., Zhou, X., Souleimanov, A. & Smith, D. L. 2005. Nod factor induces soybean resistance to powdery mildew. Plant Physiology and Biochemistry 43: 1022-1030.

[117] Eaglesham, A. R. J. & Ayanaba, A. 1984. Tropical stress ecology of rhizobia, root nodulation and legume fixation. In: Subba Rao, N. S. (ed.), Current Developments in Biological Nitrogen Fixation. Edward Arnold Ltd, London, UK, p. 1-35.

[118] Eaglesham, A., Seaman, B. & Ahmad, H. 1981. High-temperature tolerant "cowpea" rhizobia. In: Gibson, A. H. and Newton, W. E. (ed.), Current perspectives in nitrogen fixation. Elsevier/North-Holland Biomedical Press, Canberra-Amsterdam.

[119] Eberbach, P. L. & Douglas, L. A. 1991. Effect of herbicide residues in a sandy loam on the growth, nodulation and nitrogenase activity (C_2H_2/C_2H_4) of *Trifolium subterraneum*. Plant and Soil 131: 67-76.

[120] El-enany, A. E. & Abd-Alla, M. H. 1995. Cadmium resistance in *Rhizobium*-faba bean symbiosis. Synthesis of cadmium binding proteins. Phyton 35: 45-53.

[121] Elsheikh, E. A. E. & Wood, M. 1989. Response of chickpea and soybean rhizobia to salt: osmotic and specific ion effects of salts. Soil Biology and Biochemistry 21: 889-895.

[122] Elsheikh, E. A. E. & Wood, M. 1990a. Salt effects on survival and multiplication of chickpea and soybean rhizobia. Soil Biology and Biochemistry 22: 343-347.

[123] Elsheikh, E. A. E. & Wood, M. 1990b. Rhizobia and bradyrhizobia under salt stress: possible role of trehalose in osmoregulation. Letters in Applied Microbiology 10: 127-129.

[124] Estévez, J., Soria Díaz, M. E., Fernández de Córdoba, F., Móron, B., Manyan, H., Gil, A., Thomas-Oates, J., van Brussel, A. A. N., Dardanelli, M. S., Sousa, C. & Megías, M. 2009. Different andnew Nod factors produced by *Rhizobium tropici* CIAT899 following Na^+ stress. FEMS Microbiology Letters 293: 220-231.

[125] Euzéby, J. P. 1997. List of bacterial names with standing in nomenclature: a folder available on the Internet. International Journal of Systematic Bacteriology 47 (2): 590-592.

[126] FAO 2008. FAOSTAT Accessed on http://faostat.fao.org/default.aspx (accessed on April 2011).

[127] Ferreira, P. A. A., Bomfeti, C. A., Junior, R. S., Soares, B. L., Soares, C. R. F. S. & Moreira, F. M. S. 2012. Symbiotic efficiency of *Cupriavidus necator* strains tolerant to zinc, cadmium, copper and lead. Pesquisa Agropecuária Brasileira 47: 85-95.

[128] Fierer, N. & Jackson, R. B. 2006. The diversity and biogeography of soil bacterial communities. Proceedings of the National Academy of Sciences USA 103: 626-631.

[129] Fierer, N., Schimel, J. P. & Holden P. A. 2003. Influence of drying rewetting frequency on soil bacterial community structure. Microbial Ecology 45: 63-71.

[130] Firmin, J. L., Wilson, K. E., Rossen, L. & Johnston, A. W. B. 1986. Flavonoid activation of nodulation genes in *Rhizobium* reversed by other compounds present in plants. Nature 324: 90-92.

[131] Firmin, J. L., Wilson, K. E., Carlson, R. W., Davies, A. E. & Downie, J. A. 1993. Resistance to nodulation of cv. Afghanistan peas is overcome by nodX, which mediates an O-acetylation of the *Rhizobium leguminosarum* lipo-oligosaccharide nodulation factor. Molecular Microbiology 10: 351-360.

[132] Fitter, A. H. & Garbaye, J. 1994. Interactions between mycorrhizal fungi and other soil organisms. Plant and Soil 159: 123-132.

[133] Foster, J. W. 2000. Microbial responses to acid stress. In: Storz, G. and Hengge-Aronis, R. (ed.), Bacterial Stress Response. ASM Press, Washington, D. C., p. 99-115.

[134] Foth, H. D. 1990. Fundamentals of soil science. Wiley, New York.

[135] Fougere, F. & Le Rudulier, D. 1990. Uptake of glycine betaine and its analogues by bacteroids of *Rhizobium meliloti*. Journal of General Microbiology 136: 157-163.

[136] Fox, J. E., Starcevic, M., Kow, K. Y., Burow, M. E. & McLachlan, J. A. 2001. Nitrogen fixation: Endocrine disrupters and flavonoid signaling. Nature 413: 128-129.

[137] Fox, J. E., Starcevic, M., Jones, P. E., Burow, M. E. & McLachlan, J. A. 2004. Phytoestrogen signaling and symbiotic gene activation are disrupted by endocrine-disrupting chemicals. Environmental Health Perspectives 112: 672-677.

[138] Fox, J. E., Gulledge, J., Engelhaupt, E., Burow, M. E. & McLachlan, J. A. 2007. Pesticides reduce symbiotic efficiency of nitrogen-fixing rhizobia and host plants. Proceedings of the National Academy of Sciences of the United States of America 104(24): 10282-10287.

[139] Frerot, H., Lefebvre, C., Gruber, W., Collin, C., Dos Santos, A. & Escarre, J. 2006. Specific interactions between local metallicous plants improve the phyto stabilisation of mine soils. Plant and Soil 282: 53-65.

[140] Frings, J. F. J. 1976. The *Rhizobium*-pea symbiosis as affected by high temperatures. Ph.D. Thesis.Wageningen Agricultural University, Germany.

[141] Fujihara, S. & Yoneyama, T. 1993. Effects of pH and osmotic stress on cellular polyamine contents in the soybean *Rhizobia fredii* P220 and *Bradyrhizobium japonicum* A1017. Applied and Environmental Microbiology 59: 1104-1109.

[142] Fujishige, N. A., Kapadia, N. N., De Hoff, P. L. & Hirsch, A. M. 2006. Investigations of *Rhizobium* biofilm formation. FEMS Microbiology Ecology 56: 195-205.

[143] Gadd, G. M. 1992. Metals and microorganisms: a problem of definition. FEMS Microbiology Letters 100: 197-204.

[144] Gagnon, H. & Ibrahim, R. K. 1998. Aldonic acids: a novel family of nod gene inducers of *Mesorhizobium loti, Rhizobium lupini* and *Sinorhizobium meliloti*. Molecular Plant-Microbe Interactions 11: 988-998.

[145] Galinski, E. A. 1995. Osmoadaptation in bacteria. Advances in Microbial Physiology 37: 273-328.

[146] Georgiev, G. I. & Atkins, C. A. 1993. Effects of salinity on N_2 fixation, nitrogen metabolism and export and diffusive conductance of cowpea nodules. Symbiosis 15: 239-255.

[147] Gerosa-Ramos, M. L., Parsons, R., Sprent, J. I. & James, E. K. 2003. Effect of water stress on nitrogen fixation and nodule structure of common bean. Pesquisa Agropecuária Brasileira 38: 339-347.

[148] Gewaily, E. M., Khan, M. F. & Kheder, A. K. 1991. Effect of temperature and calcium carbonate contents of soil on the survival and growth of rhizobia. In: 13th North American Symbiotic Nitrogen Fixation Conference. Banff Centre for Conferences, Canada, p. 50.

[149] Ghassemi, F., Jakeman, A. J. & Nix, H. A. 1995. Salinization of land and water resources: human causes, extent, management and case studies, CAB International, p. 526.

[150] Ghittoni, N. E. & Bueno, M. A. 1996. Changes in the cellular content of trehalose in our peanut rhizobia strains cultured under hypersalinity. Symbiosis 20: 117-127.

[151] Giller, K. E., McGarth, S. P. & Hirsch, P. R. 1989. Absence of nitrogen fixation in clover grown on soil subjected to long-term contamination with heavy metals is due to survival of only ineffective *Rhizobium*. Soil Biology and Biochemistry 21: 841-848.

[152] Giller, K. E., Witter, E. & McGrath, S. P. 1998. Toxicity of heavy metals to microorganisms and microbial process in agricultural soils: a review. Soil Biology and Biochemistry 30: 1389-1414.

[153] Giller, K. E., Beare, M. H., Lavelle, P., Izac, A. M. N. & Swift, M. J. 1997. Agricultural intensification, soil biodiversity and agroecosystem function. Applied Soil Ecology 6: 3-16.

[154] Glenn, A. R. & Dilworth, M. J. 1994. The life of root nodule bacteria in the acidic underground. FEMS Microbiology Letters 123: 1-10.

[155] Gloux, K. & Le Rudulier, D. 1989. Transport and catabolism of proline betaine in salt-stressed *Rhizobium meliloti*. Archives of Microbiology 151: 143-148.

[156] Gonzalez-Gonzalez, R., Botsford, J. L. & Lewis, T. 1990. Osmoregulation in *Rhizobium meliloti*: Charaterization of enzymes involved in glutamate synthesis. Canadian Journal of Microbiology 36: 469-474.

[157] Gouffi, K. & Blanco, C. 2000. Is the accumulation of osmoprotectant the unique mechanism involved in bacterial osmoprotection? International Journal of Food Microbiology 55: 171-174.

[158] Gouffi, K., Bernard, T. & Blanco, C. 2000. Osmoprotection by pipecolic acid in *Sinorhizobium meliloti*: specific effects of D and L isomers. Applied and Environmental Microbiology 66: 2358-2364.

[159] Gouffi, K., Pica, N., Pichereau, V. & Blanco, B. 1999. Disaccharides as a new class of nonaccumulated osmoprotectants for *Sinorhizobium meliloti*. Applied and Environmental Microbiology 65: 1491-1500.

[160] Graham, P. H. & Vance, C. P. 2000. Nitrogen fixation in perspective: an overview of research and extension needs. Field Crops Research 65: 93-106.

[161] Graham, P. H., Ocampo, G., Ruiz, L. D. & Duque, A. 1980. Survival of *Rhizobium phaseoli* in contact with chemical seed protectants. Agronomy Journal 72: 625-626.

[162] Graham, P. H., Sadowsky, M. J., Keyser, H. H., Barnet, Y. M., Bradley, R. S., Cooper, J. E., De Ley, D. J., Jarvis, B. D. W., Roslycky, E. B., Stijdom, B. W., & Young, J. P. W. 1991. Proposed minimal standards for the description of new genera and species of root and Stem nodulating bacteria. International Journal of Systematic Bacteriology, 41: 582-587.

[163] Guasch-Vidal, B., Estévez, J., Dardanelli, M. S., Soria-Diaz, M. E., de Córdoba, F. F., Balog, C. I., Manyani, H., Gil-Serrano, A., Thomas-Oates, J., Hensbergen, P. J., Deelder, A. M., Megías, M. & van Brussel, A. A. 2013. High NaCl concentrations induce the nod genes of *Rhizobium tropici* CIAT899 in the absence of flavonoid inducers. Molecular Plant-Microbe Interactions 26(4): 451-460.

[164] Guerin, V., Trinchant, J. C. & Rigaud, J. 1990. Nitrogen fixation (C_2H_2 reduction) by broad bean (*Vicia faba* L.) nodules and bacteroids under water-restricted conditions. Plant Physiology 92: 595-601.

[165] Guerin, V., Pladys, D., Trinchant, J. C. & Rigaud, J. 1991. Proteolysis and nitrogen fixation in faba-bean *Vicia faba* nodules under water stress. Physiologia Plantarum 82: 360-366.

[166] Gustafson, A. M., O'Connell, K. P. & Thomashow, M. F. 2002. Regulation of *Sinorhizobium meliloti* 1021 rrnA-Reporter gene fusions in response to cold shock. Canadian Journal of Microbiology 48: 821-830.

[167] Hafeez, F. Y., Shah, N. H., Malik, K. A. 2000. Field evaluation of lentil cultivars inoculated with *R. leguminosarum* bv. *viciae* strains for nitrogen fixation using nitrogen-15 isotope dilution. Biology and Fertility of Soils 31: 65-69.

[168] Han, Y., Zhou, D., Pang, X., Zhang, L., Song, Y., Tong, Z., Bao, J., Dai, E., Wang, J., Guo, Z., Zhai, J., Du, Z., Wang, X., Wang, J., Huang, P. & Yang, R. 2005. Comparative

transcriptome analysis of *Yersinia pestis* in response to hyperosmotic and high-salinity stress. Research in Microbiology 156: 403-415.

[169] Hansen, A. P. 1994. Symbiotic N_2 fixation of crop legumes: Achievement and perspectives, Hohenheim Tropical Agricultural Series. Margraf Verlag, p. 31-36.

[170] Hatimi, A. 1999. Effect of salinity on the association between root symbionts and *Acacia cyanophylla* Lind.: growth and nutrition. Plant and Soil 216: 93-101.

[171] Hayman, D. S. 1986. Mycorrhizae of nitrogen-fixing legumes. MIRCEN Journal 2: 121-145.

[172] Hecker, M. & Völker, U. 1998. Non-specific, general and multiple stress resistance of growth restricted *Bacillus subtilis* cells by the expression of the σB Regulon. Molecular Microbiology 29: 1129-1136.

[173] Hecker, M., Schumann, W. & Völker, U. 1996. Heat-shock and general stress response in *Bacillus subtilis*. Molecular Microbiology 19: 417-428.

[174] Heckman, J. R., Angel J. S. & Chaney R. L. 1984. Residual effects of sewage sludge on soybean. II. accumulation of soil and symbiotically fixed nitrogen. Journal of Environmental Quality 16: 118-124.

[175] Hengge-Aronis, R. 1999. Interplay of global regulators and cell physiology in the general stress response of *Escherichia coli*. Current Opinions in Microbiology 2: 148-152.

[176] Herman P. L., Behrens, M., Chakraborty, S., Crastil, B. M., Barycki, J. & Weeks, D. P. 2005. A three component dicamba O-demethylase from *Pseudomonas maltiphilia* strain DI-6: Gene isolation, characterization and heterologous expression. Journal of Biological Chemistry 280: 24759-24767.

[177] Hernandez-Armenta, R., Wien, H. C. & Eaglesham, A. R. J. 1989. Maximum temperature for nitrogen fixation in common bean. Crop Science 29: 1260-1265.

[178] Hernandez-Jimenez, M. J., Lucas, M. M. & de Felipe, M. R. 2002. Antioxidant defense and damage in senescing lupin nodules. Plant Physiology and Biochemistry 40: 645-657.

[179] Hirsch, A.M. 2010. How rhizobia survive in the absence of a legume host, a stressful world indeed. In Symbioses and Stress: Joint Ventures in Biology, Cellular Origin, Life in Extreme Habitats and Astrobiology, Seckbach, J. and Gruber, M. (ed.) Springer Science and Business Media, p. 375-391.

[180] Hirsch, P. R., Jones, M. J., McGrath, S. P. & Giller, K. E. 1993. Heavy metals from past applications of sewage sludge decrease the genetic diversity of *Rhizobium leguminosarum* biovar *trifolii* populations. Soil Biology and Biochemistry 25: 1485-1490.

[181] Hoglund, J. H. & Brock, J. L. 1987. Nitrogen fixation in managed grasslands. In: Naydon, R. W. (ed.), Managed Grasslands: Analytical studies.Ecosystems of the World 17B. Elsevier, Amsterdam, p. 187-196.

[182] Howieson, J. G., Robson, A. D. & Abbot, L. K. 1992. Acid-tolerant species of *Medicago* produce root exudates at low pH which induce the expression of nodulation genes in *Rhizobium meliloti*. Australian Journal of Plant Physiology 19: 287-296.

[183] Hsiao, C. T. 1973. Plant responses to water stress. Annual Review of Plant Physiology 24: 519-570.

[184] Hua, S. S. T., Tsai, V. Y., Lichens, G. M. & Noma, A. T. 1982. Accumulation of amino acid in *Rhizobium* sp. strain WR1001 in response to sodium chloride salinity. Applied and Environmental Microbiology 44: 135-140.

[185] Huang, C. Y., Bazzaz, F. A. & Vanderhoef, L. N. 1974. The inhibition of soybean metabolism by cadmium and lead. Plant Physiology 53: 308-312.

[186] Huang, C. Y., Boyer, J. S. & Vanderhoef, L. N. 1975. Limitation of acetylene reduction (nitrogen fixation) by photosynthesis in soybean having low water potentials. Plant Physiology 56: 228-232.

[187] Hungria, M. & Franco, A. A. 1993. Effects of high temperature on nodulation and nitrogen fixation by *Phaseolus vulgaris* L. Plant and Soil 149: 95-102.

[188] Hungria, M. & Phillips, D. A. 1993. Effects of a seed color mutation on rhizobial nod-gene-inducing flavonoids and nodulation in common bean. Molecular Plant-Microbe Interactions 6: 418-422.

[189] Hungria, M. & Stacey, G. 1997. Molecular signals exchanged between host plants and rhizobia: Basic aspects and potential application in agriculture. Soil Biology and Biochemistry 29: 819-830.

[190] Hungria, M. & Vargas, M. A. T. 2000. Environmental factors affecting N_2 fixation in grain legumes in the tropics, with an emphasis on Brazil. Field Crops Research 65: 151-164.

[191] Irigoyen, J. J., Emerich, D. W. & Sanchez-Diaz, M. 1992. Water stress induced changes in concentrations of proline and total soluble sugars in nodulated alfalfa (*Medicago sativa*) plants. Physiologia Plantarum 84: 55-60.

[192] Ishaq, Y. Z., Zouhdi, L. I., El Haddad, M. E., Moawad, H., Abdelaziz, R. A. 1989. Population dynamics of introduced *R. trifolii* strains in the rhizosphere of Egyptian clover in relation to salt stress and pattern of competition. In: Abstracts of the fifth International Symposium on Microbial Ecology, Kyoto, Japan.

[193] Jiang, W., Hou, Y. & Inouye, M. 1997. CspA, the major cold-shock protein of *Escherichia coli*, is an RNA chaperone. Journal of Bacteriology 272: 196-202.

[194] Johnsen, K., Jacobsen, C. S., Torsvik, V. & Sorensen, J. 2001. Pesticide effects on bacterial diversity in agricultural soils-a review. Biology and Fertility of Soils 33: 443-453.

[195] Johnson, A. C. & Wood, M. 1990. DNA as a possible site of action of aluminium in *Rhizobium* sp. Applied and Environmental Microbiology 56: 3629-3633.

[196] Jones, F. R. & Tisdale, W. B. 1921. Effect of soil temperature upon the development of nodules on the roots of certain legumes. Journal of Agricultural Research 22: 17-37.

[197] Jurcisek, J. A. & Bakaletz, L. O. 2007. Biofilms formed by nontypeable *Haemophilus influenzae* in vivo contain both double-stranded DNA and type IV pilin protein. Journal of Bacteriology 189: 3868-3875.

[198] Karanja, N. K. & Wood, M. 1988. Selecting *Rhizobium phaseoli* strains for use with beans (*Phaseolus vulgaris* L.) in Kenya: Tolerance of high temperature and antibiotic resistance. Plant and Soil 112: 15-22.

[199] Kassem, M., Capellano, A. & Gounot, A. M. 1985. Effets du chlorure de sodium sur la croissance in vitro, l'infectivité et l'efficience de *Rhizobium meliloti*. MIRCEN Journal 1: 63-73.

[200] Kaur, C., Maini, P. & Shukla, N. P. 2007. Effect of captan and carbendazim fungicides on nodulation and biological nitrogen fixation in soybean. Asian Journal of Experimental Sciences 21 (2): 385-388.

[201] Khan, M. & Scullion, J. 2002. Effects of metal (Cd, Cu, Ni, Pb or Zn) enrichment of sewage-sludge on soil micro-organisms and their activities. Applied Soil Ecology 20: 145-155.

[202] Khan, M. S., Zaidi, A., Wani, P. A. & Oves, M. 2009. Role of plant growth promoting rhizobacteria in the remediation of metal contaminated soils. Environmental Chemistry Letters 7: 1-19.

[203] Kiss, E., Huguet, T., Poinsot, V. & Batut, J. 2004. The typA is required for stress adaptation as well as for symbiosis of *Sinorhizobium meliloti* 1021 with certain *Medicago truncatula* Lines. Molecular Plant-microbe Interactions 17: 235-244.

[204] Kluson, A. R., Kenworthy, W. J. & Weber, D. F. 1986. Soil temperature effects on competitiveness and growth of *Rhizobium japonicum* and on *Rhizobium*-induced chlorosis of soybean. Plant and Soil 95: 201-207.

[205] Kosanke, J. W., Osburn, R. M., Shuppe, G. I. & Smith, R. S. 1991. Slow rehydration improves the recovery of dried bacterial populations. Canadian Journal of Microbiology 38: 520-525.

[206] Kosslak, R. M., Bookland, R., Barkei, J., Paaren, H. & Appelbaum, E. R. 1987. Induction of *Bradyrhizobium japonicum* common nod genes by isoflavones isolated from *Glycine max*. Proceedings of the Natural Academy of Science, USA 84: 7428-7432.

[207] Kozik, A., Heidstra, R., Horvath, B., Kulikova, O., Tikhonovich, I., Ellis, T. H. N., van Kammen, A., Lie, T. A. & Bisseling, T. 1995. Pea lines carrying sym1 or sym2 can be

nodulated by *Rhizobium* strains containing nodX; sym1 and sym2 are allelic. Plant Science 108: 41-49.

[208] Krujatz, F., Haarstrick, A., Nortemann, B. & Greis, T. 2012. Assessing the toxic effects of nickel, cadmium and EDTA on growth of the plant growth-promoting rhizobacterium *Pseudomonas brassicacearum*. Water, Air and Soil Pollution 223 (3): 1281-1293.

[209] Kumar, S. 1981. Effect of simazine and prometryne on the growth and nodulation of chickpea (*Cicer arietinum* L.). Journal of Agricultural Science 97: 663-671.

[210] Kumar, S., Mukerji, K. G. & Lal, R. 1996. Molecular aspects of pesticide degradation by microorganisms. Critical Reviews in Microbiology 22: 1-26.

[211] Kurchak, O. N., Provorov, N. A. & Simarov, B. V. 2001. Plasmid pSym1-32 of *Rhizobium leguminosarum* bv. *viceae* controlling nitrogen fixation activity, effectiveness of symbiosis, competitiveness and acid tolerance. Russian Journal of Genetics 37: 1025-1031.

[212] Kust, C. A. & Struckmeyer, B. E. 1971. Effects of Trifluralin on the growth, nodulation and anatomy of soybean. Weed Science 19: 147-152.

[213] Lakzian, A., Murphy, P., Turner, A., Beynon, J. L. & Giller, K. E. 2002. *Rhizobium leguminosarum* bv. *viciae* populations in soils with increasing heavy metal contamination: abundance, plasmid profiles, diversity and metal tolerance. Soil Biology and Biochemistry 34: 519-529.

[214] Lanyon, W. E. 1995. Eastern meadowlark (*Sturnella magna*). The birds of North America, number 160. The American Ornithologists' Union, Washington D.C., USA, and The Academy of Natural Sciences, Philadelphia, Pennsylvania, USA.

[215] Laranjo, M. & Oliveira, S. 2011. Tolerance of *Mesorhizobium* type strains to different environmental stresses. Antonie Van Leeuwenhoek 99: 651-662.

[216] Lentze, N., Aquilina, J. A., Lindbauer, M., Robinson, C. V. & Narberhaus, F. 2004. Temperature and concentration-controlled dynamics of rhizobial small heat shock proteins. European Journal of Biochemistry 271: 2494-2503.

[217] Lie, T. A. 1984. Host genes in *Pisum sativum* L. conferring resistance to European *Rhizobium leguminosarum* strains. Plant and Soil 82: 415-425.

[218] Lindström, K., Terefewrok, Z., Suominen, L. & Lortet, G. 2002. Signalling and development of *Rhizobium*-legume symbioses. Biology and Environment: Proceedings of the Royal Irish Academy 1: 61-64.

[219] Liu, C. M., McLean, P. A., Sookdeo, C. C. & Cannon, F. C. 1991. Degradation of the herbicide glyphosate by members of the family Rhizobiaceae. Applied and Environmental Microbiology 57 (6): 1799-1804.

[220] Lloret, J., Wulff, B. B. H., Rubio, J. M., Downie, J. A., Bonilla, I. & Rivilla, R. 1998. Exopolysaccharide II production is regulated by salt in the halotolerant strain *Rhizobium meliloti* EFB1. Applied and Environmental Microbiology 64: 1024-1028.

[221] Long, S. R. 1989. *Rhizobium* genetics. Annual Review of Genetics 23: 483-506.

[222] Long, S. R. 2001. Genes and signals in the *Rhizobium*-legume symbiosis. Plant Physiology 125: 69-72.

[223] Lynch, D. H. & Smith, D. L. 1994. The effect of low root zone temperature stress on two soybean (*Glycine max*) genotypes when combined with *Bradyrhizobium* strains of varying geographic origin. Physiologia Plantarum 90: 105-113.

[224] Madkour, M. A., Smith, L. T. & Smith, G. M. 1990. Preferential osmolyte accumulation: a mechanism of osmotic stress adaptation in diazotrophic bacteria. Applied and Environmental Microbiology 56: 2876-2881.

[225] Mahieu, S., Frerot, H., Vidal, C., Galiana, A., Heulin, K. & Maure, L. 2011. *Anthyllis vulneraria / Mesorhizobium metallidurans*, an efficient symbiotic nitrogen fixing association able to grow in mine tailings highly contaminated by Zn, Pb and Cd. Plant and Soil 342: 405-417.

[226] Mahmoud, A. L. E. & Omar, S. A. 1995. Growth, cell wall-degrading enzymes and aflatoxin production by lemon-rotting fungi in relation to insecticide application. Microbiological Research 150: 195-200.

[227] Mallik, M. A. B. & Tesfai, K. 1983. Compatibility of *Rhizobium japonicum* with commercial pesticides in vitro. Bulletin of Environmental Contamination and Toxicology 31: 432-437.

[228] Mandal, S. M., Gouri, S. S., De, D. Das, B. K. Mondal, K. C., Bikas R & Pati, B. R. 2011. Effect of arsenic on nodulation and nitrogen fixation of blackgram (*Vigna mungo*).Indian Journal of Microbiology 51(1): 44-47.

[229] Marcar, N. E., Dart, P. & Sweeney, C. 1991. Effect of root-zone salinity on growth and chemical composition of *Acacia ampliceps* B. R. Maslin, *A. auriculiformis* A. Cunn ex Benth and *A. mangium* Willd at two nitrogen levels. New Phytologist 119: 567-573.

[230] Martin, R. B. 1988. The chemistry of aluminium as related to biology and medicine. Clinical Chemistry 23: 1797-1806.

[231] Martin, R. B. 1991. Aluminium in biological systems. In: Nicolini, M., Zatta, P. F. and Corain, B. (ed.), Aluminium in Chemistry, Biology and Medicine. Aseries of advances. Raven Press, New York, p 3-20.

[232] Martinez-Romero, E. 2003. Diversity of *Rhizobium-Phaseolus vulgaris* symbiosis: overview and perspectives. Plant and Soil 252: 11-23.

[233] Mary, P., Ochin, D. & Tailliez, R. 1985. Rates of drying and survival of *Rhizobium meliloti* strains during storage at different relative humidities. Applied and Environmental Microbiology 50: 207-211.

[234] Mary, P., Ochin, D. & Tailliez, R. 1986. Growth status of rhizobia in relation to their tolerance to low water activities and desiccation stress. Soil Biology and Biochemistry 18: 179-184.

[235] Mary, P., Dupuy, C., Biremon, C. D., Defives, C. & Tailliez, R. 1994. Differences among *Rhizobium meliloti* and *Bradyrhizobium japonicum* strains in tolerance to desiccation and storage at different relative humidities. Soil Biology and Biochemistry 26: 1125-1132.

[236] Matsukawa, M. & Greenberg E. P. 2004. Putative exopolysaccharide synthesis genes influence *Pseudomonas aeruginosa* biofilm development. Journal of Bacteriology 186: 4449-4456.

[237] Mattimore, V. & Battista, J. R. 1996. Radioresistance of *Deinococcus radiodurans*: functions necessary to survive ionizing radiation are also necessary to survive prolonged desiccation. Journal of Bacteriology 178: 633-637.

[238] McGrath, S. P. & Lane, P. W. 1989. An explanation for the apparent losses of metals in a long-term field experiment with sewage sludge. Environmental Pollution 60: 235-256.

[239] McGrath, S. P., Hirsch, P. R. & Gillar, K. E. 1988. Effects of heavy metals contamination on the genetics of nitrogen fixing populations of *Rhizobium legumonosarum* bv. *trifolii*. In: Orio, A. A. (ed.), Proceedings of the International Conference of Environmental Contamination. Venice, Italy, p. 164-166.

[240] Mcilveen, W. & Cole, H. 1974. Influence of heavy metals on nodulation of red clover. Phytopathology 64: 583-589.

[241] Mcintyre, J. J., Davies, H., Hore, T. A., Miller, S. H., Dufour, J. P.& Ronson, C. W. 2007. Trehalose biosynthesis in *Rhizobium leguminosarum* bv. *trifolii* and its role in desiccation tolerance. Applied and Environmental Microbiology 73: 3984-3992.

[242] McKay, I. A. & Djordjevic, M. A. 1993. Production and excretion of nod metabolites by *Rhizobium leguminosarum* bv. *trifolii* are disrupted by the same environmental factors that reduce nodulation in the field. Applied and Environmental Microbiology 59: 3385-3392.

[243] McLean, E. O. 1976. Chemistry of soil aluminium. Communications in Soil Science and Plant Analysis 7: 619-636.

[244] Meghvansi, M. K. & Mahna, S. K. 2009. Evaluating the symbiotic potential of *Glomus intraradices* and *Bradyrhizobium japonicum* in Vertisol with two soybean cultivars. American-Eurasian Journal of Agronomy 2(1): 21-25.

[245] Mergaert, P., van Montagu, M. & Holsters, M. 1997. Molecular mechanisms of Nod factor diversity. Molecular Microbiology 25: 811-817.

[246] Mergeay, M., Monchy, S., Vallaeys, T., Auquier, V., Benotmane, A., Bertin, P., Taghavi, S., Dunn, J., Van der Lelie, D. & Wattiez, R. 2003. *Ralstonia metalllidurans*, a bacterium specifically adapted to toxic metals: towards a catalogue of metal-responsive genes. FEMS Microbiology Reviews 27: 385-410.

[247] Meyer, D. R. & Anderson, A. J. 1959. Temperature and symbiotic nitrogen fixation. Nature 183: 161.

[248] Michiels, J., Verreth, C. & Vanderleyden, J. 1994. Effects of temperature stress on bean-nodulating *Rhizobium* strains. Applied and Environmental Microbiology 60: 1206-1212.

[249] Miettinen, P. & Echegoyen, P. E. 1996. The effect of two pesticides (Vitavax-300 and Gaucho) on rhizobia and on the nodulation of four legumes. Agriculture Food and Science Finland 5: 203-207.

[250] Miller, K. J. & Wood, J. M. 1996. Osmoadaptation by rhizosphere bacteria. Annual Reviews of Microbiology 50: 101-136.

[251] Miransari, M. & Smith, D. L. 2007. Overcoming the stressful effects of salinity and acidity on soybean [*Glycine max* (L.) Merr.] nodulation and yields using signal molecule genistein under field conditions. Journal of Plant Nutrition 30: 1967–1992.

[252] Miransari, M. & Smith, D.L. 2008. Using signal molecule genistein to alleviate the stress of suboptimal root zone temperature on soybean-Bradyrhizobium symbiosis under different soil textures. Journal of Plant Interactions 3: 287–295.

[253] Miransari, M. & Smith, D. L. 2009. Alleviating salt stress on soybean [*Glycine max* (L.) Merr] – *Bradyrhizobium japonicum* symbiosis using signal molecule genistein. European Journal of Soil Biology 45: 146-152.

[254] Mohammadi, K., Sohrabi, Y., Heidari, G., Khalesro, S. & Majidi, M. 2012. Effective factors on biological nitrogen fixation. African Journal of Agricultural Research 7(12): 1782-1788.

[255] Morón, B., Soria-Diaz, M. E., Ault, J., Verrolos, G., Noreen, S., Rodriquez-Navarro, D. N., Gil-Serrano, A., Thomas-Oates, J., Megias, M. & Sousa, C. 2005. Low pH changes the profile of nodulation factors produced by *Rhizobium tropici* CIAT899. Chemistry and Biochemistry 12: 1029-1040.

[256] Moulin, L., Munive, A., Dreyfus, B. & Boivin-Masson, C. 2001. Nodulation of legumes by members of the beta-subclass of Proteobacteria. Nature 411: 948-950.

[257] Mpepereki, S., Makonese, F. & Wollum, A. G. 1997. Physiological characterization of indigenous rhizobia nodulating *Vigna unguiculata* in Zimbabwean soils. Symbiosis 22: 275-292.

[258] Muglia, C. I., Grasso, D. H. & Aguilar, O. M. 2007. *Rhizobium tropici* response to acidity involves activation of glutathione synthesis. Microbiology 153: 1286-1296.

[259] Münchbach, M., Nocker, A. & Narberhaus, F. 1999. Multiple small heat shock proteins in rhizobia. Journal of Bacteriology 181: 83- 90.

[260] Munevar, F. & Wollum, A. G. 1981. Effect of high root temperature and *Rhizobium* strain on nodulation, nitrogen fixation, and growth of soybeans. Soil Science Society of America Journal 45: 1113-1120.

[261] Munns, D. N. 1986. Acid soils tolerance in legumes and rhizobia. Advances In Plant Nutrition 2: 63-91.

[262] Munns, R. & Termaat, A. 1986. Whole-plant responses to salinity. Australian Journal of Plant Physiology 13: 143-60.

[263] Munns, R. & Tester, M. 2008. Mechanisms of salinity tolerance. Annual Review of Plant Biology 59: 651-681.

[264] Naeem, F., Malik, K. A. & Hafeez, F. Y. 2008. *Pisum sativum–Rhizobium* interactions under different environmental stresses. Pakistan Journal of Botany 40: 2601-2612.

[265] Nandal, K., Sehrawat, A. R., Yadav, A. S., Vashishat, R. K. & Boora, K. S. 2005. High temperature-induced changes in exopolysaccharides, lipopolysaccharides and protein profile of heat-resistant mutants of *Rhizobium* sp. (Cajanus). Microbiological Research 160: 367-373.

[266] Nandasena, K. G., O'Hara, G. W., Tiwari, R. P., Willems, A. & Howieson, J. G. 2009. *Mesorhizobium australicum* sp. nov. and *Mesorhizobium opportunistum* sp. nov., isolated from *Biserrula pelecinus* L. in Australia. International Journal of Systematic and Evolutionary Microbiology 59: 2140-2147.

[267] Naya, L., Ladrera, R., Ramos, J., Gonzalez, E., Arrese-Igor, C., Minchin, F. R. & Becana, M. 2007. The response of carbon metabolism and antioxidant defenses of alfalfa nodules to drought stress and to the subsequent recovery of plants. Plant Physiology 144: 1104-1114.

[268] Netzer, W. J. & Hartl, F. U. 1998. Protein folding in the cytosol: chaperonin-dependant and -Independent mechanisms. Trends in Biochemical Science 23: 68-73.

[269] Nishiwaki, T., Sato, T., Yashima, H., Ikarashi, T., Ohyama, T., Harper, J.E., Akao, S., Kouchi, H. 1997. Changes in concentration of leghemoglobin components in hypernodulation mutants of soybean, Soil Science and Plant Nutrition, 43: 1091-1096.

[270] Nocker, A., Krstulovic, N. P., Perret, X. & Narberhaus, F. 2001. ROSE elements occur in disparate rhizobia and are funtionally interchangeable between species. Archives of Microbiology 176: 44-51.

[271] O'Connell, K., Gustafson, A. M., Lehmann, M. D. & Thomashow, M. F. 2000. Identification of cold shock gene loci in *Sinorhizobium meliloti* by using a luxAB reporter transposon. Applied and Environmental Microbiology 66: 401-405.

[272] Ogutcu, H., Algur, O. F., Elkoca, E. & Kantar, F. 2008. The determination of symbiotic effectiveness of *Rhizobium* strains isolated from wild chickpeas collected from high altitudes in erzurum. Turkish Journal of Agriculture and Forestry 32: 241-248.

[273] Ohyama, T. & Kumazawa, K. 1981. Nitrogen assimilation in soybean nodules V. Possible pathway of allantoin synthesis in soybean nodules. Soil Science and Plant Nutrition 27: 111-114.

[274] Ohyama, T., Ohtake, N., Sueyoshi, K., Tewari, K., Takahashi, Y., Ito, S., Nishiwaki, T., Nagumo, Y., Ishii, S. & Sato, T. 2009. Nitrogen fixation and metabolism in soybean plants, Nova Science Publishers, Inc., New York.

[275] Ohyama, T., Fujikake, H., Yashima, H., Tanabata, S., Ishikawa, S., Sato, T., Nishiwaki, T., Ohtake, N., Sueyoshi, K., Ishii, S. & Fujimaki, S. 2011. Effect of nitrate on nodulation and nitrogen fixation of soybean. In: El-Shemy, H. (ed.), Soybean Physiology and Biochemistry. InTech, Shanghai, China. DOI: 10.5772/17992.

[276] Oldroyd, G. E. D. 2001. Dissecting symbiosis: developments in Nod factor signal transduction. Annals of Botany (London) 87: 709-718.

[277] Oldroyd, G. E. D., Engstrom, E. M. & Long, S. R. 2001. Ethylene inhibits the Nod factor signal transduction pathway of *Medicago truncatula*. The Plant Cell 13: 1835-1849.

[278] Olsthoorn, M. M. A., Stokvis, E., Haverkamp, J., Spaink, H. P. & Thomas-Oates, J. E. 2000. Growth temperature regulation of host specific modifications of rhizobial lipochitin oligosaccharides: the function of nodX is temperature regulated. Molecular Plant-Microbe Interactions 13: 808-820.

[279] Ortiz-Hernández, M. L. & Sánchez-Salinas, E. 2010. Biodegradation of the organophosphate pesticide tetrachlorvinphos by bacteria isolated from agricultural soils in México. Revista internacional de contaminación ambiental 26: 27-38.

[280] Ovtsyna, A. O., Rademaker, G. J, Esser, E., Weinman, J., Rolfe, B. G., Tikhonovich, I. A., Lugtenberg, B. J. J., Thomas-Oates, J. E. & Spaink, H. P. 1999. Comparison of characteristics of the nodX genes from various *Rhizobium leguminosarum* strains. Molecular Plant-Microbe Interactions 12: 252-258.

[281] Paivoke, A. 1983a. The long term effects of zinc on the growth and development, chlorophyll content and nitrogen fixation of the garden pea (*Pisum sativum* cv Dipple May). Annales Botanici Fennici 20: 205-213.

[282] Paivoke, A. 1983b. The long term effects of lead and arsenate on the growth and development, chlorophyll content and nitrogen fixation of the garden pea (*Pisum sativum* cv. Dipple May). Annales Botanici Fennici 20: 297-306.

[283] Pan, B. & Smith, O. L. 1998. Genistein and daidzein concentrations and contents in seedling roots of three soybean cultivars grown under three foot zone temperatures. Journal of Agronomy and Crop Science 180: 77-82.

[284] Pankhurst, C. E. & Sprent, J. I. 1976. Effects of temperature and oxygen tension on the nitrogenase and respiratory activities of turgid and water-stressed soybeans and French bean root nodules. Journal of Experimental Botany 27: 1-9.

[285] Panoff, J. M., Corroler, D., Thammavongs, B. & Boutibonnes, P. 1997. Differentiation between cold shock proteins and cold acclimation proteins in a mesophilic Gram-positive bacterium, *Enterococcus faecalis* JH2-2. Journal of Bacteriology 179: 4451-4454.

[286] Pan-UK 2003. Current Pesticide Spectrum, Global use and Major Concerns.

[287] Paudyal, S. P., Aryal, R. R., Chauhan, S. V. S. & Maheshwari, D. K. 2007. Effect of heavy metals on growth of *Rhizobium* strains and symbiotic efficiency of two species of tropical legumes. Scientific World 5: 27-32.

[288] Paul, M. J. & Cockburn, W. 1989. Pinitol, a compatible solute in *Mesembryanthemum crvstallinum* L. Journal of Experimental Botany 40: 1093-1098.

[289] Pawlak-Sprada, S., Arasimowicz-Jelonek, M., Podgórska, M. & Deckert, J. 2011a. Activation of phenylpropanoid pathway in legume plants expose to heavy metals: Part I. Effects of cadmium and lead on phenylalanine ammonia-lyase gene expression, enzyme activity and lignin content. Acta Biochimica Polonica 58: 211-216.

[290] Pawlak-Sprada, S., Stobiecki, M. & Deckert, J. 2011b. Activation of phenylpropanoid pathway in legume plants exposed to heavy metals. Part ii. Profiling of isoflavonoids and their glycoconjugates induced in roots of lupine (*Lupinus luteus*) seedlings treated with cadmium and lead. Acta Biochimica Polonica 58: 217-223.

[291] Pena-Cabrales, J. J. & Castellanos, J. Z. 1993. Effect of water stress on N_2 fixation and grain yield of *Phaseolus vulgaris*. Plant and Soil 152: 151-155.

[292] Pereira, S. I. A., Lima, A. I. G. & Figueira, E. M. A. P. 2006. Heavy metal toxicity in *Rhizobium leguminosarum* biovar *viciae* isolated from soils subjected to different sources of heavy metal contamination: effect on protein expression. Applied Soil Ecology 33: 286-293.

[293] Perret, X., Staehelin, C. & Broughton, W. 2000. Molecular basis of symbiotic promiscuity. Microbiology and Molecular Biology Reviews 64: 180-201.

[294] Phadtare, S., Yamanaka, K. & Inouye, M. 2000. The Cold Shock Response. In: Storz, G. and Hengge-Aronis, R. (ed.), Bacterial stress response. ASM Press, Washington, D.C., p. 33-45.

[295] Phillips, D. A. 1999. Using rhizobia in the 21st century. Symbiosis 27: 83-102.

[296] Phillips, D. A. & Torrey, J. G. 1973. Research in grain legume improvement. Span 16: 9-11.

[297] Piha, M. I. & Munnus, D. N. 1987. Sensitivity of the common bean (*Phaseolus vulgaris* L.) symbiosis to high soil temperature. Plant and Soil 98: 183-194.

[298] Pladys, D. & Rigaud, J. 1985. Senescence in French-bean nodules: occurrence of different proteolytic activities. Physiologia Plantarum 63 (1): 43-48.

[299] Poolman, B., Blount, P., Folgering, J. H. A., Friesen, R. H. E., Moe, P. C. & van der Heide, T. 2002. How do Membrane Proteins Sense Water Stress? Molecular Microbiology 44: 899-902.

[300] Porcel, R., Barea, J. M. & Ruiz-Lozano, J. M. 2003. Antioxidant activities in mycorrhizal soybean plants under drought stress and their possible relationship to the process of nodule senescence. New Phytologist 157: 135-143.

[301] Postgate, J. R. 1982. The fundamentals of nitrogen fixation. Cambridge University Press, Cambridge, United Kingdom.

[302] Potts, M. 1994. Desiccation tolerance of prokaryotes. Microbiology Reviews 58: 755-805.

[303] Pueppke, G. S. & Broughton, W. J. 1999. *Rhizobium* sp. strain NGR234 and *R. fredii* USDA257 share exceptionally broad, nested host ranges. Molecular Plant-Microbe Interactions 12: 293-318.

[304] Purwantari, N. D., Date, R. A. & Dart, P. J. 1995. Nodulation and N_2 by *Calliandra calothyrsus* and *Sesbania sesban* grown at different root temperatures. Soil Biology and Biochemistry 27: 421-425.

[305] Rai, R. & Prasad, V. 1983. Studies on compatibility of nitrogen fixation by high temperature-adapted *Rhizobium* strains and *Vigna radiata* genotype at two levels in calcareous soil. Journal of Agricultural Science 101: 377-381.

[306] Ramos, J. L., Galledgos, M. T., Marquéz, S., Ramos-Gonzalez, M. I., Espinosa Uergel, M. & Segura, A. 2001. Responses of Gram-negative bacteria to certain environmental stressors. Current Opinions in Microbiology 4: 166-171.

[307] Rao, A. V. & Venkateswarlu, B. 1983. Microbial ecology of the soils of Indian desert. Agriculture, Ecosystems and Environment 10: 361-369.

[308] Ratnayake, M., Leonard, R. T. & Menge, J. A. 1978. Root exudation in relation to supply of phosphorus and its possible relevance to mycorrhizal formation. New Phytologist 81: 543 552.

[309] Redmond, J. W., Batley, M., Djordjevic, M. A., Innes, R. W., Kuempel, P. L. & Rolfe, B. G. 1986. Flavones induce expression of nodulation genes in *Rhizobium*. Nature 323: 632-635.

[310] Reeve, W. G., Tiwari, R. P., Dilworth, M. & Glenn, A. R. 1993. Calcium affects the growth and survival of *Rhizobium meliloti*. Soil Biology and Biochemistry 25: 581-586.

[311] Reeve, W. G., Tiwari, R. P., Wong, C. M., Dilworth, M. J. & Glenn, A. R. 1998. The transcriptional regulator gene phrR in *Sinorhizobium meliloti* WSM419 is gulated by Low pH and other stresses. Microbiology 144: 3335-3342.

[312] Rhoades, J. D., Chanduvi, F. & Lesch, M. 1999. Soil salinity assessment: Methods and Electrical Conductivity Measurements. M5 ISBN 92-5-104281-0, FAO, Italy.

[313] Riccillo, P. M., Muglia, C. J., de Bruijn, F. J., Roe, A. J., Booth, I. R. & Aguilar, O. M. 2000. Glutathione is involved in environmental stress responses in *Rhizobium tropici*, including acid tolerance. Journal of Bacteriology 182: 1748-1753.

[314] Rice, W. A., Olsen, P. E. & Collins, M. M. 1995. Symbiotic effectiveness of *Rhizobium meliloti* at low temperature. Plant and Soil 170: 351-358.

[315] Richardson, A. E. & Simpson, R. J. 1989. Acid-tolerance and symbiotic effectiveness of *Rhizobium trifolii* associated with a *Trifolium subterraneum* L. based pasture growing in an acid soil. Soil Biology and Biochemistry 21: 87-95.

[316] Richardson, A. E., Djordjevic, M. A., Rolfe, B. G. & Simpson, R. J. 1988a. Effects of pH, Ca and Al on the exudation from clover seedlings of compounds that induce the expression of nodulation genes in *Rhizobium trifolii*. Plant and Soil 109: 37-47.

[317] Richardson, A. E., Simpson, R. J., Djordjevic, M. A. & Rolfe, B. G. 1988b. Expression of nodulation genes in *Rhizobium leguminosarum* biovar *trifolii* is affected by low pH and by Ca and Al ions. Applied and Environmental Microbiology 54: 2541-2598.

[318] Richardson, A. E., Djordjevic, M. A., Rolfe, B. G. & Simpson, R. J. 1989. Expression of nodulation genes in *Rhizobium* and acid-sensitivity of nodule formation. Australian Journal of Plant Physiology 16 (1): 117-129.

[319] Robertson, J. G., Lyttleton, P., Bullivant, S. & Grayston, G. F. 1978. Membranes in *Lupin* root nodules. I. The role of golgi bodies in the biogenesis of infection threads and peribacteroid membranes. Journal of Cell Science 30: 129-149.

[320] Rodrigues, C., Laranjo, M. & Oliveira, S. 2006. Effect of heat and pH stress in the growth of chickpea mesorhizobia. Current Microbiology 53: 1-7.

[321] Rokitko, P. V., Romanovskaia, V. A., Malashenko, L. R., Chernaia, N. A., Gushcha, N. I. & Mikheev, A. N. 2003. Soil drying as a model for the action of stress factors on natural bacterial populations. Microbiology 72: 756-761.

[322] Roth, L. E. & Stacey, G. 1989a. Bacterium release into host cells of nitrogen-fixing soybean nodules: the symbiosome membrane comes from three sources. European Journal of Cell Biology 49: 13-23.

[323] Roth, L. E. & Stacey, G. 1989b. Cytoplasmic membrane systems involved in bacterium release into soybean nodule cells as studied with two *Bradyrhizobium japonicum* mutant strains. European Journal of Cell Biology 49: 24-32.

[324] Rother, J. A., Millbank, J. W. & Thornton, I. 1983. Nitrogen fixation by white clover (*Trifolium repens*) in grasslands on soils contaminated with cadmium, lead and zinc. Journal of Soil Science 34: 127-136.

[325] Rup, L. 1988. Pesticide and nitrogen cycle. Volume III. CRC Press, Inc, Boca Raton, Florida.

[326] Saeki, Y., Aimi, N., Tsukamoto, S., Yamakawa, T., Nagatomo, Y. & Akao, S. 2006. Diversity and geographical distribution of indigenous soybean-nodulating bradyrhizobia in Japan. Soil Science and Plant Nutrition 52: 418-426.

[327] Saeki, Y., Ozumi, S., Yamamoto, A., Umehara, Y., Hayashi, M. & Sigua G. C. 2010. Changes in population occupancy of bradyrhizobia under different temperature regimes. Microbes and Environments 25: 309-312.

[328] Sakamoto, K., Ogiwara, N. & Kaji, T. 2013. Involvement of autoregulation in the interaction between rhizobial nodulation and AM fungal colonization in soybean roots. Biology and Fertility of Soils DOI 10.1007/s00374-013-0804-8.

[329] Salema, M. P., Parker, C. A., Kirby, D. K., Chatel, D. L. & Armitage, T. M. 1981. Rupture of nodule bacteria on drying and rehydration. Soil Biology and Biochemistry 14: 15-22.

[330] Sarkar, A., Nagarajan, R., Chaphadkar, S., Pal, S. & Singbal, S. Y. S. 1997. Contamination of oraganochlorine pesticides in sediments from the Arabian Sea along the west coast of India, Water Research 31: 195-200.

[331] Sarnaik, S. S., Kanekar, P. P., Raut, V. M., Taware, S. P., Chavan, K. S. & Bhadbhade, B. J. 2006. Effect of application of different pesticides to soybean on the soil microflora. Journal of Environmental Biology 27: 423-426.

[332] Sauvage, D., Hamelin, J. & Larher, F. 1983. Glycine betaine and other structurally related compounds improve the salt tolerance of *Rhizobium meliloti*. Plant Science Letters 31: 291-302.

[333] Sawicka, A. & Selwet, M. 1998. Effect of active ingredients on *Rhizobium* and *Bradyrhizobium* legume dinitrogen fixation. Polish Journal of Environmental Studies 7 (5): 317-320.

[334] Saxena, N. P., Saxena, M. C., Johansen, C., Virmani, S. M. & Harris, H. 1996. Adaptation of chickpea in the west asia and north african region. ICRISAT, Hyderabad, India; ICARDA, Aleppo, Syria.

[335] Schmeisser, C., Liesegang, H., Krysciak, D., Bakkou, N., Le Quere, A., Wollherr, A., Heinemeyer, I., Morgenstern, B., Pommerening-Roser, A., Flores, M., Palacios, R., Brenner, S., Gottschalk, G., Schmitz, R. A., Broughton, W. J., Perret, X., Strittmatter, A. W. & Streit, W. R. 2009. *Rhizobium* sp. strain NGR234 possesses a remarkable number of secretion systems. Applied and Environmental Microbiology 75: 4035-4045.

[336] Schoot Uiterkamp, A. J. M. 1990. Nitrogen cycling and human intervention, In Gresshoff, P. M., Roth, L. E., Stacey G. & Newton W. E. (ed.), Nitrogen fixation: achievements and objectives. Chapman and Hall, New York, N. Y., p. 55-66.

[337] Schultze, M. & Kondorosi, A. 2008. Regulation of symbiotic root nodule development. Annual Review of Genetics 32: 33-57.

[338] Sen, D., Baldani, J. I. & Weaver, R. W. 1990. Expression of heat induced proteins in wild type and plasmidcured derivatives of *Rhizobium leguminosarum* by. *trifolii*. In: Gresshoff, P. M., Roth, L. E., Stacey, G., Newton, W. E. (ed.): Nitrogen Fixation Achievements and Objectives. Chapman and Hall, London, p. 200-201.

[339] Sepehri, M., Rastin, N. S., Rahmani, H. A. & Alikhani, H. 2006. Effects of soil pollution by cadmium on nodulation and nitrogen fixation ability of native strains of *Sinorhizobium meliloti*. Journal of Science and Technology of Agriculture and Natural Resources 10: 153-163.

[340] Serraj, R. & Drevon, J. J. 1998. Effects of salinity and nitrogen source on growth and nitrogen fixation in alfalfa. Journal of Plant Nutrition 21: 1805-1818.

[341] Serraj, R., Sinclair, T. R. & Purcell, L. C. 1999. Symbiotic nitrogen fixation response to drought. Journal of Experimental Botany 50: 143-155.

[342] Shamseldin, A. & Werner, D. 2004. Selection of competitive strains of *Rhizobium* nodulating *Phaseolus vulgaris* and adapted to environmental conditions in Egypt, using the gus-reporter gene technique. World Journal of Microbiology and Biotechnology 20: 377-382.

[343] Shime-Hattori, A., Iida, T., Arita, K., Park, K. S., Kodama, T. & Honda, T. 2006. Two type IV pili of *Vibrio parahaemolyticus* play different roles in biofilm formation. FEMS Microbiology Letters 264: 89-97.

[344] Shirkey, B., Kovarcik, D. P., Wright, D. J., Wilmoth, G., Prickett, T. F., Helm, R. F., Gregory, E. M. & Potts, M. 2000. Active Fe-containing superoxide dismutase and abundant SodF mRNA in *Nostoc commune* (cyanobacteria) after years of desiccation. Journal of Bacteriology 182: 189-197.

[345] Silva, K., Florentino, L. A., Silva, K. B., Brandt, E., Vandammme, P., Moreira, F. M. S. 2012. *Cupriavidus necator* isolates are able to fix nitrogen in symbiosis with different legume species. Systematic and Applied Microbiology 35: 175-182.

[346] Singh, G. & Wright, D. 2002. In vitro studies on the effects of herbicides on the growth of rhizobia. Letters in Applied Microbiology 35: 12-16.

[347] Singh, C. S., Lakshmi-Kumari, M., Biswas, A. & Subba Rao, N. S. 1973. Effect of carbonate and bicarbonate of sodium on growth of rhizobia and nodulation in lucerne (*Medicago sativa* L). Indian Journal of Microbiology 13: 125-128.

[348] Singleton, P.W. & Bohlool, B. B. 1984. Effect of salinity on nodule formation by soybean. Plant Physiology 74: 72-76.

[349] Smith, L. T. & Smith, G. M. 1989. An osmoregulated dipeptide in stressed *Rhizobium meliloti*. Journal of Bacteriology 171: 4714-4717.

[350] Smith, L. T., Smith, G. M. & Madkour, M. A. 1990. Osmoregulation in *Agrobacterium*

tumefaciens: accumulation of a novel disaccharide is controlled by osmotic strength and glycine betaine. Journal of Bacteriology 172: 6849-6955.

[351] Smith, L. T., Pocard, J. A., Bernard, T. & Le Rudulier, D. 1988. Osmotic control of glycine betaine biosynthesis and degradation in *Rhizobium meliloti*. Journal of Bacteriology 170: 3142-3149.

[352] Smith, L. T., Smith, G. M., Dsouza, M. R., Pocard, J. A., Le Rudulier, D. & Madkour, M. A. 1994. Osmoregulation in *Rhizobium meliloti*: mechanism and control by other environmental signals. Journal of Experimental Zoology 268: 162-165.

[353] Soria, R. S., Correa, N. S. & Rosas, S. B. 1996. Effect of water stress on the Guar-*Bradyrhizobium* system using PEG 6000. Phyton-International Journal of Experimental Botany 58: 97-106.

[354] Soussi, M., Ocaña, A. & Lluch, C. 1998. Effects of salt stress on growth, photosynthesis and nitrogen fixation in chickpea (*Cicer arietinum* L.). Journal of Experimental Botany 49: 1329-1337.

[355] Soussi, M., Santamaria, M., Ocaña, A. & Lluch, C. 2001. Effect of salinity on protein and lipoplysaccharide pattern in a salt-tolerant strain of *Mesorhizobium ciceri*. Journal of Applied Microbiology 90: 476-481.

[356] Spaink, H. P. 2000. Root nodulation and infection factors produced by rhizobial bacteria. Annual Review of Microbiology 54: 257-288.

[357] Sprent, J. I. & Sprent P. 1990. Nitrogen fixing organisms. Pure and Applied Aspects. Chapman and Hall, Cambridge, p. 256.

[358] Staddon, W. J., Trevors, J. T., Duchesne, L. C. & Colombo, C. A. 1998. Soil microbial diversity and community structure across a climatic gradient in Western Canada. Biodiversity and Conservation 7: 1081-1092.

[359] Stead, D. & Park, S. F. 2000. Roles of Fe superoxide dismutase and catalase in resistance of *Campylobacter coli* to freeze-thaw stress. Applied and Environmental Microbiology 66: 3110 3112.

[360] Steinborn, J. & Roughley, R. J. 1975. Toxicity of sodium and chloride ions to *Rhizobium* spp. in broth and peat culture. Journal of Applied Bacteriology 39: 133-138.

[361] Streeter, J. G. 1985. Accumulation of alpha, alpha-trehalose by *Rhizobium* bacteria and bacteroids. Journal of Bacteriology 164: 78-84.

[362] Streeter, J. G. 2003. Effect of trehalose on survival of *Bradyrhizobium japonicum* during desiccation. Journal of Applied Microbiology 95: 484-491.

[363] Streeter, J. G. & Bhagwat, A. 1999. Biosynthesis of trehalose from maltooligosaccharides in rhizobia. Canadian Journal of Microbiology 45: 716-721.

[364] Streeter, J. G. & Gomez, M. L. 2006. Three enzymes for trehalose synthesis in *Bradyr-*

hizobium cultured bacteria and in bacteroids from soybean nodules. Applied and Environmental Microbiology 72: 4250-4255.

[365] Talibart, R., Jebbar, M., Gouesbet, G., Himdi-Kabbab, S., Wróblewski, H., Blanco, C. & Bernard, T. 1994. Osmoadaptation in rhizobia: Ectoine-induced salt tolerance. Journal of Bacteriology 176: 5210-5217.

[366] Talibart, R., Jebbar, M., Gouffi, K., Pichereau, V., Gouesbet, G., Blanco, C., Bernard, T. & Pocard, J. A. 1997. Transient accumulation of glycine betaine and dynamics of endogenous osmolytes in salt-stressed cultures of *Sinorhizobium meliloti*. Applied and Environmental Microbiology 63: 4657-4663.

[367] Tanabe, S., Iwata, H. & Tatsukawa, R. 1994. Global contamination by persistent organochlorine and their ecotoxiological impact on marine mammals. Science of the Total Environment 154: 163-177.

[368] Tester, M. & Davenport, R. 2003. Na^+ tolerance and Na^+ transport in higher plants. Annals of Botany 91: 503-27.

[369] Thornton, F. C. & Davey, C. B. 1983. Acid dolerance of *Rhizobium trifolii* in culture media. Soil Science Society of America Journal 47: 496-501.

[370] Tindall, B. J. 2008. The genus name *Sinorhizobium* Chen et al. 1988 is a later synonym of *Ensifer* Casida 1982 and is not conserved over the latter genus name, and the species name '*Sinorhizobium adhaerens*' is not validly published. Opinion 84 - Judicial Commission of the International Committee on Systematics of Prokaryotes. International Journal of Systematic and Evolutionary Microbiology 58: 1973-1973.

[371] Tiwari, R. P., Reeve, W. G., Dilworth, M. J. & Glenn, A. R. 1996a. An essential role for actA in acid tolerance of *Rhizobium meliloti*. Microbiology 142: 601-610.

[372] Tiwari, R. P., Reeve, W. G., Dilworth, M. J. & Glenn, A. R. 1996b. Acid tolerance in *Rhizobium meliloti* strain WSM419 involves a two-component sensor-regulator system. Microbiology 142: 1693-1704.

[373] Tong, Z. & Sadowsky, M. J. 1994. A selective medium for the isolation and quantification of *Bradyrhizobium japonicum* and *Bradyrhizobium elkanii* strains from soils and inoculants. Applied and Environmental Microbiology 60: 581-586.

[374] Truchet, G., Roche, P., Lerouge, P., Vasse, J., Caumt, S., de Billy, F., Prome, J. & Denarie, J. 1991. Sulphated lipo-oligosaccharide signals of *Rhizobium meliloti* elicit root nodule organogenesis in alfalfa. Nature 351: 670-673.

[375] Tu, J. C. 1981. Effect of salinity on *Rhizobium*- root hair interaction, nodulation and growth of soyabean. Canadian Journal of Plant Science 61 (2): 231-239.

[376] Valverde, A., Velázquez, E., Gutiérrez, C., Cervantes, E., Ventosa A. & Igual, J. M. 2003. *Herbaspirillum lusitanum* sp. nov., a novel nitrogen-fixing bacterium associated

with root nodules of *Phaseolus vulgaris*. International Journal of Systematic and Evolutionary Microbiology 53: 1979-1983.

[377] Vance, C. P., Heichel, G. H. & Phillips, D. A. 1988. Nodulation and symbiotic nitrogen fixation. In: Hanson, A. A., Barnes D. K. and Hill, R. R. (Eds), Alfalfa and Alfalfa Improvement. Madison: American Society of Agronomy, p. 229-257.

[378] Vanderlinde, E. M., Harrison, J. J., Muszyński, A., Carlson, R. W., Turner, R. J. & Yost, C. K. 2010. Identification of a novel ABC transporter required for desiccation tolerance, and biofilm formation in *Rhizobium leguminosarum* bv. *viciae* 3841. FEMS Microbiology Ecology 71 (3): 327-340.

[379] Vaneechoutte, M., Kampfer, P., De Baere, T., Falsen, E. & Verschraegen, G. 2004. *Wautersia* gen. nov., a novel genus accommodating the phylogenetic lineage including *Ralstonia eutropha* and related species, and proposal of *Ralstonia* (*Pseudomonas*) *syzygii* (Roberts et al. 1990) comb. Nov. International Journal of Systematic and Evolutionary Microbiology 54: 317-327.

[380] Vincent, J. M., Thompson, J. A. & Donovan, K. O. 1961. Death of root nodule bacteria on drying. Australian Journal of Agricultural Research 13: 258-270.

[381] Vinuesa, P., Silva, C., Lorite, M. J., Izaguirre-Mayoral, M. L., Bedmar, E. J. & Martínez-Romero, E. 2005. Molecular systematics of rhizobia based on maximum likelihood and Bayesian phylogenies inferred from rrs, atpD, recA and nifH sequences, and their use in the classification of *Sesbania* microsymbionts from Venezuelan wetlands. Systematic and Applied Microbiology 28: 702-716.

[382] Vriezen, J. A. C., de Bruijn, F. J. & Nüsslein, K. 2007. Responses of rhizobia to desiccation in relation to osmotic stress, oxygen and temperature. Applied and Environmental Microbiology 73 (11): 3451-3459.

[383] Vriezen, J. A. C., de Bruijn, F. J. & Nusslein, K. 2006. Desiccation responses and survival of *Sinorhizobium meliloti* USDA 1021 in relation to growth-phase, temperature, chloride and sulfate availability. Letters in Applied Microbiology 42: 172-178.

[384] Wang, P., Zhong, Z., Zhou, J., Cai, T. & Zhu, J. 2008. Exopolysaccharide biosynthesis is importantfor *Mesorhizobium tianshanense*: plant host interaction. Archives of Microbiology 189: 525-530.

[385] Wang, X., Pan, Q., Chen, F., Yan, X. & Liao, H. 2011. Effects of coinoculation with arbuscular mycorrhizal fungi and rhizobia on soybean growth as related to root architecture and availability of N and P. Mycorrhiza 21: 173-181.

[386] Wani, P. A., Khan, M. S. & Zaidi, A. 2007a. Effect of metal tolerant plant growth promoting *Bradyrhizobium* sp. (vigna) on growth, symbiosis, seed yield and metal uptake by green gram plants. Chemosphere 70: 36-45.

[387] Wani, P. A., Khan, M. S. & Zaidi, A. 2007b. Impact of heavy metal toxicity on plant growth, symbiosis, seed yield and nitrogen and metal uptake in chickpea. Australian Journal of Experimental Agriculture 47: 712-720.

[388] Wani, P. A., Khan, M. S. & Zaidi, A. 2008a. Effect of metal-tolerant plant growth-promoting *Rhizobium* on the performance of pea grown in metal-amended soil. Archives of Environmental Contamination and Toxicology 55: 33-42.

[389] Wani, P. A., Khan, M. S. & Zaidi, A. 2008b. Impact of zinc-tolerant plant growth promoting rhizobacteria on lentil grown in zinc-amended soil. Agronomy for Sustainable Development 28: 449-455.

[390] Weaver, R. W. & Holt, E. C. 1990. Short-term survival of rhizobia on arrow leaf clover seed at different depths. Plant and Soil 122: 147-150.

[391] Weaver, R.W., Materon, L.A., Krautmann, M.E. & Rouquette, F.M. 1985. Survival of *Rhizobium trifolii* in soil following inoculation of arrow leaf clover. MIRCEN Journal of Applied Microbiology and Biotechnology 1: 311-318.

[392] Weisz, P. R., Denison, R. F. & Sinclair, T. R. 1985. Response to drought stress of nitrogen fixation (acetylene reduction) rates by field-grown soybean. Plant Physiology 78: 525-530.

[393] Welsh, D. T. 2000. Ecological significance of compatible solute accumulation by micro-organisms: from single cells to global climate. FEMS Microbiology Reviews 24: 263-290.

[394] Werner, M., Uehlein, N., Proksch, P. & Kaldenhoff, R. 2001, Characterization of two tomato aquaporins and expression during the incompatible interaction of tomato with the plant parasite *Cuscuta reflexa*. Planta, 213: 550-555.

[395] Wetzel, A. & Werner, D. 1995. Ecotoxicological evaluation of contaminated soil using the legume root nodule symbiosis as effect parameter. Environmental Toxicology and Water Quality 10: 127-134.

[396] Whelan, A. M. & Alexander, M. 1986. Effects of low pH and high Al, Mn, and Fe, levels on the survival of *Rhizobium trifolii* and the nodulation of subterranean colver. Plant and Soil 92: 363-371.

[397] Whiteley, M., Bangera, M. G., Bumgarner, R. E., Parsek, M. R., Teitzel, G. M., Lory, S. & Greenberg, E. P. 2001. Gene expression in *Pseudomonas aeruginosa* biofilms. Nature 413: 860-864.

[398] Wood, M. & Cooper, J. E. 1988. Acidity, aluminium and multiplication of *Rhizobium trifolii* possible mechanisms of aluminium toxicity. Soil Biology and Biochemistry 20: 95-99.

[399] Wood, M., Cooper, J. E. & Bjourson, A. J. 1988. Response of *Lotus* rhizobia to acidity and aluminium in liquid culture and in soil. Plant and Soil 107: 227-231.

[400] Xi, C., Schoeters, E., van der Leyden, J. & Michels, J. 2000. Symbiosis-specific expression of *Rhizobium etli* casA encoding a secreted calmodulin-related protein. Proceedings of the National Academy of Sciences of the United States of America 97: 11114-11119.

[401] Yakoleva, Z. M. 1984. The effect of metal oxides on synthesis between *Rhizobium trifolii* and clover plants. Microbiologia 53: 308-312.

[402] Yao, P. Y. & Vincent, J. M. 1969. Host specificity in the root hair "curling factor" of *Rhizobium* sp. Australian Journal of Biological Sciences 22: 413-423.

[403] Yildiz, F. H. & Schoolnik, G. K. 1999. *Vibrio cholerae* O1 El Tor, identification of a gene cluster required for the rugose colony type, exopolysaccharide production, chlorine resistance, and biofilm formation. Proceedings of the National Academy of Sciences of the United States of America 96: 4028-4033.

[404] Young, J. M., Kuykendall, L. D., Martinez-Romero, E., Kerr, A. & Sawada, H. 2001. A revision of *Rhizobium* Frank 1889, with an emended description of the genus, and the inclusion of all species of *Agrobacterium* Conn 1942 and *Allorhizobium undicola* de Lajudie et al. 1998 as new combinations: *Rhizobium radiobacter, R. rhizogenes, R. rubi, R. undicola* and *R. vitis*. International Journal of Systematic and Evolutionary Microbiology 51: 89-103.

[405] Young, J. P., Crossman, L., Johnston, A., Thomson, N., Ghazoui, Z., Hull, K., Wexler, M., Curson, A., Todd, J., Poole, P., Mauchline, T., East, A., Quail, M., Churcher, C., Arrowsmith, C., Cherevach, I., Chillingworth, T., Clarke, K., Cronin, A., Davis, P., Fraser, A., Hance, Z., Hauser, H., Jagels, K., Moule, S., Mungall, K., Norbertczak, H., Rabbinowitsch, E., Sanders, M., Simmonds, M., Whitehead, S. & Parkhill, J. 2006. The genome of *Rhizobium leguminosarum* has recognizable core and accessory components. Genome Biology 7: 34. doi:10.1186/gb-2006-7-4-r34.

[406] Younis, M. 2007. Responses of *Lablab purpureus-Rhizobium* symbiosis to heavy metals in pot and field experiments. World Journal of Agricultural Sciences 3(1): 111-122.

[407] Yousef, A. N. & Sprent, J. I. 1983. Effects of NaCl on growth, nitrogen incorporation and chemical composition of inoculated and NH_4NO_3 fertilized *Vicia faba* (L.) plants. Journal of Experimental Botany 34: 941-950.

[408] Yura, T., Kanemori, M. & Morita, M. T. 2000. The Heat shock response: regulation and function: In Storz G. and Hengge-Aronis R. (ed.), Bacterial stress response, ASM Press, Washington, D.C., p. 3-18.

[409] Zablotowicz, R. M. & Reddy, K. N. 2004. Impact of glyphosate on the *Bradyrhizobium japonicum* symbiosis with glyphosate-resistant transgenic soybean: A minireview. Journal of Environmental Quality 33 (3): 825-831.

[410] Zablotowicz, R. M., Eskew, D. L. & Focht, D. D. 1978. Denitrification in *Rhizobium*. Canadian Journal of Microbiology 24: 757-776.

[411] Zahran, H. H. 1999. *Rhizobium*-legume symbiosis and nitrogen fixation under severe conditions and in an arid climate. Microbiology and Molecular Biology Reviews 63: 968-989.

[412] Zahran, H. H. & Sprent, J. I. 1986. Effects of sodium chloride and polyethylene glycol

on root hair infection and nodulation of *Vicia faba* L. plants by *Rhizobium leguminosarum*. Planta 167: 303-309.

[413] Zahran, H. H., Räsänen, L. A., Karsisto, M. & Lindström, K. 1994. Alteration of lipopolysaccharide and protein profiles in SDS-PAGE of rhizobia by osmotic and heat stress. World Journal of Microbiology and Biotechnology. 10: 100-105.

[414] Zawoznik, M. S. & Tomaro, M. L. 2005. Effect of chlorimuron–ethyl on *Bradyrhizobium japonicum* and its symbiosis with soybean. Pest Management Science 61 (10): 1003-1008.

[415] Zhang, F., Smith, D. L. 1994. Effects of low root zone temperature on the early stages of symbiosis establishment between soybean (*Glycine max* L. Merr) and *Bradyrhizobium japonicum*. Journal of Experimental Botany 45: 1467-1473.

[416] Zhang, F., Smith, D. L. 1995. Preincubation of *Bradyrhizobium japonicum* with genistein accelerates nodule development of soybean at suboptimal root zone temperatures. Plant Physiology 108: 961-968.

[417] Zhang, F. & Smith, D. L. 1996. Genistein accumulation in soybean (*Glycine max* L Merr) root systems under suboptimal root zone temperatures. Journal of Experimental Botany 47: 785-792.

[418] Zhang, F. & Smith, D. L. 1997. Application of genistein to inocula and soil to overcome low spring soil temperature inhibition of soybean nodulation and nitrogen fixation. Plant and Soil 192: 141-151.

[419] Zhang, F. & Smith, D. L. 2002. Interorganismal signaling in suboptimum environments: The legume rhizobia symbiosis. Advances in Agronomy 76: 125-157.

[420] Zhang, J., Hamill, A. S. & Weaver, S. E. 1995. Antagonism and synergism between herbicides: trends from previous studies. Weed Technology 9: 86-90.

9

Autoregulation of Nodulation in Soybean Plants

Sayuri Tanabata and Takuji Ohyama

1. Introduction

Soybean (*Glycine max* [L.] Merr.) seeds contain a high concentration of protein and oil. Therefore, soybean is an important source of protein and calories for humans and livestock in the world. Although to cultivate soybean is in need of a large amount of nitrogen, soybean plants can form root nodules which are symbiotic organs with soil bacteria bradyrhizobia. The partner bradyrhizobia fix atmospheric nitrogen in nodules, and then the plants can use the fixed nitrogen. So soybean plants can grow well even in the absence of soil nitrogen. On the other hand, the plants give photosynthate to the bradyrhizobia. Substantial amounts of photosynthate are required for nitrogen fixation activity in matured nodules. The formation of excess nodules might be a disadvantage because of decreasing the carbon supply each nodule. Therefore, the number of nodule is strictly regulated by the host soybean plant. This system is referred to as the autoregulation of nodulation. The nodule growth of later infection site is suppressed by the rapid response to the earlier rhizobial infection and subsequent nodule initiation. In this chapter, we discuss the autoregulation of nodulation in soybean plants.

2. Research of autoregulation of nodulation in soybean plants

Using a split root technique, it was shown that nodule formation of the latter infected root was systemically suppressed by a prior infected root [1]. Host plants systemically suppress excess nodulation through communications between shoots and roots, by using unknown signals [2]. It is postulated that when soybean root is infected with bradyrhizobia, infection signal is synthesized in root and is transported toward shoot. And then shoot-derived autoregulation signal is synthesized in shoot and is transported toward root. Differentiation of nodule at later infected root is suppressed by the shoot-derived autoregulation signal.

Several hypernodulation mutant lines of soybean were isolated using chemical mutagens such as ethyl-methane-sulfonate (EMS) or N-nitroso-methyl-urea (NMU) since the 1980s; nts lines (from cv. Bragg) [3], NOD lines (from cv. Williams) [4] and En6500 line (from cv. Enrei) [5] were obtained. The mutant lines can form profuse nodulation in the presence or absence of nitrate compared with their parents. These mutants are thought to lack a part of the autoregulation of nodulation. Reciprocal grafting experiments between the hypernodulation mutant and the wild type soybean showed that the hypernodulation phenotype depends on the shoot [6], especially mature leaves [7, 8]. Hypernodulation mutant supplied a large amount of photosynthate to the nodule than the roots [9]. The specific nitrogen-fixation activity and the concentration of leghemoglobin in hypernoduation mutant lines are lower than in the parent line [10, 11, 12]. Therefore, it is reasonable that the number of nodules should be regulated to be optimum by the host plant.

Autoregulatory response is induced during after nodule meristem formation but before nodule emergence [13, 14], perhaps there are some suppressing points [2]. The control has been proposed to operate 2-4 days after inoculation in soybean [15, 16]. About 7-8 days after inoculation, nodules initially appear on roots, only which has not suppressed their growth by shoot-derived autoregulation signal. These time course is a point in the study of autoregulation of nodulation.

3. Photosynthate allocation and regulation of nodulation

One of the idea, plants might suppress the translocation of photosynthate to underground part to control the excessive nodulation. It is shown that availability of photosynthate might be involved in the control of nodule number, with approach grafting [17], and light enhancement and CO_2 enrichment study [18]. From the split root experiment, the inoculation treatment of bradyrhizobia appeared to stimulate the allocation of photosynthate to the inoculated root [19, 20]. However, photosynthate requirement during nodule initiation was not known. In general, growth point has a strong localized sink activity of photosynthate, which is the energy source as well as structural resources. So we investigated whether a requirement of carbon source is increased or not in early stages of nodule formation. The early stage of nodule formation (a period of 2-8 days after inoculation in this study) is important, because autoregulation of nodulation is already activated in this period.

Current photosynthate allocation of soybean cv. Williams was conducted in relation to the nodule initiation [21]. Whole shoots were exposed to $^{14}CO_2$ for 120 min, and the distribution of radioactivity in each organ was determined. During the early stage of nodule formation (4, 6, 8 days after inoculation) the ^{14}C distribution to the inoculated roots did not increase in comparison with uninoculated control roots. ^{14}C respired by underground parts was also similar between the inoculated and control roots. Visualized Images of the distribution of ^{14}C indicated that the shoot apex and root apex, which were growing point in above and underground parts, showed a high radioactivity, but the intense signals were not seen in the expected nodulating parts of roots. These results indicate that current photosynthate allocated to the inoculated roots did not increase in comparison with uninoculated roots, in the early stage of nodule formation.

After the emergence of nodules (10 and 12 days after inoculation), the inoculated roots gradually had priority of photosynthate allocation compared with the uninoculated control roots. At 12 days after inoculation, ^{14}C distribution of inoculated root was statistically increased in comparison with the uninoculated control roots. Also, consumption of current photosynthate by the respiration of underground parts increased at day 12 after inoculation. The radioactivity per dry weight was higher in the nodules than that of the growing point of shoots (i.e. shoot apex). Thus, the underground part after emergence of the nodule is already a high consumer of energy (Figure 1).

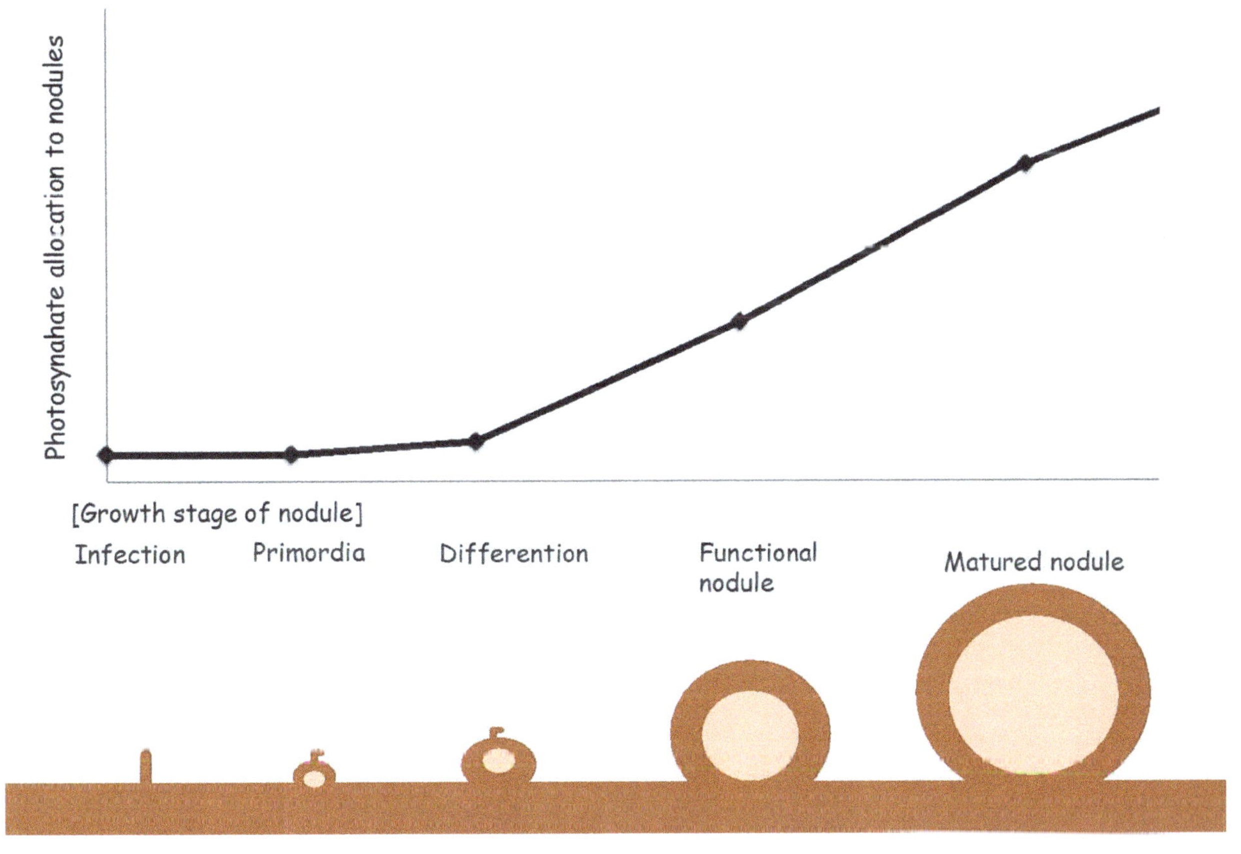

Figure 1. A model for the requirement of photosynthate of nodules during initial stages of nodule formation.

It was shown that an appreciable amount of photoassimilate is not required for nodule initiation before emergence in Williams. What about in hypernodulation mutant lines? The availability of photosynthate might be involved in autoregulation of nodulation. For example, if photosynthate allocation to nodulation in the early stages of nodule formation is markedly increased in NOD1-3, autoregulation of nodulation might be involved in the availability of photosynthate.

To demonstrate this idea, a current photosynthate allocation of hypernodulation mutant of soybean NOD1-3 (isolated from Williams) was examined at 8 days after inoculation using

$^{14}CO_2$ as shown above [22]. The results showed that the ^{14}C distribution in the roots on 8 days after inoculation did not increase when compared with uninoculated control plants in NOD1-3. In visualized images of radioactivity by an imaging plate, the nodules were observed as the strong signal spots in the underground organ. It was concluded that appreciable amount of photoassimilate is not required for the nodule initiation in NOD1-3 in 8 days after inoculation. It was also indicated that the growth of nodule in NOD1-3 is slightly early, compared with Williams; because nodule at early stage had already high sink activity in NOD1-3.

These results indicate that an appreciable amount of photoassimilate is not required for nodule initiation, irrespective of wild type and hypernodulation mutant. It is considered that photosynthate allocation to the nodulated root is not related to autoregulation of nodulation. However, it is unclear whether photosynthate allocation to root in the uninoculated NOD1-3 is similar with the uninoculated Williams or not. If photosynthate allocation to root in uninoculated NOD1-3 is higher than that in uninoculated Williams, cause of hypernodulating trait in NOD1-3 might be attributed to the photosyanthate allocation. So we compared ^{14}C partitioning in the uninoculated plants between Williams and NOD1-3 (unpublished data). Results show that there was no difference between NOD1-3 and Williams in ^{14}C distribution per plant parts (Figure 2). It is strongly suggested that control of nodule formation may be independent of allocation of photosynthetic product.

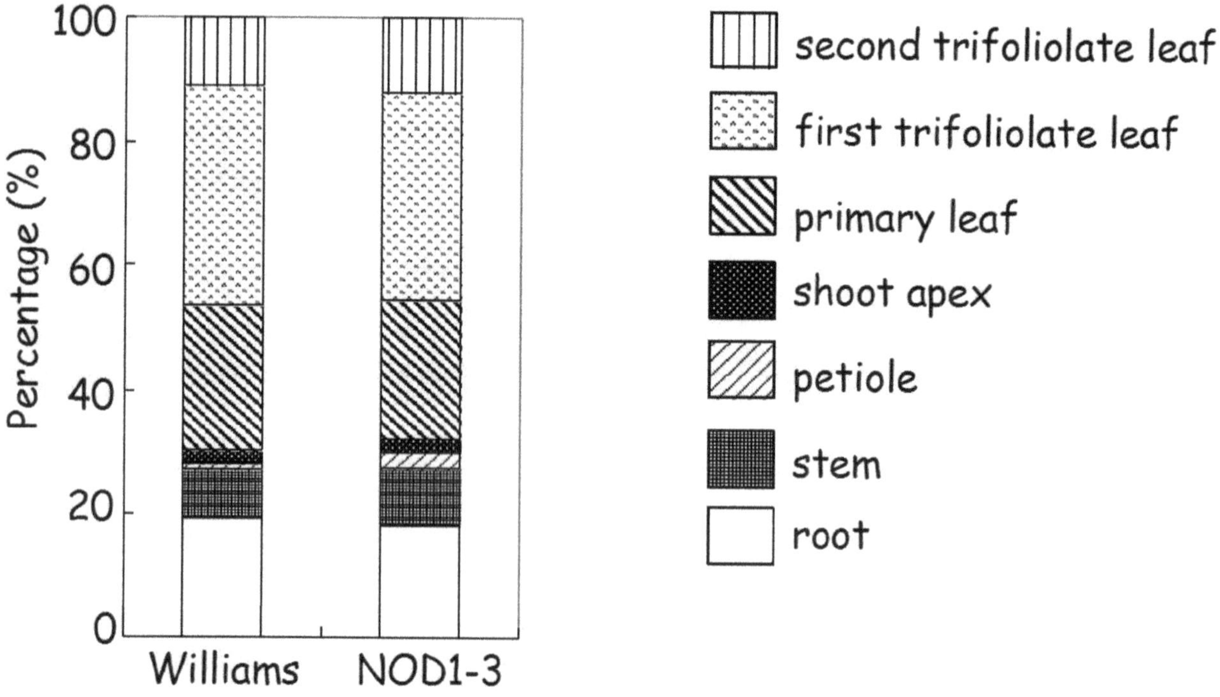

They were cultured without inoculation of bradyrhizobium under –N condition. At 18 days after sowing, whole shoots were exposed to $^{14}CO_2$ for 120min, and the distribution of radioactivity in each organ was determined.

Figure 2. Percentage distribution of radioactivity of ^{14}C in whole plants in uninoculated Williams and NOD1-3.

4. Role of autoregulation system

While it is known that substantial amounts of photosynthate are required for the nitrogen-fixing activity in mature nodules, real-time analysis of photosynthate allocation in hypernodulation has not been reported. Time course study was performed with the aim to clarify the real-time allocation of photosynthetic products in relation to excess nodulation in soybean plants [23]. Allocation of photosynthates to underground part in soybean plants was analyzed using $^{11}CO_2$ and positron-emitting tracer imaging system (PETIS). PETIS can monitor 2D-distribution of positron-emitting tracer like ^{11}C, and a PETIS imaging provides an animation movies of radioactivity non-invasively. Soybean plants were inoculated with *B. japonicum* when they were sown and were grown hydroponically. Whole shoots of the plants at 35 days after sowing were exposed to $^{11}CO_2$, and the ^{11}C imaging was performed for 180 min. In this study, the distribution of ^{11}C-photosynthates in the nodule of hypernodulation NOD1-3 was characterized by comparison to wild type Williams. Results showed that both in the NOD1-3 and Williams, ^{11}C-photosynthates were transported to the root base within about 20 min after feeding of $^{11}CO_2$ and to the root tips within one hour. Most of ^{11}C-photosynthates in the underground part were localized to the root base where many nodules are formed. It was shown that a larger amount of ^{11}C-photosynthates was transported into the nodules on the root base than into those on distal root regions, both per nodule and per volume of nodule (Figure 3). This suggested that the basal nodules may have higher activity for nitrogen fixation both in the mutant and wild type, and such position of a nodule may be a dominant determining factor for the activity. Surprisingly, there was no difference between the mutant and wild type in the amount of ^{11}C-photosynthates accumulated into the nodules per volume of nodule, both in basal and distal regions (Figure 3). These results suggested that the reduced activity of nitrogen fixation in the mutant might be generally caused by the increased proportion of the distal nodules which are poorly fed with photosynthates. So regulation of nodulation on distal region of root is thought important for efficient activity of nitrogen fixation. It might be that shoot-derived autoregulation signal regulate nodulation of distal region of root (Figure 4).

Generally, hypernodulation mutant lines tend to show an inferior growth and seed yield compared with the parent. The peculiar supernodulating variety of soybean "Sakukei 4" showed the improved growth [24], so it was expected high yield. However, Sakukei 4 could not produce higher seed yield compared with the parental varieties Enrei or Tamahomare, because N use efficiency of Sakukei4 was low [25]. One of the reason of low N efficiency might be nodule formation on distal region of root in Sakukei 4.

5. The growth of hypernodulation mutant

Hypernodulation mutants form profuse nodules compared with their wild type parents. Physiological characterization of hypernodulation mutants showed some features of plant growth in addition to the hypernodulation trait. Especially, less vigorous plant growth had

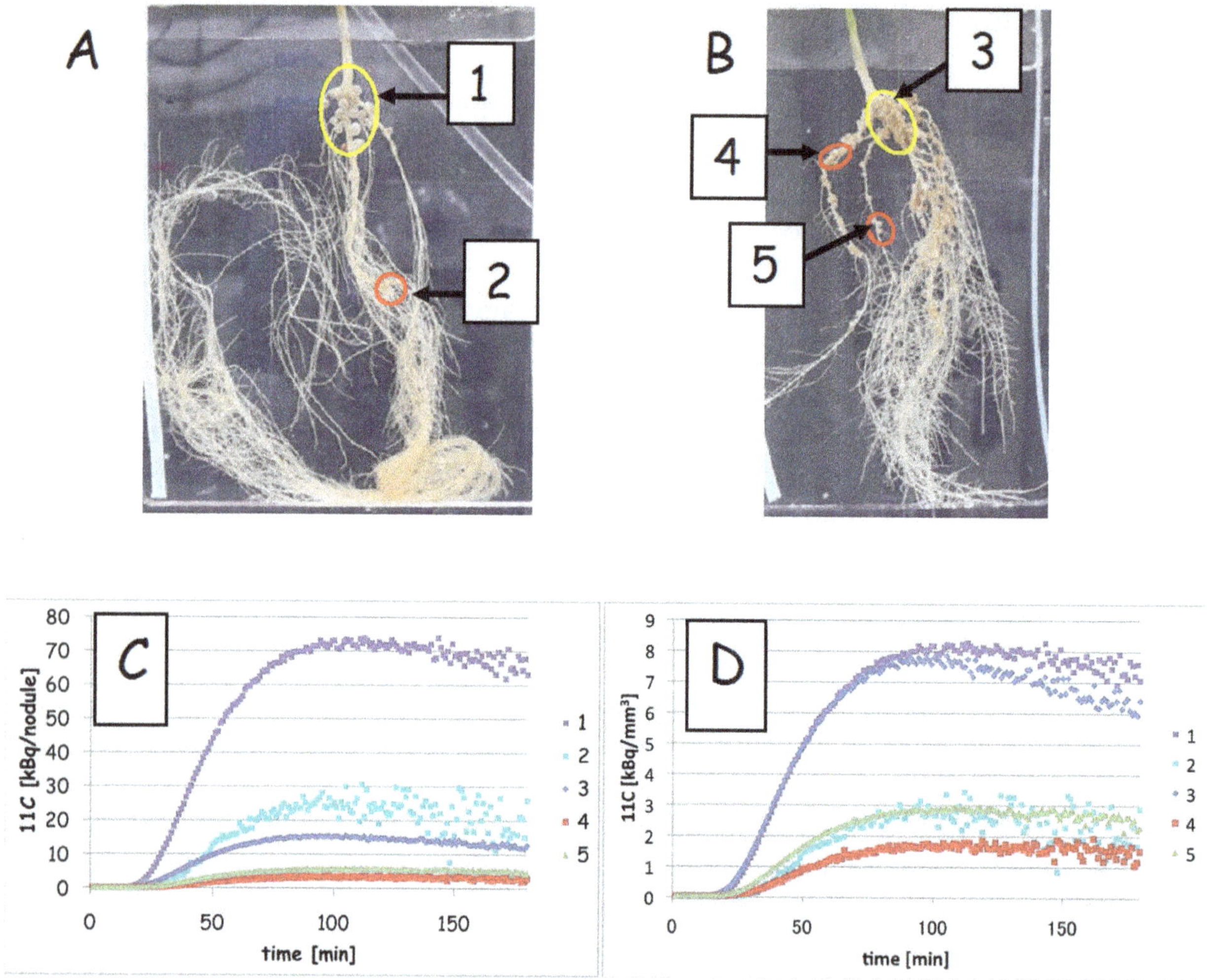

Figure 3. Accumulation of ^{11}C-photosynthates into nodules. (from reference [23]) (A, B) Photographs of underground part in wild-type (A) and NOD1-3 (B). Yellow ellipses indicate regions of interest (ROIs) for the basal nodules and red ellipses indicate ROIs for distal nodules. (C) Time activity curves (TACs) per nodule. (D) TACs per volume of nodules.

been reported [3, 4, 5]. In general, it is because of strong requirement for carbon source of nodules. But it was unclear whether inferiority of growth occurs as a secondary effect of hypernodulation trait or directly due to gene defects by mutation. The details of the growth characteristics of hypernodulation mutant lines may be important to understand the systemic features of the autoregulation mechanism.

GmNARK, which plays an important role in autoregulation of nodulation, was identified in soybean [26, 27]. The soybean hypernodulation NOD mutant lines were isolated from the Williams parent, NOD1-3 and NOD3-7 by N-nitroso-N-methylurea and NOD2-4 by ethyl methanesulfonate [4]. Allelism analysis showed that hypernodulation of all NOD mutant lines from Williams and the En6500 mutant from Enrei is controlled by a single recessive allele [28, 29]. The mutation site of En6500 and NOD3-7 has been identified in *GmNARK* [26, 30]. Therefore, the mutation site of all NOD mutant lines is thought to be in *GmNARK*. There are some different phenotypes between these NOD mutant lines [4, 31], although the reason is unclear so far. Our previous study showed that NOD1-3 had larger nodule compared

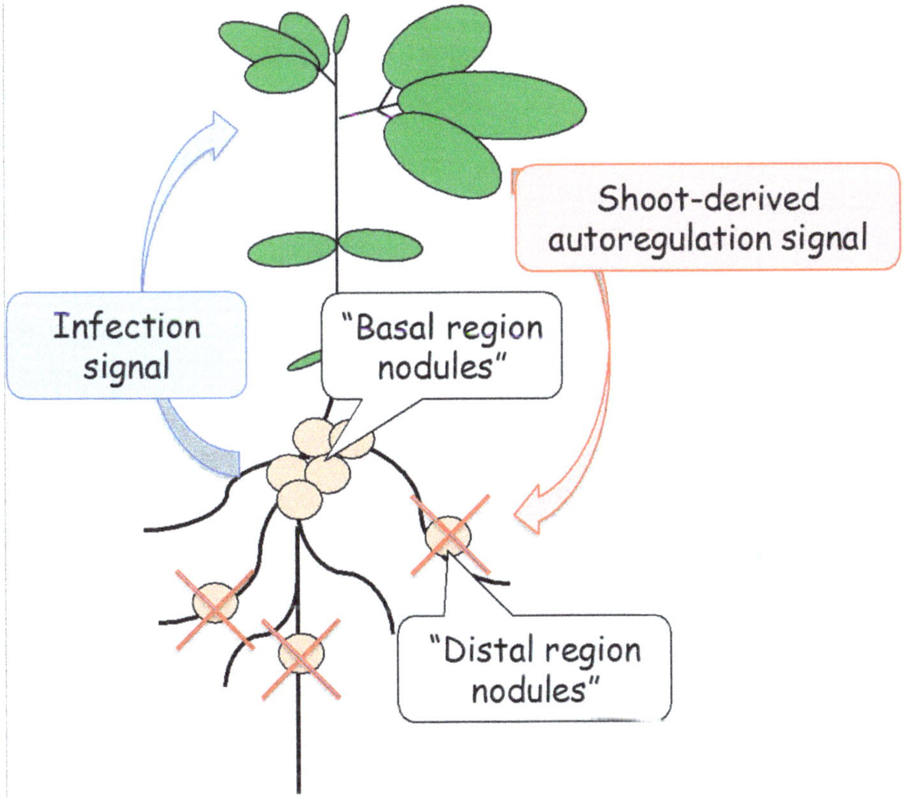

Figure 4. A model for autoregulation of nodulation in soybean plants. Nodule formation of distal region might be regulated by shoot-derived autoregulation signal.

with Williams in 8 days after inoculation [21, 22]. And images analyzed using imaging plate shows that nodules can be observed as partially strong signal in the underground organ at 8 days after inoculation, whereas strong signal of nodule in Williams was narrowly observed at 10 days after inoculation. This observation was consistent with in the literature [32], they reported that initial nodule growth of NOD1-3 was earlier than that of Williams. Also another mutant line, reference [33] suggested that supernodulating mutant nts382 growth in early stages was faster than its wild-type cv.Bragg, especially with early lateral root formation. These observations may indicate that the start of new underground organogenesis of mutant is partially faster than its parents. However, the shoot growth of the hypernodulation mutant had not been well investigated.

The objective of our study was to investigate the characteristics of the initial growth of the NOD mutant lines with the dry matter weight of each part and a whole plant. The experiment was designed within the period when the plants could grow without nitrogen supply. We investigated the phenotypes of not only inoculated plants but also uninoculated plants, and clarified whether phenotypes appeared by a secondary effect of a hypernodulation trait or not; secondary effect means the phenomenon which is caused by excess number of nodules, while a hypernodulation trait is primary effect of mutation gene. The point of this study is that soybean seeds of hypernodulation mutant lines, NOD1-3, NOD2-4, and NOD3-7, and of the Williams parent were carefully selected by uniform seed weight. If seeds were selected randomly, seeds of NOD1-3 and NOD3-7 tend to be smaller than that of Williams and NOD2-4. Difference of seed weight affects initial growth of plants.

First, characteristics of the initial growth of hypernodulation NOD mutant lines were compared with that of Williams with or without inoculation of *B. japonicum* at 7 or 8 days after sowing [34]. When plants were grown without inoculation of bradyrhizobia, the total dry weight of each hypernodulation mutant seedling was not significantly different from that of Williams. Also in inoculated condition, the plant dry weight was not different between hypernodulation mutants and Williams at this stage. The decrease in the growth of hypernodulation mutant is thought to become evident after this growth stage. Some characteristics of each mutant line were observed in this study. Nodule number of NOD1-3 was the largest in all lines in inoculated condition. On the other hand, nodule number of NOD3-7 was similar with that of Williams in this growth stage. It might be that hypernodulation trait of NOD3-7 is not appeared when they are seedling at 8 days after sowing. It is also characterized that stem length of NOD3-7 seedling was shorter than other lines, so the shoot growth of NOD3-7 might be different from other lines. Seedling growth of NOD2-4 was very similar with that of Williams, except for nodule number.

Next, characteristics of the initial growth of NOD mutant lines were compared with Williams at 17 or 18 days after sowing [35]. The plants were grown with or without seed-inoculation of bradyrhizobia, and in the absence or presence of nitrate in the culture solution. When the plants were grown without inoculation, the total dry weight of all mutant lines was not different statistically from Williams, both in the absence and presence of nitrate. When they were grown with inoculation of bradyrhizobia, however, the total dry weight of each mutant line was significantly lower than that of Williams, both in the absence and presence of nitrate. These results indicated that less total dry matter accumulation of hypernodulation mutant lines than the wild type may be the secondary effect due to the large number of nodules, while the hypernodulation trait is a primary effect of the mutated gene. When the plants were grown with inoculation, the nodule number was decreased by the presence of nitrate in Williams, NOD1-3 and NOD2-4, but not in NOD3-7. NOD3-7 may be the most tolerant to nitrate inhibition of nodulation among NOD mutant lines. Growth of each leaf of NOD3-7 and NOD1-3 was different from the wild type and NOD2-4; the expanded leaf was smaller but the new leaf was larger compared with Williams under all conditions (Figure 5). This indicates that NOD3-7 and NOD1-3 might decrease the ability for leaf expansion and have a faster leaf emergence rate. In order to compare the growth rate of the leaves, study was conducted to measure leaf length of the plants daily between 6 and 18 days after sowing (unpublished data). Plants were grown in bradyrhizobium-free nutrient solution without N supply. Figure 6 shows the increase in the length of each leaf blades with time, of Williams and NOD mutant lines. In Williams, an expansion of new leaf started after growth of the previous leaf had been completed. The leaf growth pattern of NOD2-4 was similar with the Williams. In NOD3-7, expansion of new leaf was started earlier before the previous leaf finished expansion. The graph of NOD1-3 shows intermediate pattern between Williams and NOD3-7. The emergence rate of new leaf of NOD3-7 was fastest in all lines. In addition, leaf shape of NOD1-3 and of NOD3-7 was narrower compared with wild type; the leaf index (the ratio of the leaf length to the leaf width) of NOD1-3 and of NOD3-7 was increased than the wild type. It was indicated that NOD3-7 and NOD1-3 had a rapid emergence rate of leaves, while the final size of their expanded leaves was smaller than that of Williams and NOD2-4.

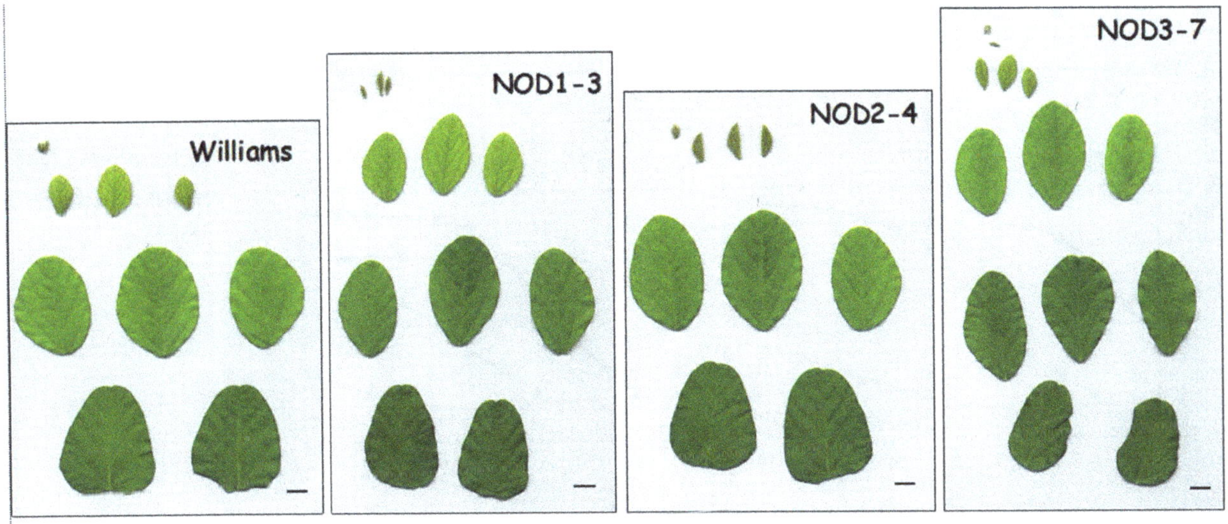

Figure 5. Photos of leaves of hypernodulation mutant lines, NOD1-3, NOD2-4 and NOD3-7, and their parent Williams. Plants were grown without inoculation of bradyrhizobium and under -N condition. In order of primary leaf, 1st trifoliolate leaf, 2nd trifoliolate leaf, and 3rd trifoliolate leaf are indicated from the bottom in each line. The bar indicates 1 cm.

Figure 7 shows summary of initial growth of NOD mutant lines, compared with Williams. When they were inoculated, all NOD mutant lines form great number of nodules than does Williams. But the individual nodule size of NOD mutant line was smaller than that of Williams. In the leaf phenotype, individual leaf size of NOD1-3 and NOD3-7 was smaller than that of Williams. But they had faster rate of new-leaf emergence and their leaf number per plant had tendency to increase. In NOD1-3 and NOD3-7, leaf phenotype was similar with nodule phenotype. On the other hand, leaf phenotype of NOD2-4 was similar with Williams.

6. Leaf phenotype of hypernodulation mutant

We investigated the difference of fully expanded leaf size in cell level using light microscopy when the plants were not inoculated without N supply [36]. Results showed that the cell number of NOD1-3 and NOD3-7 was significantly lower than that of Williams. Cell area of all lines was similar and there were not significant difference between each line. The cell area of NOD3-7 was a little higher than that of other lines, it maybe because of compensation effect of decreased cell number. Compensation effect is characterized by cell enragement triggered by the decrease in cellular proliferation of leaf [37]. It was indicated that NOD1-3 and NOD3-7 produced small-size leaves due to the smaller number of leaf cells,

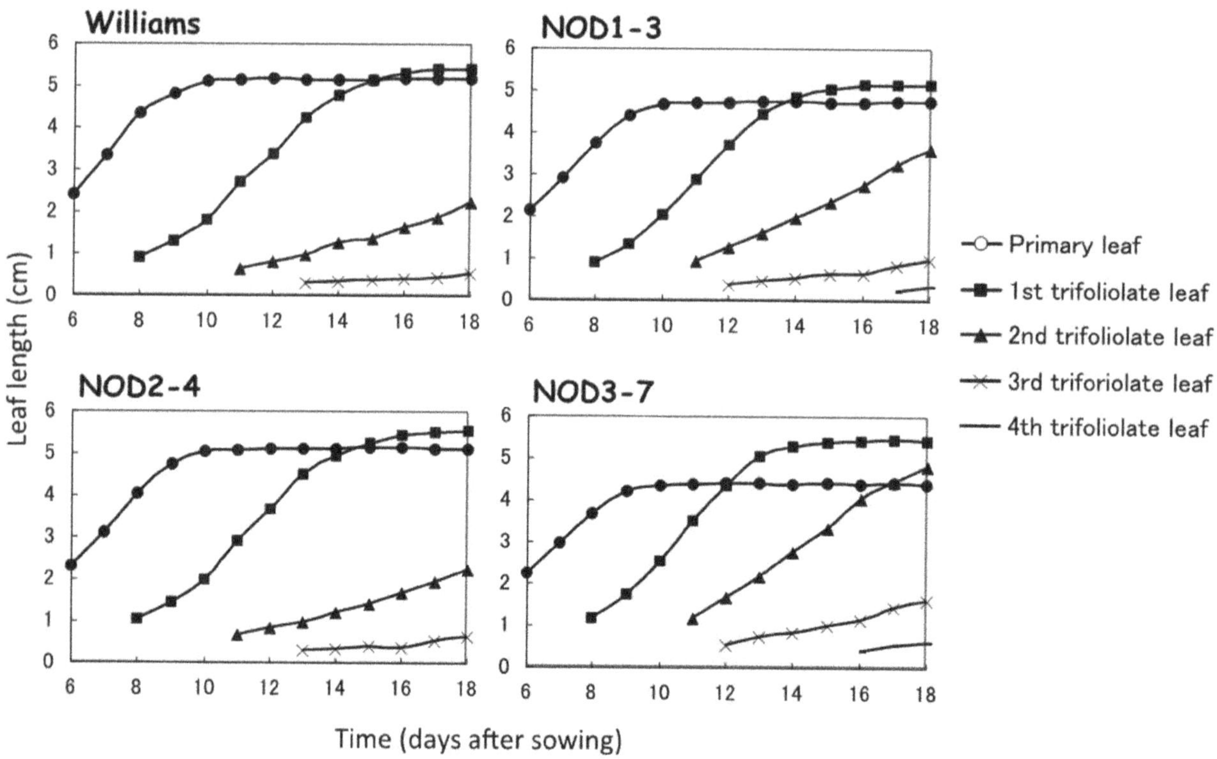

Figure 6. The increase in the length of each leaf blades with time of Williams, NOD1-3, NOD2-4 and NOD3-7. Plants were grown without inoculation of bradyrhizobium and under -N condition. Length of each leaf blade was measured daily between 6 and 18 days after sowing.

	Nodule		Leaf	
	Size	No. / plant	Size	No. / plant
NOD1-3	↘	↗	↘	↗
NOD2-4	↘	↗	similar	similar
NOD3-7	↘	↗	↘	↗

Figure 7. Summary of initial growth of NOD mutant lines, compared with Williams.

compared to Williams parent. These phenotypes were not affected by the inoculation with bradyrhizobia or nitrate supply.

Also in *Lotus japonicus,* leaf growth of hypernodulation mutant har1-4 was compared with their wild type Gifu [38]. Plants were grown in rhizobium-free nutrient solution with 5 mM nitrate supply. Sampling was done at day 14 after sowing. The total dry weight of har1-4 was similar to wild type. Leaf area of primary and 1st and 2nd trifoliolate leaf on har1-4 was similar with that of wild type (Figure 8A). Leaf area of 3rd trifoliolate leaf of har1-4 was slightly smaller than that of wild type. A microscopic study showed that cell number per leaf of 1st trifoliolate leaf of har1-4 was tended to be lower than that of wild type, but it was not significantly difference (Figure 8B). On the other hand, cell area of har1-4 tended to be larger than that of wild type (Figure 8C). Increased cell area of har1-4 might be caused by "compensation effect" of decreased cell number, so leaf area of har1-4 was similar that of wild type. A part of autoregulation system might be related to the control system of leaf-cell proliferation also in *Lotus japonicus*, although could not be clarified in this study.

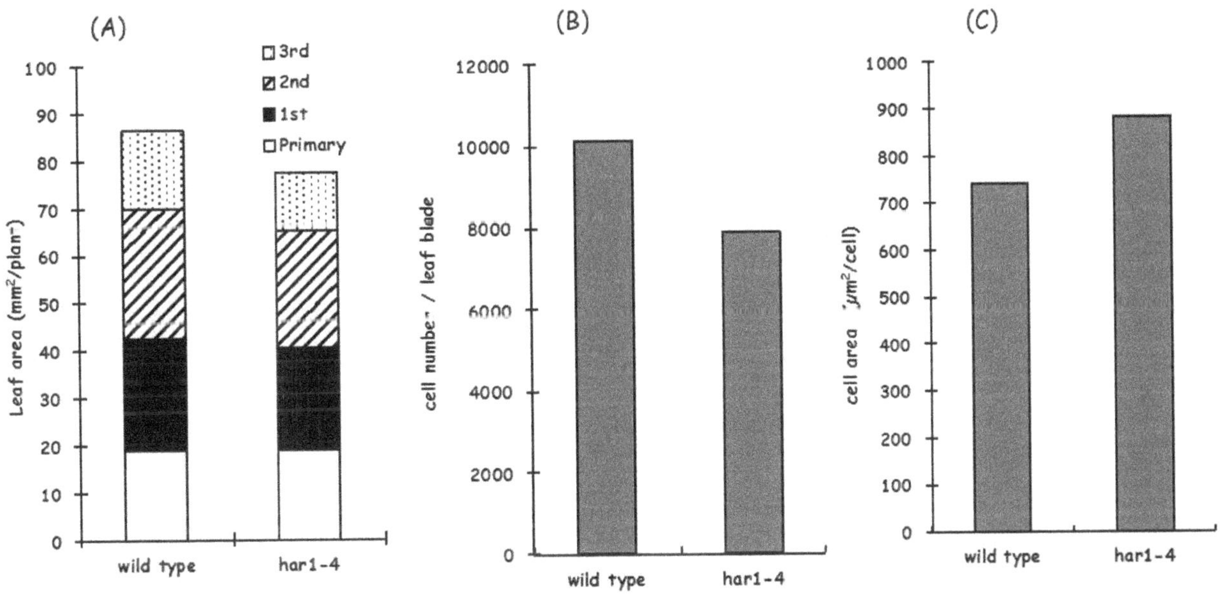

Plants were grown without inoculation under 5 mM NO_3^- supply. Comparison of leaf area (A), palisade cell number per leaf (B), and calculated cell area (C) of the first trifoliolate leaves.

Figure 8. Leaf phenotypes of hypernodulation mutant har1-4 compared with wild type (Gifu) in *Lotus japonicus*. (from reference [38])

To investigate the relationship between nodule and leaf phenotypes, the leaf growth of seed-inoculation plants (active-autoregulation) was compared with the uninoculated plants (basal-autoregulation) in Williams [36]. Results showed that the leaf area of inoculated plants was significantly larger than that of uninoculated plants. The cell number of primary leaves of active-autoregulation plants was significantly higher than that of basal-autoregulation plants. So we concluded that the recognition of infection signal in the shoot might stimulate the cell proliferation of leaf blade. The autoregulation of nodulation system might be active at the basal level when plants are grown without inoculation. Symbiotic nitrogen fixation in root nodules requires a large amount of photosynthates. To activate cell proliferation of leaves along with the autoregulation mechanism by infection of bradyrhizobia would be reasonable.

7. Conclusion

In order to prevent a decrease in growth due to excess nodulation, especially of distal region of root, the control mechanism plays an important role. In the autoregulation of nodulation, yet-unknown signal molecules are used in the communication between shoot and root. Our study showed that control of nodule formation may be independent of allocation of photosynthetic product [21, 22]. Recently, it was indicated that the shoot-derived autoregulation signal is a downregulator of nodulation and is produced by the wild type, rather than activator produced by the mutants [39]. They also indicated that both bradyrhizobia inoculation and visible nodule formation are not essential for shoot-derived autoregulation signal synthesis [39]. Our study showed that the autoregulation of nodulation might be related to the control system of leaf-cell proliferation. Universal regulation system of plant growth might closely link to the autoregulation of nodulation. So it might be reasonable that plant hormones regulate nodule development [40]. For example, salicylic acid [41], abscisic acid [42], methyl jasmonic acid [43] and polyamine [44] are proposed to be involved in autoregulation of nodulation in leguminous plants. The autoregulation of nodulation might be regulated through interplay of several signaling compounds. The non-symbiotic phenotype of the mutants might be helpful to isolate signal molecules of autoregulation of nodulation.

Author details

Sayuri Tanabata and Takuji Ohyama

Niigata University, Japan

References

[1] Kosslak R.M. and Bohlool B.B. Suppression of nodule development of one side of a split-root system of soybeans caused by prior inoculation of the other side. Plant Physiology 1984; 75: 125-130

[2] Caetano-anolles G. and Gresshoff P.M. Plant genetic control of nodulation. Annual Review of Microbiology 1991; 45: 345-382

[3] Carroll B.J., McNeil D.L. and Gresshoff P.M. Isolation and properties of soybean [*Glycine max* (L.) Merr.] mutants that nodulate in the presence of high nitrate concentrations. Proceedings of National Academy of Science 1985; 82: 4162-4166

[4] Gremaud M.F. and Harper J.E. Selection and initial characterization of partially nitrate tolerant nodulation mutants of soybean. Plant Physiology 1989; 89: 169-173

[5] Akao S. and Kouchi H. A supernodulating mutant isolated from soybean cultivar Enrei. Soil Science and Plant Nutrition 1992; 38: 183-187

[6] Delves A.C., Mathews A., Day D.A., Carter A.S., Carroll B.J. and Gresshoff P.M. Regulation of the soybean-*Rhyzobium* nodule symbiosis by shoot and root factors. Plant Physiology 1986; 82: 588-590

[7] Francisco P.B. Jr. and Harper J.E. Translocatable leaf signal autoregulates soybean nodulation. Plant Science 1995; 107: 167-176

[8] Sato T., Yashima H., Harper J.E., Akao S. and Ohyama T. Nodule formation and N_2 fixation traits of the rooted-single leaf isolated from hypernodulating mutants and the wild type. Japanese Journal of Soil Science and Plant Nutrition 1997; 68: 444-447 (in Japanese)

[9] Fujikake H., Yashima H., Sato T., Ohtake N., Sueyoshi K. and Ohyama T. Rapid and reversible nitrate inhibition of nodule growth and N_2 fixation activity in soybean (*Glycine max* (L.) Merr.). Soil Science and Plant Nutrition 2002; 48: 211-217

[10] Nishiwaki T., Sato T., Yashima H., Ikarashi T., Ohyama T., Harper J.E., Akao S. and Kouchi H. Changes in concentration of leghemoglobin components in hypernodulation mutants of soybean. Soil Science and Plant Nutrition 1997; 43: 1091-1096

[11] Sato T., Yashima H., Ohtake N., Sueyoshi K., Akao S., Harper J.E. and Ohyama T. Determination of leghemoglobin components and xylem sap composition by capillary electrophoresis in hypernodulation soybean mutants cultivated in the field. Soil Science and Plant Nutrition 1998; 44: 635-6445

[12] Sato T., Yashima H., Ohtake N., Sueyoshi K., Akao S. and Ohyama T. Possible involvement of photosynthetic supply in changes of nodule characteristics of hypernodulating soybeans. Soil Science and Plant Nutrition 1999; 45: 187-196

[13] Calvert H.E., Pence M.K., Piece M., Malik N.S.A. and Bauer W.D. Anatomical analysis of the development and distribution of *Rhizobium* infections in soybean roots. Canadian Journal of Botany 1984; 62: 2375-2384

[14] Caetano-anolles G. and Gresshoff P.M. Early induction of feedback regulatory responses governing nodulation in soybean. Plant Science 1990; 71: 69-81

[15] Garahty N., Caetano-Anolles G., Joshi P.A. and Gresshoff P.M. Anatomical analysis of nodule development in soybean reveals an additional autoregulatory control point. Plant Science 1992; 58: 1-7

[16] Bano A. and Harper J.E. Plant growth regulators and phloem exudates modulate root nodulation of soybean. Functional Plant Biology 2002; 29: 1299-1307

[17] Francisco P.B. Jr and Harper J.E. Tranlocatable leaf signal autoregulates soybean nodulation. Plant Science 1995; 107:167-176

[18] Bacanamwo M. and Harper J.E. Response of a hypernodulating soybean mutant to increased photosyathate supply. *Plant Science* 1997; 124: 119-129

[19] Singleton P.W. and Kessel C.V. Effect of localized nitrogen availability to soybean half-root systems on photosynthate partitioning to roots and nodules. Plant Physiology 1987; 83: 552-556

[20] Olsson J.E., Nakao P., Bohlool B.B. and Gresshoff P.M. Lack of systemic suppression of nodulation in split root system of supernodulating soybean (*Glycine max* [L.] Merr.) mutants. Plant Physiology 1989; 90: 1347-1352

[21] Ito S., Ohtake N., Sueyoshi K. and Ohyama T. Allocation of photosynthetic products in soybean during the early stages of nodule formation. Soil Science and Plant Nutrition 2006; 52: 438-443

[22] Ito S., Ohtake N., Sueyoshi K. and Ohyama T. Allocation of photosynthetic products in hypernodulation mutant of soybean NOD1-3 in the early stage of nodule formation. Bulletin of Faculty of Agriculture Niigata University 2006; 59: 33-38

[23] Ito S., Suzui N., Kawachi N., Ishii S., Ishioka N.S. and Fujimaki S. Real time analysis of translocation of photosynthate to nodules in soybean plants using $^{11}CO_2$ with a positron-emitting tracer imaging system (PETIS). Radioisotopes 2010; 59: 145-154 (in Japanese)

[24] Takahashi, M., Arihara, J., Nakayama, N., Kokubun, M., Shimada, S., Takahashi, K. and Hajika, M. Breeding of supernodulating soybean cultivar "Sakukei 4". Bulletin of National Institute of Crop Science 2003; 4: 17-28 (in Japanese)

[25] Nakamura T., Nakayama N., Yamamoto R., Shimamura S., Kim Y., Hiraga S., Ohyama T., Komatsu S. and Shimada S. Nitrogen utilization in the supernodulating soybean variety "Sakukei 4" and its parental varieties, "Enrei" and "Tamahomare." Plant Production Science 2010; 13: 123-131

[26] Nishimura R., Hayashi M., Wu G.J., Kouchi H., Imaizumi-Anraku H., Murakami Y., Kawasaki S., Akao S., Ohmori M., Nagasawa M., Harada K. and Kawaguchi M. HAR1 mediates systemic regulation of symbiotic organ development. Nature 2002; 420: 426-429

[27] Searle I.R., Men A.E., Laniya T.S., Buzas D.M., Iturbe-Ormaetxe I., Carroll B.J. and Gresshoff P.M. Long-distance signaling in nodulation directed by a CLAVATA1-like receptor kinase. Science 2003; 299: 109-112

[28] Vuong T.D., Nickell C.D. and Harper J.E. Genetic and allelism analyses of hypernodulation soybean mutants from two genetic backgrounds. Crop Science 1996; 36: 1153-1158

[29] Vuong T.D. and Harper J.E. Inheritance and allelism analysis of hypernodulating genes in the NOD3-7 and NOD2-4 soybean mutants. Crop Science 2000; 40: 700-703

[30] Reid D.E., Ferguson B.J, and Gresshoff P.M. Inoculation- and nitrate-induced CLE peptides of soybean control NARK-dependent nodule formation. Molecular Plant-Microbe Interactions 2011; 24: 606-618

[31] Ohyama T., Nicholas J.C. and Harper J.E. Assimilation of $^{15}N_2$ and $^{15}NO_3^-$ by partially nitrate-tolerant nodulation mutants of soybean. Journal of Experimental Botany 1993; 44: 1739-1747

[32] Sato. T., Onoma N., Fujikake H., Ohtake N., Sueyoshi K. and Ohyama T. Changes in four leghemoglobin components in nodules of hypernodulating soybean (Glycine max [L] Merr.) mutant and its parent in the early nodule developmental stage. Plant Soil 2001; 237: 129-135

[33] Day D.A., Lambers H., Bateman J., Carroll B.J. and Gresshoff P.M. Growth comparisons of a supernodulating soybean (Glycine max) mutant and its wild-type parent. Physiologia Plantarum 1986; 68: 375-382

[34] Ito S., Ohtake N., Sueyoshi K. and Ohyama T. Comparison of initial growth of hyper-nodulation soybean mutants, NOD1-3, NOD2-4 and NOD3-7, and their parent cv. Williams. Bulletin of Faculty of Agriculture Niigata University 2006; 59: 39-43

[35] Ito S., Ohtake N., Sueyoshi K. and Ohyama T. Characteristics of initial growth of hypernodulation soybean mutants, NOD1-3, NOD2-4 and NOD3-7, affected by inoculation of bradyrhizobia and nitrate supply. Soil Science and Plant Nutrition 2007; 53: 66-71

[36] Ito S., Kato T., Ohtake N., Sueyoshi K. and Ohyama T. The autoregulation of nodulation mechanism is related to leaf development. Plant and Cell Physiology 2008; 49: 121-125

[37] Tsukaya H. Controlling size in multicellular organs: focus on the leaf. PLOS Biology. 2008; 6: 1373-1376

[38] Tanabata S., Ohtake N., Sueyoshi K., Kawaguchi M. and Ohyama T. Leaf growth of *Lotus japonicus* hypernodulation mutant *har1-4*. Bulletin of Faculty of Agriculture Niigata University 2013; 66(2): (in press)

[39] Yamaya H. and Arima Y. Evidence that a shoot-derived substance is involved in regulation of the super-nodulation trait in soybean. Soil Science and Plant Nutrition 2010; 56: 115-122

[40] Ferguson B.J. and Mathesius U. Signaling interactions during nodule development. Journal of Plant Growth Regulation 2003; 22: 47-72

[41] Sato T., Fujikake H., Ohtake N., Sueyoshi K., Takahashi T., Sato A. and Ohyama T. Effect of exogenous salicylic acid supply on nodule formation hypernodulating mutant and wild type of soybean. Soil Science and Plant Nutrition 2002; 48: 413-420

[42] Suzuki A., Akune M., Kogiso M., Imagama Y., Osuki K., Uchiumi T., Higashi S., Han S.M.,Yshida S., Asami T. and Abe M. Control of nodule number by the phytohor-

mone abscisic acid in the roots of two leguminous species. Plant and Cell Physiology 2004; 45: 914-922

[43] Nakagawa T. and Kawaguchi M. Shoot-applied MeJA suppresses root nodulation in Lotus japonicus. Plant and Cell Physiology 2006; 47: 176-180

[44] Terakado-Tonooka J. and Fujihara S. Involvement of polyamines in the root nodule regulation of soybeans (Glycine max). Plant Root 2008; 2: 46-53

10

Nitrogen Fixation in Sugarcane

Takuji Ohyama, Atsushi Momose, Norikuni Ohtake,
Kuni Sueyoshi, Takashi Sato, Yasuhiro Nakanishi,
Constancio A. Asis Jr., Soraya Ruamsungsri and
Shotaro Ando

1. Introduction

1.1. Biological nitrogen fixation by endophytic diazotrophic bacteria

Nitrogen (N) is a major essential element for all organisms, and generally the amount of available N (mainly inorganic nitrogen such as nitrate or ammonia) in soil is limiting factor for natural and agricultural plant production [40]. Biological nitrogen fixation (BNF) is a process by which atmospheric dinitrogen (N_2) is reduced into 2 molecules of ammonia (NH_3) by the enzyme nitrogenase with $8H^+$, $8e^-$ and 16 Mg ATP. BNF have important role in N cycle in both global ecosystem and agro-ecosystem. Based on the data compiled by Bezdicek and Kennedy in 1988 [11], about 175 million metric tons of nitrogen per year is estimated to be fixed in global ecosystems, in which 90 million metric tones in agricultural land, 50 million metric tones in forest and non-agricultural land, and 35 million metric tones in sea. At that time, non-biological nitrogen fixation was estimated about 50 million metric tones per year by industrial nitrogen fixation mainly for the synthesis of ammonia fertilizer, and about 20 million metric tones by combustion, and about 10 million metric tones by lightening. In 2009, the production of N fertilizers increased to 106 million metric tones (FAOSTAT), but the amount of BNF still exceeds over non-biological nitrogen fixation.

Historically, the first experimental data to show the BNF by legumes was obtained by Boussingault (from [24]). He began crop rotation experiment including clover (legumes) in 1837, and the result clearly indicated the increase in nitrogen content was associated with cultivation of clover. Hellriegel and Wilfarth demonstrated the bacterial nitrogen fixation in association with legumes in 1888 (from [24]). Beijerinck isolated and cultivated the bacteria from legume nodules, and then he succeeded to inoculate the isolated bacteria to the uninfected host legume plant. In 1893, Winogradsky found that the free-living strictly anaerobic bacterium, *Clostridium pasteurianum,* could fix N_2 (from [24]). Beijerinck showed that the aerobic bacteria, *Azotobacter agilis* and *A. chroococcum,* could also fix N_2. Now a vast genus of bacteria (Green sulfur bacteria, Thallobacteria, Heliobacteria, Cynobacteria, Campylobacter, Proteobacteria), and Archaea can fix N_2 [56].

Several types of nitrogen fixing bacteria are recognized. First is "symbiotic bacteria", such as *Rhizobium,* associated with leguminous plants, and frankia or cyanobacteria with non-legume plants. The second type is "free-living" (non-symbiotic) bacteria, including the cyanobacteria (or blue-green algae) *Anabaena* and *Nostoc* and such genera as *Azotobacter, Beijerinckia,* and *Clostridium* living in soil or water. The third type resides around the plant roots (rhizosphere) and provides fixed nitrogen to the plant. This is called "associative nitrogen fixation". Fourth is "endophytic nitrogen fixation" associated with cereal grasses such as sugarcane.

2. Characteristics of sugarcane

Sugarcane (*Saccharum* spp. hybrid) is a tall, perennial grass (family Poaceae, subfamily Panicoide), and is cultivated in tropical and warm-temperate regions between 35°N and 35°S and from sea level to altitudes of 1,000 m in a wide variety of soil types [47]. Most of commercial sugarcane varieties are hybrids with *Saccharum officinarum* [1]. The optimal temperature for sugarcane cultivation is between 20 and 35 °C and the minimum rainfall requirement is 1,200mm per year [1]. The stalks (stems) of sugarcane are harvested at 9 to 18 months after planting the mother stem cutting (setts). Once planted, sugarcane can be harvested several times, because new stalks, called ratoons, repeatedly grow from the stubble. For many years, sugarcane has been used for sugar and an alcoholic drink production. Recently, the use of sugarcane alcohol (ethanol) as an automotive fuel to replace gasoline has rapidly increased [16, 35]. At the moment, sugarcane is the most economically and environmentally advantageous crop for bio-ethanol production.

In 2011, world production of sugarcane was 1,794 million tons (FAOSTAT). This is much greater than for the other major crops such as maize (883 million tons), paddy rice (723 million tons), wheat (704 million tons) and potatoes (374 million tons). Sugarcane production is highest in Brazil (734 million tons), followed by India (342 million tons), and China (115 million tons). In 2011, sugarcane was cropped over an area of 25 million hectares. Sugarcane is a C4 plant, which has an efficient photosynthetic system, and it can convert up to 2% of incident solar energy into biomass. In 2011, the average yield was 70.5 tons per hectare. It grows up to 2-6 m in height (Figure 1) and the thick stalks (stems) store a high concentration of sucrose accumulated in stalk internodes, and the expressed stalk juice contains sucrose concentration at between 12 and 20% (W/V).

Figure 1. Growth of sugarcane (left), and sugarcane mother stalk for planting (right).

3. Contribution of biological nitrogen fixation in sugarcane cultivation

In Brazil, sugarcane crops accumulate N between 100 and 200 kg N per hectare per year, while N fertilization rates are relatively low, usually less than 60 kg N per hectare [47]. Also, the response of sugarcane crops to N fertilizers is usually very weak [12,47]. The use of low N fertilizer input was justified by the results of 135 field experiments in all of main cane-growing areas of Brazil, with only 19% of the studies showing a significant increase in yield owing to the application of N fertilizer [6]. In some sites of Brazil, sugarcane has been grown continuously for more than 100 years without any N fertilizer being applied at all [20]. This circumstantial evidence suggests a high potential for BNF in sugarcane.

4. Estimation of contribution of biological nitrogen fixation by ^{15}N dilution technique and ^{15}N natural abundance method

Using a ^{15}N dilution technique involving the supply of a ^{15}N-labeled fertilizer, Lima *et al.* [33] evaluated the contribution of BNF in four sugarcane varieties planted in large-size pots and concrete tanks. The result showed that 40-60% of plant N was derived from BNF in the variety CB 47-89. Urquiaga *et al.* [54] also calculated the contribution of BNF in several cultivars of sugarcane, and found it to be about 70% for the most promising genotypes. Again, using the ^{15}N natural-abundance method, Boddey *et al.* [13] showed that 25-60% of the N assimilated in sugarcane at various sites in Brazil was derived from BNF. The analysis showed very large BNF-inputs to several sugarcane varieties, especially the wild non-commercial species Krakatau (*S. spontaneum*) used in plant breeding in Brazil, as well as the commercial varieties SP 70-1143 and CB 45-3 in low-fertility soils. Yoneyama *et al.* [57] examined the contribution of BNF using a ^{15}N natural-abundance method in Brazil, the Philippines and Japan, comparing the abundance of ^{15}N in sugarcane with that in neighboring weeds as control plants. At many but not in all of the sites in Brazil, a contribution from BNF was indicated.

Asis *et al.* [5] estimated the contribution of nitrogen fixation of Japanese sugarcane cultivar NiF8 by ^{15}N dilution and natural ^{15}N abundance techniques, and total %Ndfa (percentage of N derived from atmospheric dinitrogen) was estimated to be 27-38%. They also reported that the estimated the %Ndfa was 26% for the roots, 14.1% for the stem and 20.5% for the leaves. Nishiguchi *et al.* [38] estimated the contribution of BNF using the ^{15}N dilution in Japanese varieties of sugarcane, and found that the percentage between 10% and 40% of sugarcane N was derived from BNF depending on the cultivar and also on the availability of mineral N.

In Niigata, a pot experiment was conducted to estimate the contribution of nitrogen fixation in sugarcane (*Saccharum* spp. var. NiF8) plants using ^{15}N dilution method in relation to N supply period [26]. Sugarcane plants were grown from a cut stalk in water for 20 days (Figure 1) and a young shoot was transplanted to the 1/5000 a Wagner pot filled with vermiculite. Three fertilizer treatments were applied; 1) N0: Nitrogen free culture solution was supplied, 2) N100: ^{15}N labeled ammonium sulfate was continuously supplied at the rate of 100 mgN per pot a week, 3) N100N0: ^{15}N labeled ammonium sulfate was supplied at the rate of 100 mgN per pot a week until 6 weeks after transplanting (WAT), then the plants were cultivated with N free solution. The growth of the plants was measured every week, and plants were harvested at 12 WAT and 20 WAT. Total N content and ^{15}N abundance in each part were determined.

Figure 2 shows the changes in the shoot length and leaf number of each treatment for 12 weeks after transplanting (WAT). The N0 plants grew very poor, and sole nitrogen fixation is not enough to support vigorous growth of sugarcane NiF8 cultivated in Niigata. The N100 plants and N100N0 plants showed relatively similar shoot length and leaf number, but the N100N0 plants leaves were pale compared with the N100 plants. At 12 WAT, total N content was 408 mgN and 286 mgN per plant in N100 treatment and N100N0 treatment, respectively. At 20 WAT, total N content was 569 mgN (N100) and 292mgN (N100N0).

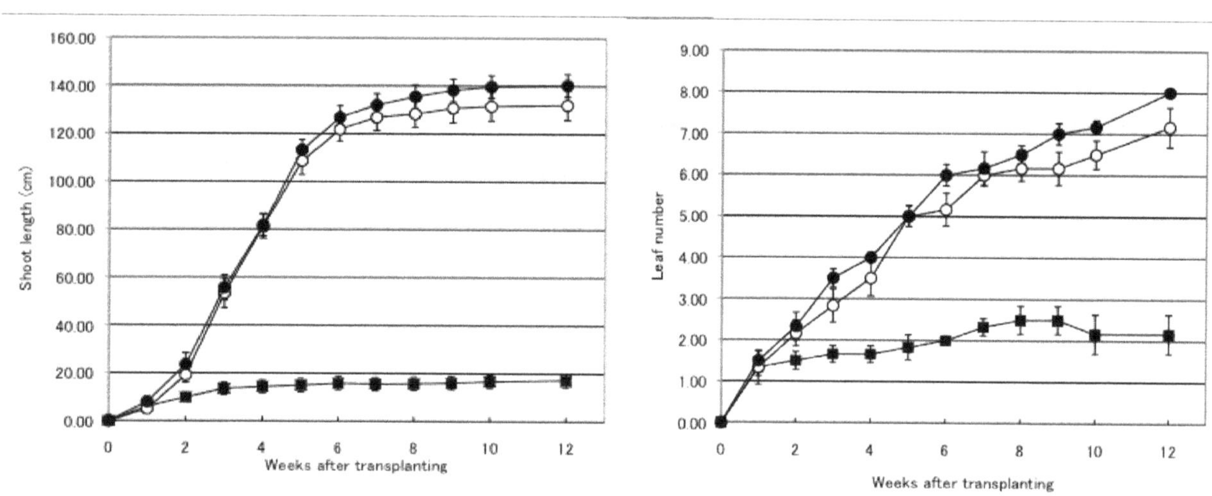

Figure 2. Shoot length and leaf number of sugarcane with N0 (■), N100 (●), and N100N0 (○) fertilizer treatments.

Figure 3 shows the amount of N derived from three sources of N; Ndfa (N derived from nitrogen fixation), Ndfs (N derived from a mother stalk), and NDff (N derived from fertilizer). The amount of Ndfa in N100 and N100N0 was 87 mgN (21%Ndfa) and 48 mgN (17%Ndfa) per plant respectively. At 20 WAT, the amount of Ndfa was 87 mg (15%Ndfa) in N100 and 57 mgN (20%Ndfa) in N100N0 treatment. Among organs, the estimated %Ndfa tended to be higher in old leaves and stalk, and lower in green leaves and stems (Table 1). From this experiment, the continuous supply of N fertilizer did not inhibit nitrogen fixation in sugarcane compared with N deficient plants.

12WAT		Percentage of Nitrogen (%)			20WAT		Percentage of Nitrogen (%)		
		N dff	Ndfs	Ndfa			N dff	Ndfs	Ndfa
Green Leaves	N100N0	66	21	13	Green Leaves	N100N0	62	20	18
	N100	73	10	17		N100	74	10	16
Old Leaves	N100N0	43	22	35	Old Leaves	N100N0	52	21	28
	N100	39	10	51		N100	65	10	25
Stem	N100N0	68	22	10	Stem	N100N0	66	21	13
	N100	80	8	12		N100	81	10	10
Stalk	N100N0	39	22	39	Stalk	N100N0	34	30	35
	N100	43	9	48		N100	66	9	24
Root	N100N0	57	21	21	Root	N100N0	55	19	26
	N100	60	9	31		N100	74	10	15
Total	N100N0	62	21	17	Total	N100N0	59	21	20
	N100	70	9	21		N100	75	10	15

Table 1. Percentage of nitrogen derived from fertilizer (%Ndff), from stalk (%NDfs), and atmospheric nitrogen (%Ndfa) at 12 WAT and 20WAT.

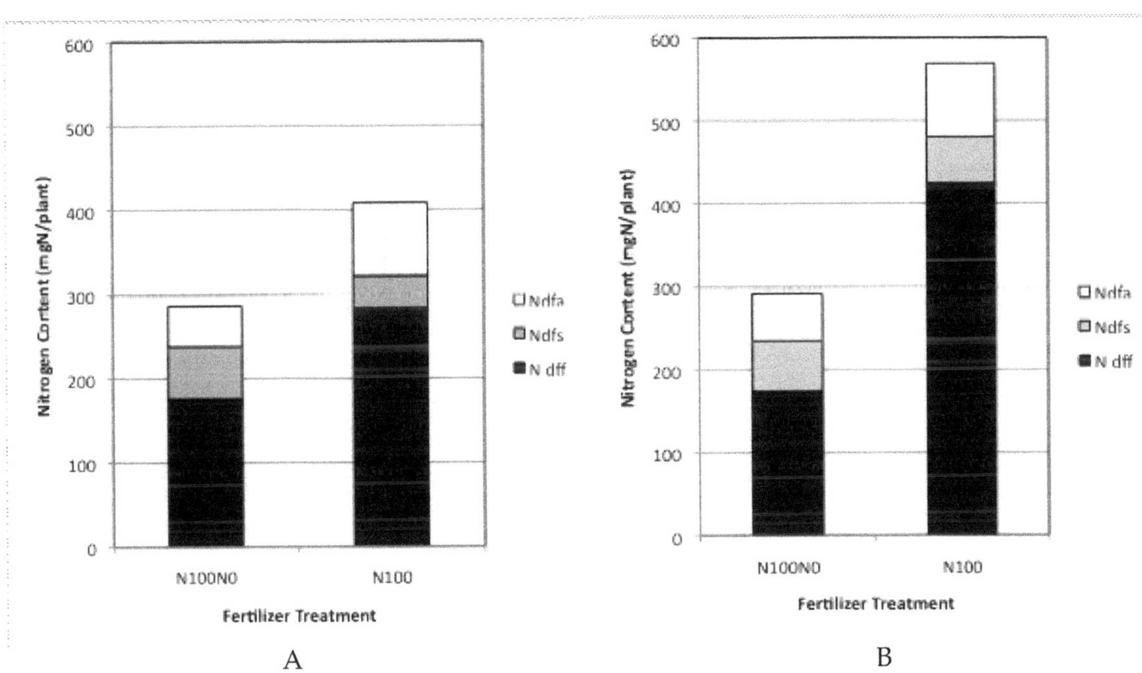

Figure 3. Nitrogen content in sugarcane plants at 12 WAP (A) and 20WAP (B).

5. $^{15}N_2$ fixation studies in sugarcane

Both ^{15}N dilution method and ^{15}N natural-abundance method are indirect methods for estimating nitrogen fixation. Direct evidence can be obtained by $^{15}N_2$ tracer experiment. In earlier studies, Ruschel *et al.* [48] exposed soil-grown 60 or 90-day-old sugarcane plants to $^{15}N_2$ under a low oxygen (O_2) concentration (2%), and near normal O_2 concentration (16%). Significant incorporation of ^{15}N was detected in roots and shoots of intact plants only under the lower O_2 concentration (2%). In addition, the detached roots fixed $^{15}N_2$ under 2% O_2, but $^{15}N_2$ fixation was not detected in the detached shoots. Based on these results, they suggested that the site of fixation is within the roots and not in the shoots but that the fixed N is rapidly translocated to the shoot within 40 hours. Again working with sugarcane, Sevilla *et al.* [50] compared plants inoculated with wild-type *Acetobacter diazotrophicus* strain PA15, with plants inoculated with a Nif$^-$ mutant of PA15, and with un-inoculated (sterile) plants. Fixation of $^{15}N_2$ was observed only in the shoots and roots of the plants inoculated with the wild-type Nif$^+$ PA15.

Figure 4. Exposure of a sugarcane plant to $^{15}N_2$. (Left) $_{15}N_2$ gas is introduced to the cylinder displacing the nutrient solution. (Right) The plant is exposed to $_{15}N_2$ under controlled light and temperature conditions.

Momose *et al.* reported on the nitrogen fixation and translocation in young sugarcane plants associated with endophytic nitrogen-fixing bacteria [36]. The tracer $^{15}N_2$ was used to investigate the sites of N_2 fixation and the possible translocation of the fixed N. Young sugarcane plants from a stem cutting were exposed to $^{15}N_2$-labeled air in a 500 mL plastic cylinder (Figure 4). After 3 days of $^{15}N_2$ feeding, the percentage of N derived from $^{15}N_2$ was higher in the roots (2.22 %Ndfa) and stem cutting (0.271 % Ndfa) than the shoot (0.027% Ndfa) (Figure 5). At day-3, the stem cutting showed the highest quantity of fixed nitrogen per plant (about 38 $\mu g^{15}N$), followed by the roots (13 $\mu g^{15}N$) and shoots (2 $\mu g^{15}N$). The large amount of ^{15}N in the stem cutting is due to the much greater N content of the cutting than young shoot and roots in the young plants at this stage.

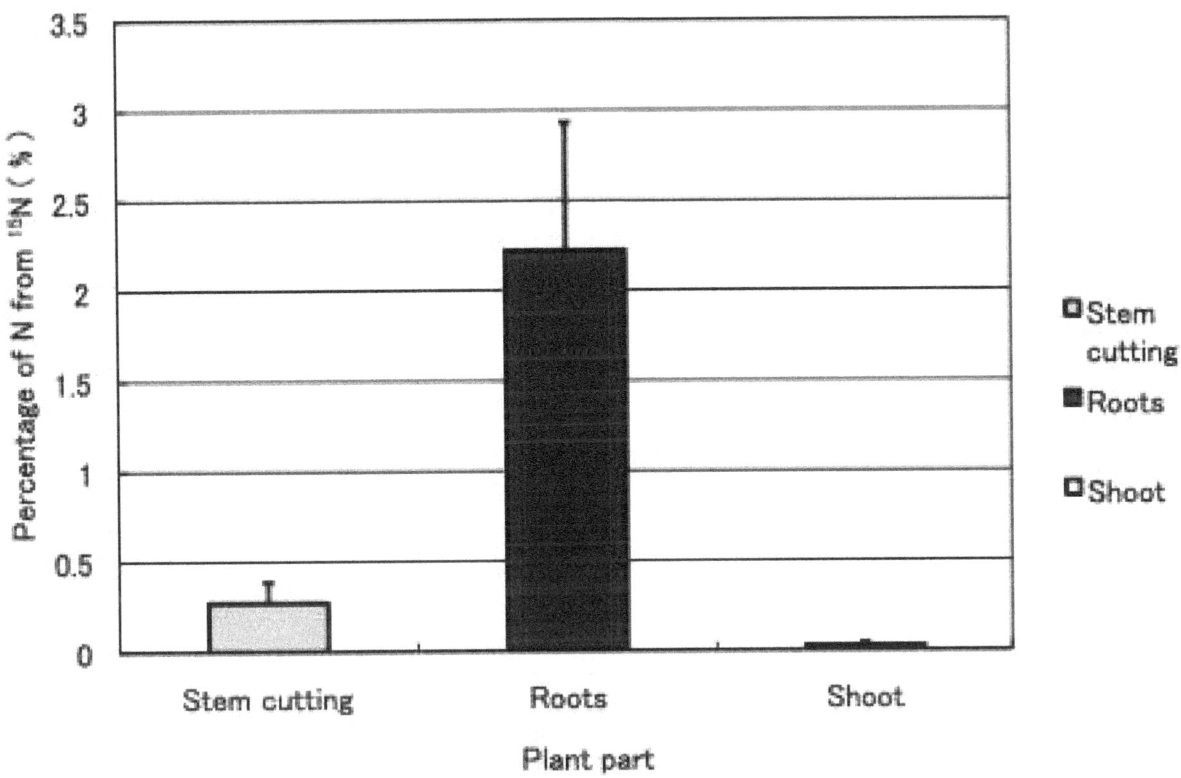

Figure 5. The fraction of ^{15}N (expressed as a % of total N) fixed from $_{15}N$-labeled gas in the stem cutting, roots and shoot of sugarcane plants after three days of exposure (average with standard deviation).

The plants exposed to $^{15}N_2$ for 7 days were grown in normal air for a further chase period. After 21 days, about half of the N originating in the stem cutting had been transported to the shoot and roots, suggesting that the cutting played a role in supplying N for the growth of shoot and roots (Figure 7). Most of the fixed N was distributed in the 80 % ethanol-insoluble fractions in each plant part, and the ^{15}N fixed either in the roots or in the stem cutting remained there and was not appreciably transported to the shoot (Figure 8). The results were quite different from the fate of fixed N in soybean nodules, which is rapidly transported from nodules to roots and shoots.

From the results obtained in this experiment with young sugarcane plants, it is confirmed that the roots are the most active site of N_2 fixation followed by the stem cutting. The sugarcane

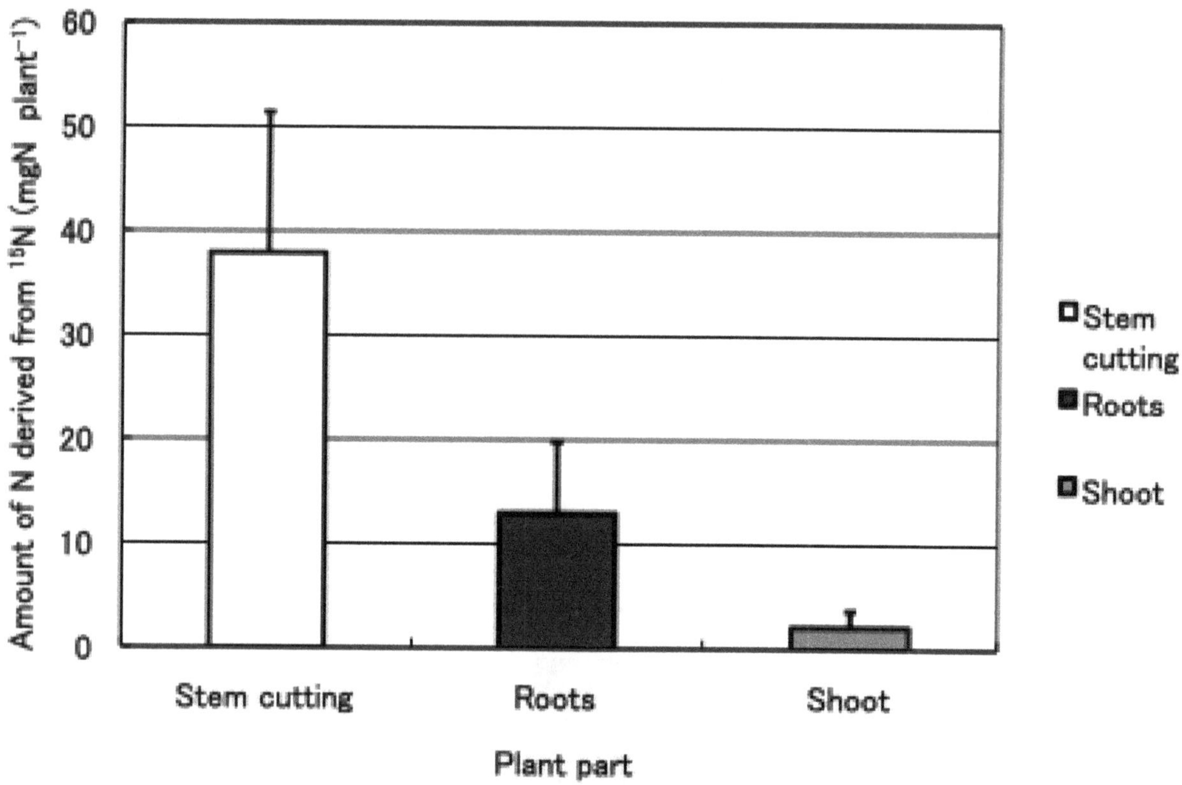

Figure 6. The mass of ^{15}N (μg) in the stem cutting, roots and shoot of sugarcane plants exposed to $^{15}N_2$ for three days (average with standard deviation).

cuttings were initially cultured in water not in soil, so the N_2-fixing endophytes in the roots might originate from the stem cutting or root primordia. If this is the case, to support active N_2 fixation, nitrogen-fixing bacteria may move into the developing roots, and colonize the intercellular space in the roots.

In an earlier study with NiF8 sugarcane using a ^{15}N dilution technique [5], after 5 months of cultivation in a pot supplied with ^{15}N-labeled mineral fertilizer, it was estimated that the roots contributed greater proportions of BNF (26%) than the stem (14%) and leaves (21%). Compared with the stem and shoot, the roots offer certain advantages as sites of N_2 fixation. These are first that the host plant provides carbohydrates to the root endophytes, and second that oxygen concentrations are usually lower in the roots and soil than in the atmosphere (pO_2 0.21). Low pO_2 is beneficial to protect nitrogenase from oxygen damage, and for optimum nitrogen fixation activity.

The fixation of nitrogen in stem cuttings of sugarcane is probably due to the activity of endophytes. Because very young sugarcane plants were used in our experiments, the finding of extremely limited N_2 fixing activity in the shoots should not necessarily be taken to imply an insignificant stem endophytic contribution to the N economies of mature field-grown sugarcane plants. Many sugarcane endophytes (e.g. *Gluconacetobacter diazotrophicus*) are adapted to high sucrose concentrations while our very young shoots did not have mature stems that offered these conditions. However, there are some evidences that BNF in sugarcane stems

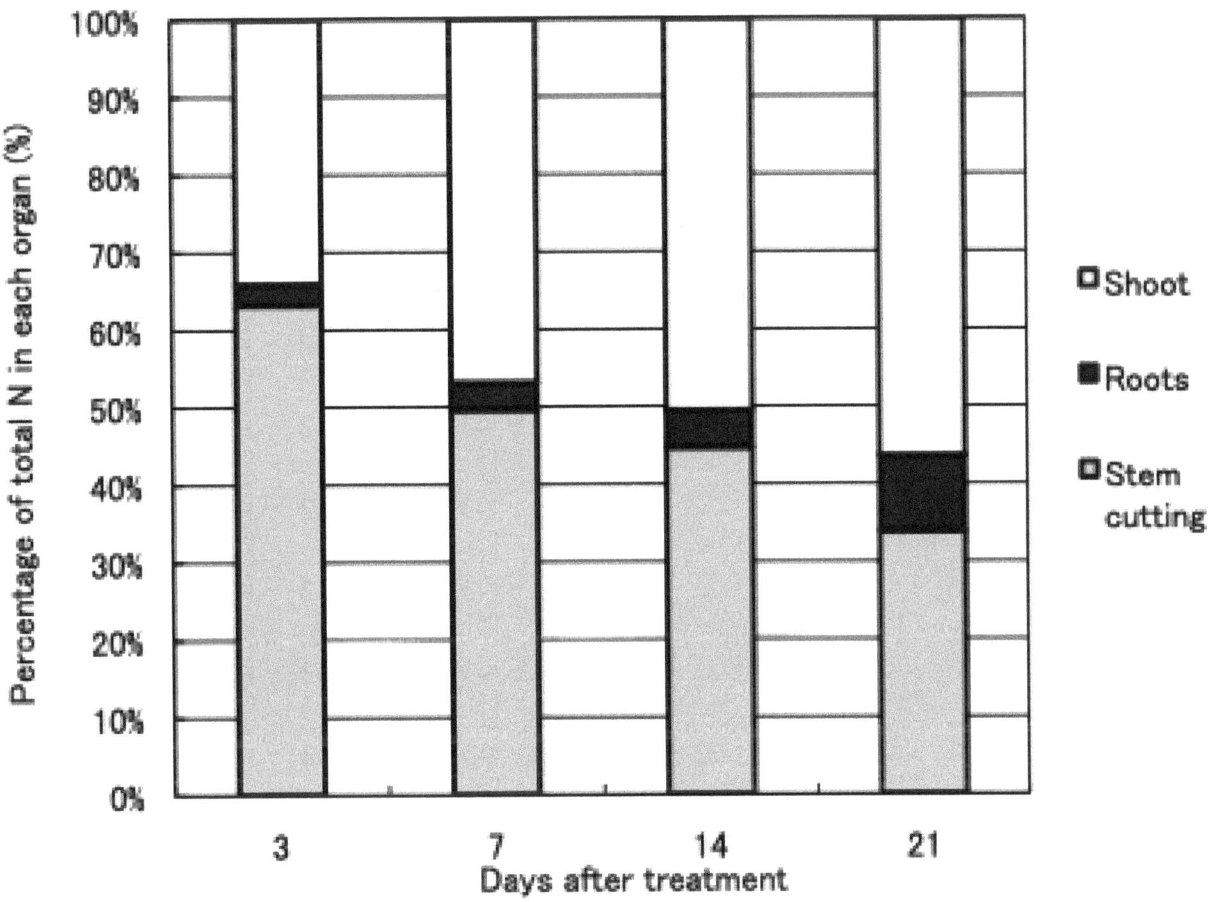

Figure 7. Changes with time in the fractional (%) distribution of total N (^{14}N+^{15}N) in the stem, roots and shoot of sugarcane plants exposed to $^{15}N_2$ for 7 days.

are very low. Zakria *et al.* reported that $^{15}N_2$ was exposed to matured sugarcane stem with vaccum infiltiration, the ^{15}N incorporation was not observed in non-inoculated stems [59]. Ando et al. [2] analyzed *nifH* genes, and they could not detect the *nifH* gene with *G. diazotrophicus.*

The absence of any significant translocation of fixed ^{15}N, most of it remaining in the 80% ethanol-insoluble fraction, in young sugarcane plants supports the possibility that fixed N may be used after the disintegration of dead endophytes, rather than being rapidly transported to the host plant as in the case of soybean nodules [44, 45]. However, it does not mean that there is no contribution of N_2 fixation in the roots to shoot growth. When the N_2-fixing bacteria in the roots eventually die, the products of their decomposition may well contribute to the growth of the roots and shoots of the plant. Even if this does not occur, then the N fixed in the roots will at least contribute to soil fertility in the field after the natural processes of root turnover and decomposition. Ando *et al.* [3] calculated N_2 fixation rates, fertilizer efficiency and the turnover of organic matter in sugarcane production in Thailand. Using the ^{15}N natural-abundance method, they estimated that the contribution of crop N_2 fixation to the overall N economy of many of the plantations was about 0-30%. Meanwhile, the contributions of N from

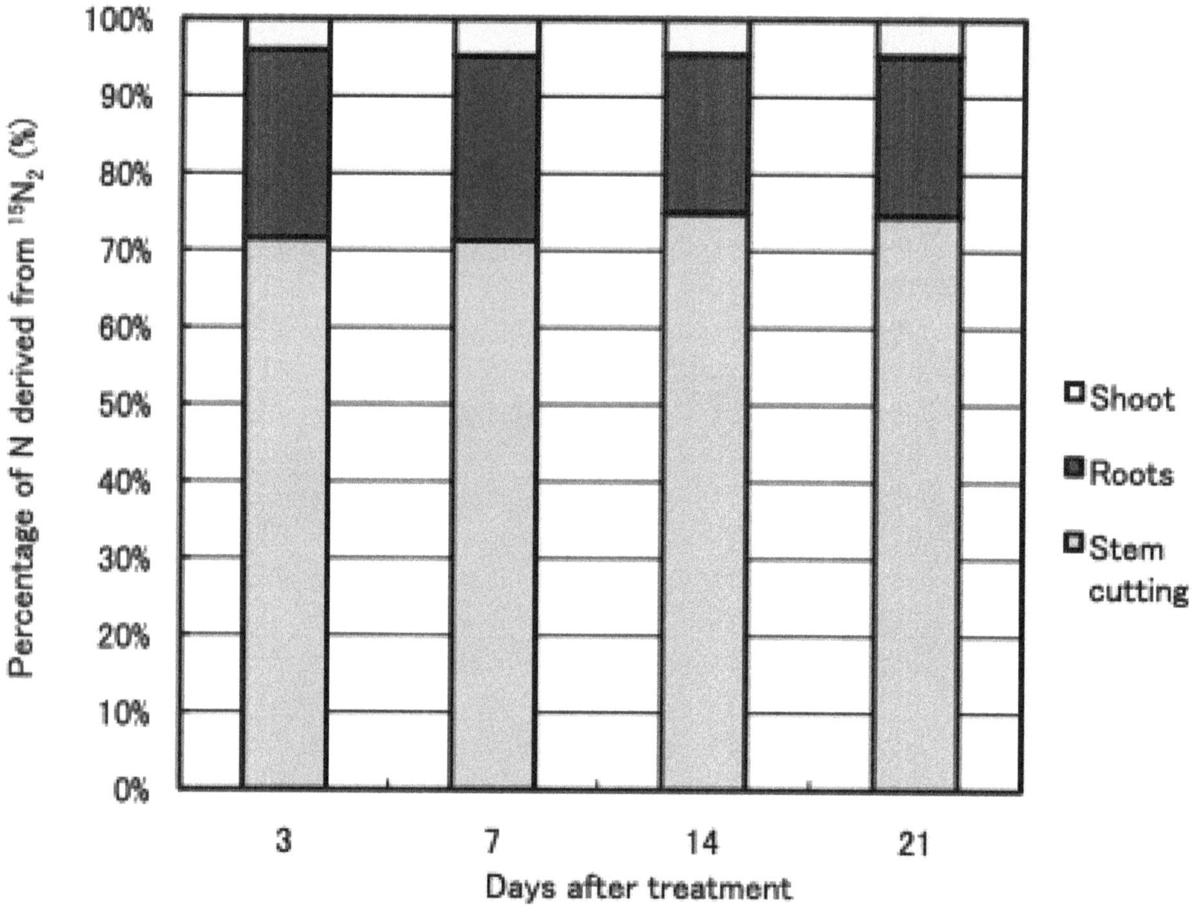

Figure 8. Changes with time in the fractional distribution (%) of ^{15}N in the stem, roots and shoot of sugarcane plants exposed to $^{15}N_2$ for 7 days.

applied mineral fertilizers were about 18-31%. This implies that N supplied from other soil-N sources such as from decomposing crop organic matter is important for supporting the fertility of the soil under sugarcane production.

6. Diazotrophic endophytes in sugarcane

As for the presence of N_2-fixing bacteria in sugarcane, diazotrophic bacteria belonging to the *Beijerinckia* genera have been found in large numbers in the rhizosphere (the soil volume adjacent to, or within a few millimeters from the root surface) and in the rhizoplane (the soil:root interface) of sugarcane [19]. Other diazotrophs, such as *Bacillus, Azotobacter, Derxia, Enterobacter, Erwinia, Klebsiella,* and *Azospirillum,* have also been isolated from the sugarcane rhizosphere [47]. In 1988, a new species of *Acetobacter* was found inside the sugarcane stem and named *Acetobacter diazotrophicus* [15], though this was later renamed to *Gluconacetobacter diazotrophicus* [55]. These microorganisms are called "endophytes" as they live inside host plant tissues without eliciting any symptoms of disease [8]. Recently, Saito *et al.* [49] reported a broad distribution and phylogeny of anaerobic endophytes of cluster XIVa clostridia in various plant species including the leaves, stems, stem cuttings and roots of sugarcane. The fixation of N_2 by endophytic bacteria has also been suggested in other crops, eg. rice (*Ozyza sativa*) [34,58], sweet potatoes (*Ipomoea batatas* L.) [51,1]. Recently, endophytic bacteria associated with nitrogen fixation and indole acetic acid synthesis were identified in ornamental bulb flower, curcuma (*Curcuma alismatifolia* Gagnep.) to promote plant growth, and the endophytes were corresponded to *Bacillus drentensis, Sphingomonas pseudosanguinis,* and *Bacillus methylotrophics* using 16S rDNA sequence analysis [53].

G. diazotrophicus is considered to be a major diazotrophic endophyte in sugarcane and has been isolated from leaves, stems and roots of sugarcane plants collected from a number of sites in Brazil and also in other countries [15]. *G. diazotrophicus* is clustered in the alpha subclass of Proteobacteria, and *Herbaspirillum seropedicae, Herbaspirullum rubrisubalbicans* and *Burkholderia* sp, clustered in the beta subclass of the Proteobacteria [14,8]. Rhizobia, which are symbiotic nitrogen fixing bacteria with legumes, survive as a free living state in soil after symbiotic state in root nodules. Different from rhizobia, *G. diazotrophicus* does not survive free in the soil, and it is thought that it is mainly transmitted in the course of vegetative propagation, which is usually done from stem cuttings or 'setts' [46].

Recently, Ando *et al.* [2] detected the *nifH* gene sequences, encoding the homodimer Fe protein of nitrogenase, in sugarcane stem and pineapple leaves to detect unculturable strains of bacteria as well as culturable strains. Their result showed that the sequences of the *nif H* clones were homologous to those of bacteria in the genera *Bradyrhizobium, Seratia,* and *Kebsiella*. On the other hand, no *nifH* sequence related to *G. diazotrophicus* was detected in sugarcane. This result indicates the absence or the presence of few *G. diazotrophicus* in the stems of the sugarcane plants used in their study. The expression of *nifH* gene in the stems and roots of sugarcane plants was investigated and the *nifH* RNA sequences similar to those of *Bradyrhizobium* sp. and *Azorhizobium caulinodans* were detected [52]. The *nifH* expression of *Bradyrhizobium* sp. and *Rhizobium* sp. in roots of field-grown Brazilian sugarcane was also found [23].

It is known that endophytic diazotrophic bacteria colonize in the vascular tissues or intercellular spaces and of sugarcane organs. For the presence of endophytic diazotrophs in sugarcane juice, and Bellone and Bellone [9] concluded that in the mature region of the sugarcane stem *G. diazotrophicus* grows more abundantly than *H. seropedicae* or *Azospirillum brasilense*. However, the sites of colonization and the movement through xylem vessels of *G. diazotrophicus* within sugarcane plants are controversial [20,21,28]. James *et al.* [28] reported that *G. diazotrophicus* may reside in xylem of sugarcane stem, however, Dong *et al.* [21] pointed out that xylem is not suitable habitat and the intercellular space appoplast is the probable location of the bacteria.

Recently, complete genome sequence of the sugarcane nitrogen-fixing endophyte *G. diazotrophicus* Pal5 was reported [10]. *G. diazotrophicus* Pal5 was the third diazotrophic endophytic bacterium to be completely sequenced, followed by *Azoarcus* sp. strain BH72 [30] and *Klebsiella pneumoniae* strain 342 [25]. Its genome is composed of a 3.9 Mb chromosome and 2 plasmids of 16.6 and 38.8 Kb, respectively. 3,938 coding sequences are annotated and those are related to the endophytic lifestyle such as nitrogen fixation, plant growth promotion, sugar metabolism, transport systems, synthesis of auxin and the occurrence of bacteriocins [10]. Gene clusters for gum-like polysaccharide biosynthesis, tad pilus, quorum sensing, for modulation of plant growth by indole acetic acid and mechanisms involved in tolerance to acidic conditions were identified and may be related to the sugarcane endophytic and plant-growth promoting traits of *G. diazotrophicus*.

A broad proteomic description of *G. diazotrophicus* identified 583 proteins, and potential metabolic pathways for nucleotides, amino acids, carbohydrates, lipids, cofactors and energy production were described [31]. A differential protein expression analysis was carried out to study *G. diazotrophicus* interaction with sugarcane [22].

The mechanisms for the association between sugarcane plants and diazotrophic endophytes are as yet poorly understood.

De Carvalho *et al.* [18] reviewed sugarcane-endophytes association. Besides biological nitrogen fixation, these endophytes exhibit growth promoting traits by mechanisms involving nutrient solubilization, plant hormone production and pathogens antagonistic activity. Cavalcante *et al.* [16] suggested that the ethylene signaling pathway may play a role in the establishment of the association between sugarcane and endophytic diazotrophic bacteria.

7. Mechanism by which N is transferred to the host plant from endophytic nitrogen-fixing bacteria

To promote sugarcane growth and high yield of sugar, the transport of N from diazotrophic endophytes to the host plant is important in addition to the occurrence of high nitrogen fixation activity. The mechanism by which N is transferred to the host sugarcane plant from N_2-fixing endophytes has not yet been fully elucidated. There are two possible ways for this transfer to occur. The first is that living bacteria actively excrete fixed N into the apoplast of the host tissue and the plant cells then absorb the released N compounds. This is an analogous to legume-rhizobia symbiosis, in which fixed ammonia is rapidly excreted from bacteroid (a symbiotic state of rhizobia) to cytosol of infected cells in soybean root nodules [45]. The second is that bacteria proliferate and colonize in the host tissue and the fixed N is released to the host cells only after their death and disintegration. No direct evidence has yet been obtained.

There is little direct evidence on how N_2 fixed is supplied from endophyte to sugarcane plants. When an amylolytic yeast was used to mimic the plant, *G. diazotrophicus* was capable to excrete part of the fixed N into the medium [17]. There are also evidences that ammonium is the product excreted by this bacterium under N_2 fixation conditions [7]. Data obtained on sugarcane EST expression analysis indicated that nitrogen metabolism was active in plants colonized by diazotrophic endophytes [39]. It suggests that the sugarcane N assimilation apparatus can play a role on the incorporation of the N compound released by the diazotrophic bacteria during association [39].

The characteristics of nitrogen fixation and transport in endophytic bacteria isolated from sugarcane stem were investigated [37]. The strains JA1 and JA2 were putatively identified as

G. diazotrophicus [4]. The cultures of strain JA1 and JA2 in N free LGIP medium were exposed to the ^{15}N labeled air with different O_2 concentrations using vacuum system (Figure 9). The result showed $^{15}N_2$ fixation activity and acetylene reduction activity (ARA) in both culture, although optimum O_2 concentration were different (Figure 10, 11). These activities in JA1 were highest at 0.4% O_2 on solid agar culture (Figure 10), and at 0% O_2 in liquid culture (Figure 11). In both culture conditions, $^{15}N_2$ fixation activity and ARA decreased with increasing O_2 concentrations up to 20%.

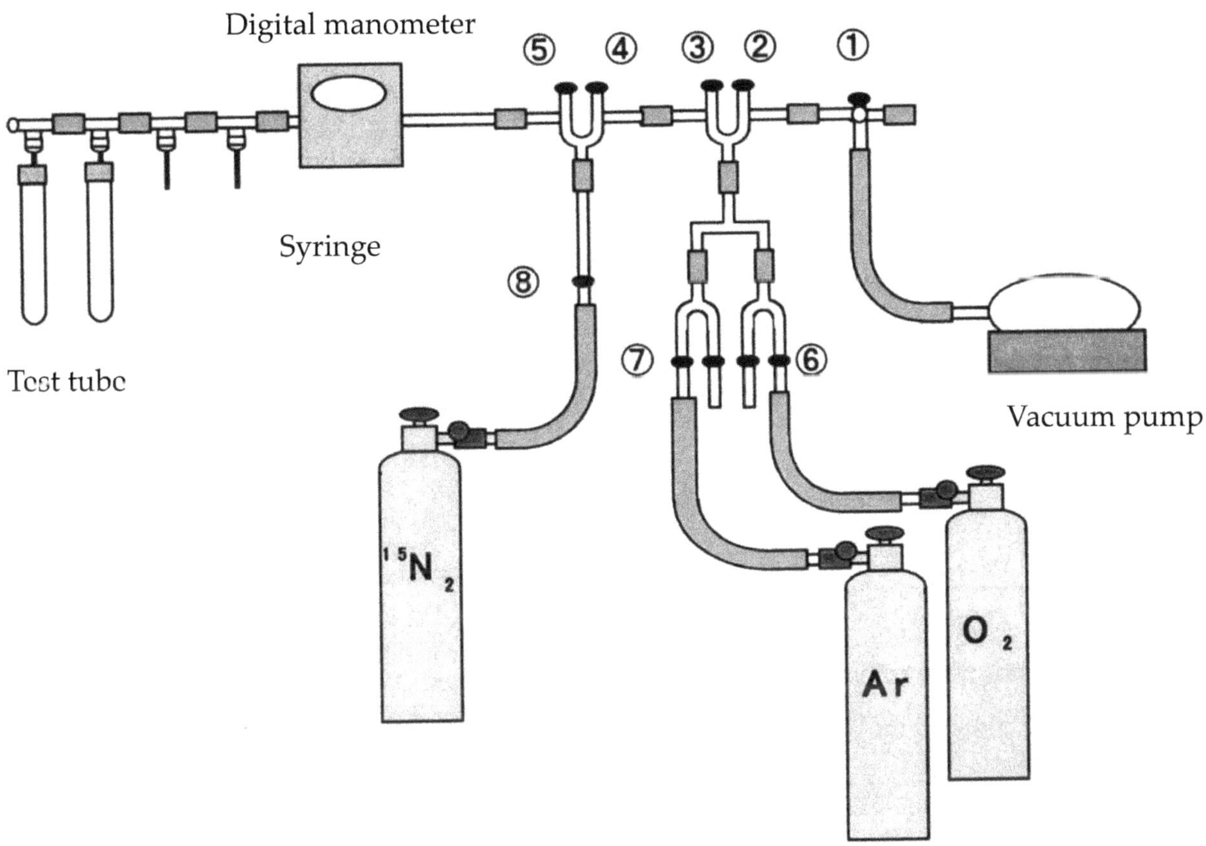

Figure 9. Vacuum system for exchanging the gas inside the test tubes.

The growth and ARA of the strain JA1 cultivated with liquid LGIP medium were measured every day for 10 days (Figure12). Bacteria growth increased rapidly at day-1 after inoculation, and continued to increase until day-6, then the increase was almost stopped thereafter. The strain might regulate bacterial density, possibly by quorum sensing mechanism. The ARA increased rapidly from day-1 to day-5, but it decreased rapidly after day-6. Very low ARA was detected after day-7 to day-10.

This result suggests that nitrogen fixation is active only during early stage of proliferation of JA1. After the bacterium growth stops, nitrogen fixation activity is inactivated. If the situation is the same inside the sugarcane organs, the continuous proliferation should be essential to keep nitrogen fixation activity of the diazotrophic endophyte.

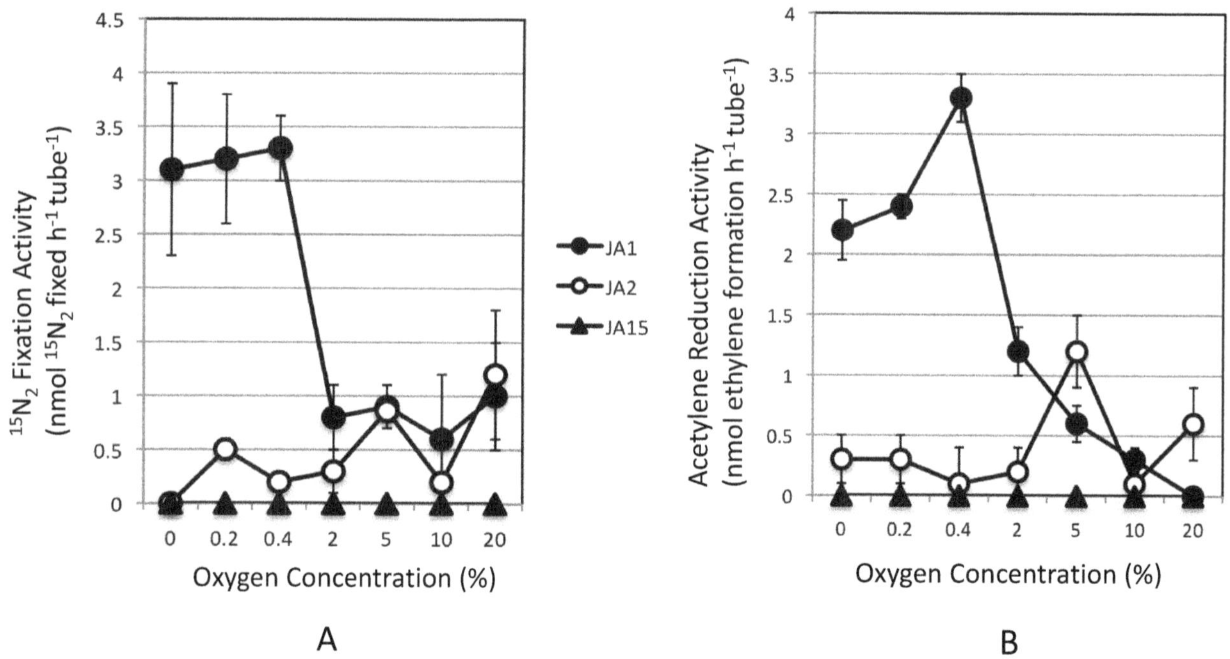

Figure 10. $^{15}N_2$ fixation activity (A) and acetylene reduction activity (B) of isolated strains from sugarcane endophytes cultivated on agar medium under various concentration of O_2.(average with standard deviation)

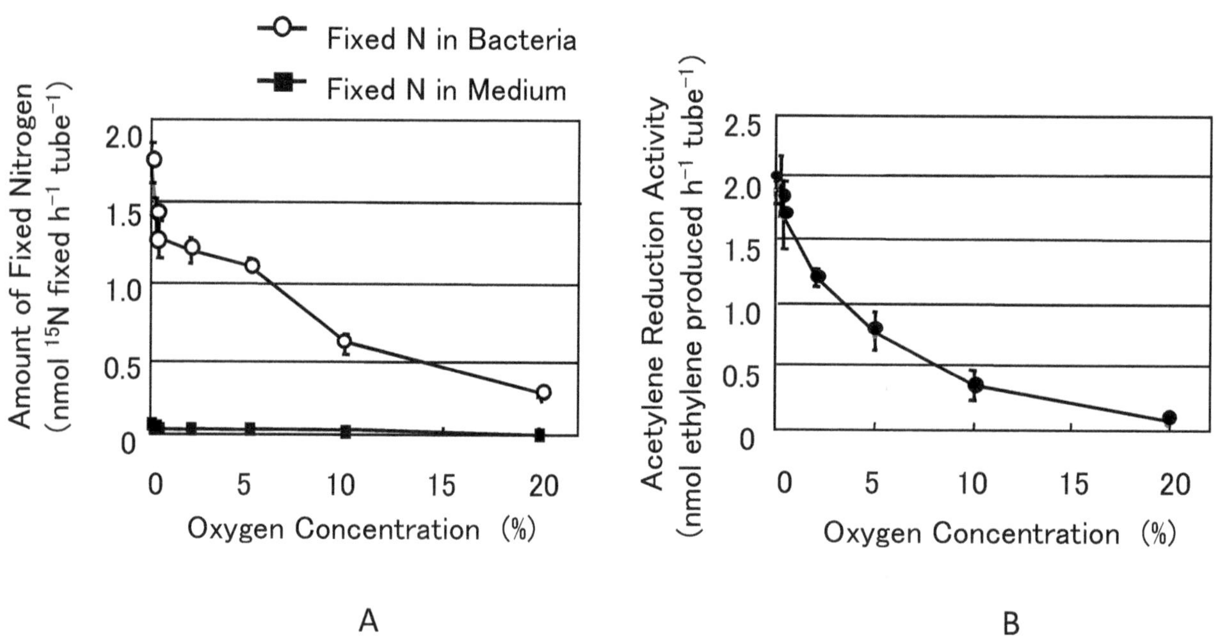

Figure 11. $^{15}N_2$ fixation activity (A) and acetylene reduction activity (B) of isolated strain JA1 cultivated in liquid medium under various concentration of O_2. (average with standard deviation)

The N release mechanism from endophytes to sugarcane plant is very important to support N nutrition for sugarcane growth. From the experiment with JA1 strain, ^{15}N fixed during 24 hours was mainly distributed in bacteria fraction and only a little portion (about 4%) was released

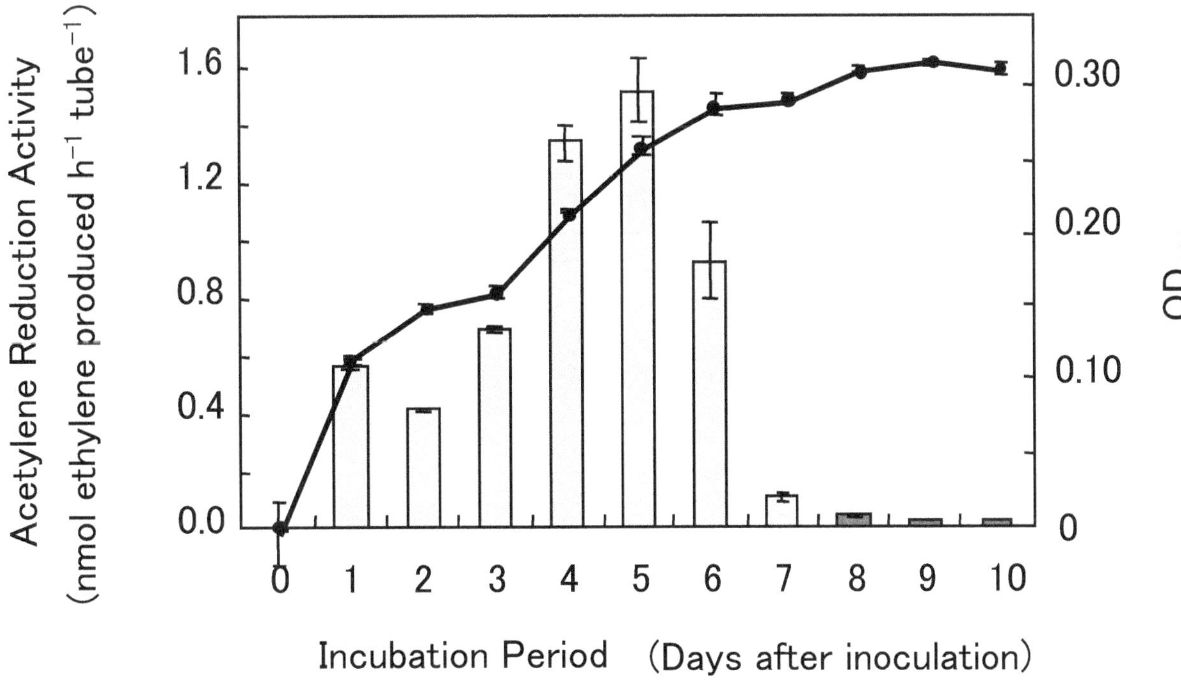

Figure 12. Growth (OD_{560}, line graph) and acetylene reduction activity (bar graph) of isolated strain JA1 cultivated in liquid medium. (average with standard deviation)

to the medium. However, at 10 days after one day feeding of $^{15}N_2$, a significant portion of fixed ^{15}N was distributed to the medium, especially the percentage of released ^{15}N was highest at 40% under 20% O_2 conditions (Figure 13). This result indicates that the cultured JA1 released N after stopping growth and nitrogen fixation, possibly by their death and degradation. Lethbridge and Davidson [32] suggested that endophytic bacteria only transferred fixed N to the plant when they died and were eventually decomposed.

8. Conclusion

In our studies, a $^{15}N_2$ tracer experiment was conducted to investigate the sites of N_2 fixation and the possible translocation of the fixed N in young sugarcane plants. Most active nitrogen fixation was observed in the roots, followed by planted stalk. The young shoot showed very little nitrogen fixation. Although many diazotrophic endophytic bacteria are reported in all parts of sugarcane plants (Figure 14), the roots may be a principal part of nitrogen fixation.

The fixed ^{15}N either in roots or stalk were not readily transported to the shoots. In addition, the fixed N was mainly located in the 80% ethanol insoluble fractions, which contains high molecular weight compounds such as protein. The isolated diazotrophic endophyte fixed $^{15}N_2$ in N free liquid and solid cultures. They fixed $^{15}N_2$ only during their active growth period, and nitrogen fixation stopped thereafter. It is possible that the endophytic bacteria can fix N during proliferation stage, and they will release N after their death and decomposition.

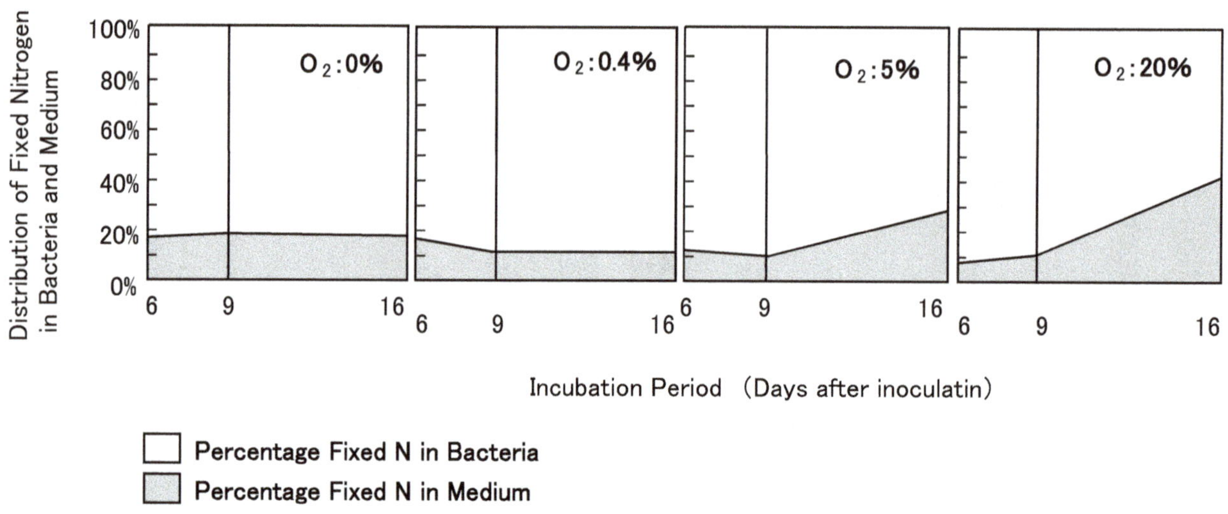

Figure 13. Distribution of fixed N in bacteria and medium fractions after $^{15}N_2$ exposure at day-6 of incubation period under different O_2 concentrations.

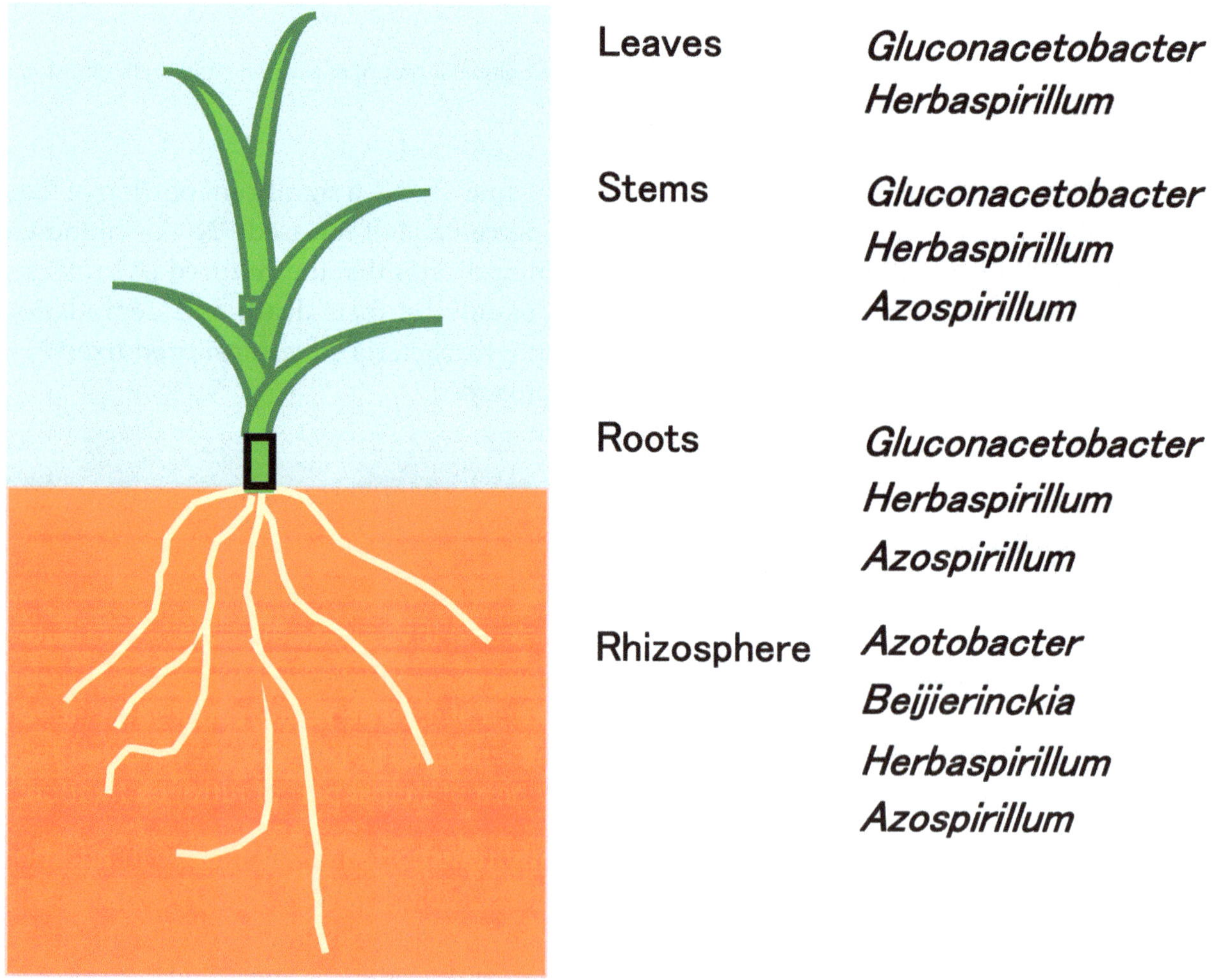

Figure 14. Presence of major diazotrophic endophytes in sugarcane

The translocation of fixed N is quite different from legume-rhizobia symbiosis. In soybean nodules, N_2 is fixed by bacteroids (N_2-fixing rhizobia) in the infected cells and ammonia or ammonium is readily excreted from bacteroid to the cytosol of the infected cells. On short-term (5 min) exposure to $^{15}N_2$, 97% of the fixed ^{15}N in the 80% ethanol-soluble fraction, which contain low molecular weight compounds such as amino acids, in the nodules was distributed in the cytosol of the nodule plant cells, while only 3% remained within the bacteroids [44]. The ammonium is then assimilated via the glutamine synthetase/glutamate synthase (GS/GOGAT) system [41,43], and used mainly to produce ureides, allantoin and allantoic acids, and the ureides are transported to the various plant organs via the xylem vessels [42,45].

Other than in Brazil, 150-250 kg of urea-N per hectare per year is usually applied to sugarcane, the actual amount depending on soil fertility, on genotype and on target yield [29]. By promoting BNF through endophytic or associative diazotrophs, the cost associated with N fertilizer usage in sugarcane production can be reduced and environmental problems such as NO_3^- leaching or N_2O gas emission, consequent upon the use of excessive chemical fertilizers, can be avoided. Further research will be important for promoting more efficient N_2 fixation rates in sugarcane production.

Acknowledgements

We greatly appreciate the late Prof. Dr. Shoichiro Akao for giving us the chance to do the experiment on sugarcane-endophyte association. This research was supported by a grant from a Research Project of MAFF, Japan.

Author details

Takuji Ohyama[1,2*], Atsushi Momose[2], Norikuni Ohtake[1,2], Kuni Sueyoshi[1,2], Takashi Sato[3], Yasuhiro Nakanishi[4], Constancio A. Asis Jr.[5], Soraya Ruamsungsri[6] and Shotaro Ando[7]

*Address all correspondence to: ohyama@agr.niigata-u.ac.jp

1 Faculty of Agriculture, Niigata University, Niigata, Japan

2 Graduate School of Science and Technology, Niigata University, Niigata, Japan

3 Faculty of Bioresource Sciences, Akita Prefectural University, Akita, Japan

4 Faculty of International Agriculture and Food Studies, Tokyo University of Agriculture, Tokyo, Japan

5 Philippine Rice Research Institute, Science City of Muñoz, Nueva Ecija, Philippines

6 Faculty of Agriculture, Chiang Mai University, Chiang Mai, Thailand

7 Japan International Research Center for Agricultural Sciences, Okinawa, Japan

References

[1] Ando, S. 2010. Nitrogen fixation associated with endophytic bacteria. In *Nitrogen Assimilation in Plants*, Ed. Ohyama, T. and Sueyoshi, K. (Research Signpost, Kerala, India), pp. 215-231.

[2] Ando, S., Goto, M., Meunchang, S., Thongra-ar, P., Fujiwara, T., Hayashi, H., and Yoneyama T., 2005. Detection of *nifH* sequences in sugarcane (*Saccharum officinarum* L.) and Pineapple (*Ananas comosus* (L.) Merr.), *Soil Sci. Plant Nutr.*, 51: 303-308.

[3] Ando, S., Meunchang, S., Thippayarugs, S., Prasertsak, P., Matsumoto, N., and Yoneyama, T. 2002. Evaluation of sustainability of sugarcane production in Thailand based on nitrogen fixation, efficiency of nitrogen fertilizer and flow of organic matters. *JIRCAS Working Report*, 30:61-64.

[4] Asis, C.A. Jr., Kubota, M., Ohta, H., Arima, Y., Chebotar, V.K., Tsuchiya, K., and Akao, S. 2000. Isolation and partial characterization of endophytic diazotrophs associated with Japanese sugarcane cultivar. *Soil Sci. Plant Nutr.*, 46:759-765.

[5] Asis, C.A. Jr., Kubota, M., Ohta, H., Arima, Y., Ohwaki, Y., Yoneyama, T., Tsuchiya, K., Hayashi, N., Nakanishi, Y., and Akao, S. 2002. Estimation of the nitrogen fixation by sugarcane cultivar NiF-8 using ^{15}N dilution and natural ^{15}N abundance techniques. *Soil Sci. Plant Nutr.*, 48:283-285.

[6] Azeredo, D.F, Bolsanello, J., Wever, M., and Vieira, J.R., 1986. Nitrogenio em canaplanta, doses e fracionameento. *STAB-Acucar Alcool e Subproductos*, 4: 26-32.

[7] Baldani, J.I., Caruso, L., Baldani, V.L.D., Goi, S.R., and Döbereiner, J. 1997. Recent advances in BNF with non-legume plants. *Soil Biol. Biochem.*, 29: 911-922.

[8] Barnett, M.J., Toman C.J., Fisher R.F., and Long S.R. (2004) A dual-genome symbiosis chip for coordinate study of signal exchange and development in a prokaryote–host interaction. *Proc. Natl. Acad. Sci. USA*, 101:16636–16641.

[9] Bellone de, S.C., and Bellone, C.H. 2006. Presence of endophytic diazotrophs in sugarcane juice. *World J. Microbiol. Biotech.*, 22, 1065-1068.

[10] Bertalan, M., et al., 2009, Complete genome sequence of the sugarcane nitrogen-fixing endophyte *Gluconacetobacter diazotrophicus* Pal5. *BMC Genomics*, 10, 450.

[11] Bezdicek, D.F., and Kennedy, A.C. 1988. Symbiotic nitrogen fixation and nitrogen cycling in terrestrial environments, In *Micro-organisms in Action; Concepts and Applications in Microbiol Ecoloby*, Ed. Lynch, J.M. and Hobbie, J.E., Blackwell Scientific Publications, Oxford, UK. 241-260

[12] Boddey, R.M. 1995. Biological nitrogen fixation in sugar cane: A key to energetically viable biofuel production. *CRC Crit. Rev. Plant Sci.*, 14:263-279.

[13] Boddey, R.M., Urquiaga, S., Reis, V., and Döbereiner. J. 1991. Biological nitrogen fixation associated with sugar cane. *Plant Soil,* 137:111-117.

[14] Boddey, R.M., Urquiaga, S., Alves, B.J.R., and Reis, V. 2003. Endophytic nitrogen fixation in sugarcane: present knowledge and future applications. *Plant Soil,* 252:139–149.

[15] Cavalcante, V.A., and Döbereiner, J. 1988. A new acid-tolerant nitrogen-fixing bacterium associated with sugarcane. *Plant Soil,* 108:23-31.

[16] Cavalcante, J.J.V., Vargas, C., Nogueira, E.M., Vinagre, F., Schwarcz, K., Baldani, J.I., Ferreira, P.C.G., and Hemerly, A.S. 2007. Members of the ethylene signalling pathway are regulated in sugarcane during the association with nitrogen-fixing endophytic bacteria. *J. Exp. Bot.,* 58:673-686.

[17] Cohjo, E.H., Reis, V.M., Schenberg, A.C.G., and Döbereiner, J. 1993. Interactions of *Acetobacter diazotrophicus* with an amylolytic yeast in nitrogen–free batch culture. *FEMS Microbiol. Lett.,* 106:341-346.

[18] De Carvalho, T.L.G., Ferreira, P.C.G, Hemerly, A.S. 2011. Sugarcane genetic controls involved in the association with beneficial endophytic nitrogen fixing bacteria, *Tropical Plant Biol.* 4:31–41.

[19] Döbereiner, J. 1961. Nitrogen-fixing bacteria of the genus *Beijerinckia* Derk in the rhizosphere of sugar cane. *Plant Soil,* 15:211-216.

[20] Dong, Z., Canny, M.J., McCully, M.E., Roboredo, M.R., Cabadilla, C.F., Ortega, E., and Rodes, R. 1994. A nitrogen-fixing endophyte of sugarcane stems. *Plant Physiol.,* 105:1139-1147.

[21] Dong, Z., McCully, M.E., and Canny, M.J. 1997. Does *Acetobacter diazotrophicus* live and move in the xylem of sugarcane stems? Anatomical and physiological data. *Ann. Bot.,* 80:147-158.

[22] Dos Santos, M.F., De Pádua, V.L.M., Nogueira, E.M., Hemerly, A.S., and Domont, G.B. 2010. Proteome of *Gluconacetobacter diazotrophicus* co-cultivated with sugarcane plantlets. *J. Proteomics,* 73:917–931.

[23] Fischer, D., Pfitzner, B., Schmid, M., Simões-Araújo, J.L., Reis, V.M., Pereira, W., Ormeño-Orrillo, E., Hai, B., Hofmann, A., Schloter, M., Martinez-Romero, E., Baldani, J.I., and Hartmann, A., 2012. Molecular characterization of the diazotrophic bacterial community in uninoculated and inoculated field-grown sugarcane (*Saccharum* sp.), *Plant Soil,* 356, 83-99.

[24] Fisher, K., and Newton, W.E. 2004. Nitrogen fixation: An historical perspective. In *Canalystes for nitrogen fixation: Nitrogenases, relevant chemical models, and commercial processes.* Ed Smith B.E., Richard, R.L., Newton, W.E., (Kluwer Academic Publisher, Dordrecht.) pp. 1-31.

[25] Fouts, D.E., Tyler, H.L., De Boy, R.T., Daugherty, S., Ren, Q., Badger, J.H., Durkin, A.S., Huot, H., Shrivastava, S., Kothari, S., et al. 2008. Complete genome sequence of the N_2-fixing broad host range endophyte *Klebsiella pneumoniae* 342 and virulence predictions verified in mice. *PLoS Genet.*, 47:e1000141.

[26] Hiyama, T., Momose, A., Nishimura, K., Ishizaki, N., Ishikawa, S., Yamamoto, M., Hung, N.V.P., Chamaiporn, A., Ruamrungsri, S., Ohtake, N., Sueyoshi, K., Ohyama, T. 2013. Estimation of contribution of nitrogen fixation in sugarcane (*Saccharum officinarum*) plants with ^{15}N dilution method in relation to nitrogen supply period. Bull. Facul. Agric. Niigata Univ., 66:11-19

[27] James, E.K., and Olivares, F.L.1997. Infection and colonization of sugar cane and other graminaceous plants by endophytic diazotrophs. *CRC Crit. Rev. in Plant Sci.*, 17:77-119.

[28] James, E.K., Olivares, F.L., de Oliveira, A.L.M., dos Reis, Jr. F.B., da Silva, L.G. and Reis, V.M. 2001. Further observations on the interaction between sugar cane and *Gluconacetobacter diazotrophicus* under laboratory and greenhouse conditions. *J. Exp. Bot.*, 52:747-760.

[29] Kennedy, I.R., Choudhury, A.T.M.A., and Kecskes, M.L. 2004. Non-symbiotic bacterial diazotrophs in crop-farming systems: can their potential for plant growth promotion be better exploited? *Soil Biol. Biochem.*, 36:1229-1244.

[30] Krause, A., Ramakumar, A., Bartels, D., Battistoni, F., Bekel, T., Boch, J., Boehm, M., Friedrich, F., Hurek, T., Krause, L., et al. 2008. Complete genome of the mutualistic, N_2-fixing grass endophyte Azoarcus sp. strain BH72. *Nat Biotechnol.*, 24:1385-1391.

[31] Lery, L.M.S., Coelho, A., von Kruger, V.M.A. *et al.* 2008. Protein expression profile of *Gluconacetobacter diazotrophicus* PAL5, a sugarcane endophytic plant growth-promoting bacterium. *Proteomics,* 8:1631–1644

[32] Lethbridge, G., and M.S. Davidson. 1983. Root-associated nitrogen-fixing bacteria and their role in the nitrogen nutrition of wheat estimated by ^{15}N isotope dilution. *Soil Biol. Biochem.*, 15:365-374.

[33] Lima, E., Boddey, R.M., and Döbereiner, J., 1987. Quantification of biological nitrogen fixation associated with sugar cane using a ^{15}N aided nitrogen balance. *Soil Biol. Biochem.*, 19:165-170.

[34] Mano, H., and Morisaki, H. 2008. Endophytic bacteria in the rice plant. *Microbes Environ.*, 23:109-117.

[35] Marris, E. 2006. Drink the best and drive the rest. Brazil's sugar-cane ethanol industry is the world's best and able to get better. *Nature,* 444:670-672.

[36] Momose, A., Ohtake, N., Sueyoshi, K., Sato, T., Nakanishi, Y., Akao, S., and Ohyama, T. 2009. Nitrogen fixation and translocation in young sugarcane (*Saccharum officina-*

rum L.) plants associated with endophytic nitrogen-fixing bacteria. *Microbes Environ.*, 24, 224-230

[37] Momose, A., Hiyama, T., Nishimura, K., Ishizaki, N., Ishikawa, S., Yamamoto, M., Hung, N.V.P., Ohtake, N., Sueyoshi, K., and Ohyama, T. 2013. Characteristics of nitrogen fixation and nitrogen release from diazotrophic endophytes isolated from sugarcane (*Saccharum officinarum* L.) stem. *Bull. Facul. Agric. Niigata Univ.*, 66 (1) 66:1-9

[38] Nishiguchi, T., Shimizu, T., Njoloma, J., Oota, M., Saeki, Y., and Akao, S. 2005. The estimation of the amount of nitrogen fixation in the sugarcane by ^{15}N dilution technique. *Bull. Facul. Agric. Univ. Miyazaki*, 51:53-62.

[39] Nogueira, E.M., Vinagre, F., Masuda, H.P., et al. 2001. Expression of sugarcane genes induced by inoculation with *Gluconacetobacter diazotrophicus* and *Herbaspirillum rubrisubalbicans*. *Gen. Mol. Biol.*, 24:199-206.

[40] Ohyama, T. 2010. Nitrogen as a major essential element of plants. In *Nitrogen Assimilation in Plants*, Ed Ohyama, T., Sueyoshi, K. Research Signpost, Kerala India, pp. 1-17.

[41] Ohyama, T., and Kumazawa. K., 1978. Incorporation of ^{15}N into various nitrogenous compounds in intact soybean nodules after exposure to $^{15}N_2$ gas. *Soil Sci. Plant Nutr.*, 24:525-533.

[42] Ohyama, T., and Kumazawa. K. 1979. Assimilation and transport of nitrogenous compounds originated from $^{15}N_2$ fixation and $^{15}NO_3$ absorption. *Soil Sci. Plant Nutr.*, 25:9-19.

[43] Ohyama, T., and Kumazawa. K. 1980. Nitrogen assimilation in soybean nodules I. The role of GS/GOGAT system in the assimilation of ammonia produced by N_2-fixation. *Soil Sci. Plant Nutr.*, 26:109-115.

[44] Ohyama, T., and Kumazawa. K. 1980. Nitrogen assimilation in soybean nodules II. $^{15}N_2$ assimilation in bacteroid and cytosol fractions of soybean nodules. *Soil Sci. Plant Nutr.*, 26:205-213.

[45] Ohyama, T., Ohtake, N., Sueyoshi, K., Tewari, K., Takahashi, Y., Ito, S., Nishiwaki, T., Nagumo, Y., Ishii, S., Sato, T. 2009. Nitrogen fixation and metabolism in soybean plants. Nova Science Publishers, Inc., New York, pp.1-131.

[46] Reis, V.M., Olivares, F.L., and Döbereiner, J. 1994. Improved methodology for isolation of *Acetobacter diazotrophicus* and confirmation of its endophytic habitat. *World J. Microbiol. Biotechnol.*10:101-104.

[47] Reis, V., Lee, S., and Kennedy, C. 2007. Biological nitrogen fixation in sugarcane. p. 213-232. *In*: *Associative and Endophytic Nitrogen-fixing Bacteria and Cyanobacterial Associations*, Ed. Emerich, C. and Newton W.E., Springer, Dordrecht, The Netherlands. pp. 213-232.

[48] Ruschel, A.P., Henis, Y., and Salati, E. 1975. Nitrogen-15 tracing of N-fixation with soil-grown sugarcane seedlings. *Soil Biol. Biochem.*, 7:181-182.

[49] Saito A., Kawahara, M., Ikeda, S., Ishimine, M., Akao, S., and Minamisawa, K. 2008. Broad distribution and phylogeny of anaerobic endophytes of cluster XIVa clostridia in plant species including crops. *Microbes Environ.*, 23:73-80.

[50] Sevilla, M., Burris, R.H., Gunapala, N., and Kennedy, C. 2001. Comparison of benefit to sugarcane plant growth and $^{15}N_2$ incorporation following inoculation of sterile plants with *Acetobacter diazotrophicus* wild-type and Nif$^-$ mutant strains. *Mol. Plant-Microbe Int.*, 14:358-366.

[51] Terakado-Tonooka, J., Ohwaki, Y., Yamakawa, H., Tanaka, F., Yoneyama, T., and Fujiwara, S. 2008. Expressed *nifH* genes of endophytic bacteria detected in field-grown sweet potatoes (*Ipomoea batatas* L.). *Microbes Environ.*, 23:89-93

[52] Thaweenut, N., Hachisuka, Y., Anso, S., Yanagisawa, S., Yoneyama, T. 2011. Two seasons' study on *nifH* gene expression and nitrogen fixation by diazotrophic endophytes in sugarcane (*Saccharum* spp. hybrids): expression of *nifH* genes similar to those in rhizobia. *Plant Soil*, 338, 435-449.

[53] Thepsukhon, A., Choonluchanon, S., Tajima, S., Nomura, M., and Ruamrungsri, S., 2013, Journal of Plant Nutrition, 36, 1424-1438

[54] Urquiaga, S., Cruz, K.H.S., and Boddey, R.M. 1992. Contribution of nitrogen fixation to sugarcane: Nitrogen-15 and nitrogen-balance estimates. *Soil Sci. Soc. Am. J.*, 56:105-114.

[55] Yamada, Y., K. Hoshino, and T. Ishikawa. 1997. The phylogeny of acetic acid bacteria based on the partial sequence of 16S ribosomal RNA: The elevation of the subgenus *Gluconoacetobacter* to the generic level. *Biosci. Biotech. Biochem.*, 61:1244-1251.

[56] Yokoyama, T. 2010 Biological nitrogen fixation: Nitrogen cycle in agriculture. In *Nitrogen Assimilation in Plants*, Ed. Ohyama, T., Sueyoshi, K. Research Signpost. Kerala, India, pp. 169-174.

[57] Yoneyama, T., Muraoka, T., Kim, T.H., Dacanay, E.V., and Nakanishi, Y. 1997. The natural ^{15}N abundance of sugarcane and neighbouring plants in Brazil, the Phillipines and Miyako (Japan). *Plant Soil*, 189:239-244.

[58] Zakria, M., Njoloma, J., Saeki, Y., and Akao, S. 2007. Colonization and nitrogen-fixing ability of *Herbaspirillum* sp. strain B501*gfp*1 and assessment of its growth-promoting ability in cultivated rice. *Microbes Environ.*, 22:197-206.

[59] Zakria, M., Udonishi, K., Saeki, Y., Yamamoto, A., and Akao, S. 2008. Infection, multiplication and evaluation of the nitrogen-fixing ability of *Herbaspirillum* sp. strain B501 *gfp*1 in sugarcane stems inoculated by the vacuum infiltration method. *Microbes Environ.*, 23:128-133.

Permissions

All chapters in this book were first published in ABENF, by InTech Open; hereby published with permission under the Creative Commons Attribution License or equivalent. Every chapter published in this book has been scrutinized by our experts. Their significance has been extensively debated. The topics covered herein carry significant findings which will fuel the growth of the discipline. They may even be implemented as practical applications or may be referred to as a beginning point for another development.

The contributors of this book come from diverse backgrounds, making this book a truly international effort. This book will bring forth new frontiers with its revolutionizing research information and detailed analysis of the nascent developments around the world.

We would like to thank all the contributing authors for lending their expertise to make the book truly unique. They have played a crucial role in the development of this book. Without their invaluable contributions this book wouldn't have been possible. They have made vital efforts to compile up to date information on the varied aspects of this subject to make this book a valuable addition to the collection of many professionals and students.

This book was conceptualized with the vision of imparting up-to-date information and advanced data in this field. To ensure the same, a matchless editorial board was set up. Every individual on the board went through rigorous rounds of assessment to prove their worth. After which they invested a large part of their time researching and compiling the most relevant data for our readers.

The editorial board has been involved in producing this book since its inception. They have spent rigorous hours researching and exploring the diverse topics which have resulted in the successful publishing of this book. They have passed on their knowledge of decades through this book. To expedite this challenging task, the publisher supported the team at every step. A small team of assistant editors was also appointed to further simplify the editing procedure and attain best results for the readers.

Apart from the editorial board, the designing team has also invested a significant amount of their time in understanding the subject and creating the most relevant covers. They scrutinized every image to scout for the most suitable representation of the subject and create an appropriate cover for the book.

The publishing team has been an ardent support to the editorial, designing and production team. Their endless efforts to recruit the best for this project, has resulted in the accomplishment of this book. They are a veteran in the field of academics and their pool of knowledge is as vast as their experience in printing. Their expertise and guidance has proved useful at every step. Their uncompromising quality standards have made this book an exceptional effort. Their encouragement from time to time has been an inspiration for everyone.

The publisher and the editorial board hope that this book will prove to be a valuable piece of knowledge for researchers, students, practitioners and scholars across the globe.

List of Contributors

Takashi Soyano and Masayoshi Kawaguchi
Division of Symbiotic Systems, National Institute for Basic Biology, Okazaki, Aichi, Japan

Maki Nagata
Department of Agricultural Sciences, Faculty of Agriculture, Saga University, Honjyomachi, Saga, Japan

Akihiro Suzuki
United Graduate School of Agricultural Sciences, Kagoshima University, Korimoto, Kagoshima, Japan

Nacer Bellaloui
USDA-ARS, Crop Genetics Research Unit, Stoneville, USA

Alemu Mengistu
USDA-ARS, Crop Genetics Research Unit, Jackson, TN, USA

My Abdelmajid Kassem
Plant Genomics and Biotechnology Lab, Department of Biological Sciences, Fayetteville State University, Fayetteville, NC, USA

Craig A. Abel
Corn Insects and Crop Genetics Research, Iowa State University, Ames, Iowa, USA

L.H.S. Zobiole
Crop Protection, R&D Dow Agro Sciences Brazil, Cascavel, Paraná, Brazil

C.P. Chanway and H. Yang
Dept of Forest and Conservation Sciences, University of British Columbia, Vancouver, Canada

R. Anand
Children's Hospital, Vancouver, Canada

Nisa Rachmania Mubarik and Tedja-Imas Sunatmo
Department of Biology, Faculty of Mathematics and Natural Sciences, Bogor Agricultural University, Jalan Agatis, IPB Dermaga, Bogor, Indonesia

Ahmed A. Issa and Mohamed Hemida Abd-Alla
Botany and Microbiology Department, Faculty of Science, Assiut University, Assiut, Egypt

Takuji Ohyama
Biological chemistry Department, Faculty of Agriculture, Niigata University, Niigata, Japan

Sayuri Tanabata
Niigata University, Japan

Norikuni Ohtake and Kuni Sueyoshi
Faculty of Agriculture, Niigata University, Niigata, Japan
Graduate School of Science and Technology, Niigata University, Niigata, Japan

Atsushi Momose
Graduate School of Science and Technology, Niigata University, Niigata, Japan

Yasuhiro Nakanishi
Faculty of International Agriculture and Food Studies, Tokyo University of Agriculture, Tokyo, Japan

Constancio A. Asis Jr.
Philippine Rice Research Institute, Science City of Muñoz, Nueva Ecija, Philippines

Soraya Ruamsungsri
Faculty of Agriculture, Chiang Mai University, Chiang Mai, Thailand

Shotaro Ando
Japan International Research Center for Agricultural Sciences, Okinawa, Japan

Index

www.ingramcontent.com/pod-product-compliance
Lightning Source LLC
Chambersburg PA
CBHW081214180926
51613CB00003B/238
* 9 7 8 1 6 3 9 8 7 3 9 7 5 *